I0071971

New Paradigms in Agroecology

New Paradigms in Agroecology

Edited by Bella Harper

SYRAWOOD
PUBLISHING HOUSE

New York

Published by Syrawood Publishing House,
750 Third Avenue, 9th Floor,
New York, NY 10017, USA
www.syrawoodpublishinghouse.com

New Paradigms in Agroecology
Edited by Bella Harper

© 2022 Syrawood Publishing House

International Standard Book Number: 978-1-64740-055-2 (Hardback)

This book contains information obtained from authentic and highly regarded sources. Copyright for all individual chapters remain with the respective authors as indicated. All chapters are published with permission under the Creative Commons Attribution License or equivalent. A wide variety of references are listed. Permission and sources are indicated; for detailed attributions, please refer to the permissions page and list of contributors. Reasonable efforts have been made to publish reliable data and information, but the authors, editors and publisher cannot assume any responsibility for the validity of all materials or the consequences of their use.

Trademark Notice: Registered trademark of products or corporate names are used only for explanation and identification without intent to infringe.

Cataloging-in-Publication Data

New paradigms in agroecology / edited by Bella Harper.
     p. cm.
Includes bibliographical references and index.
ISBN 978-1-64740-055-2
1. Agricultural ecology. 2. Agriculture--Environmental aspects. 3. Ecology. I. Harper, Bella.
S589.7 .N49 2022
577.55--dc23

TABLE OF CONTENTS

Permissions

List of Contributors

Index

PREFACE

This book has been an outcome of determined endeavour from a group of educationists in the field. The primary objective was to involve a broad spectrum of professionals from diverse cultural background involved in the field for developing new researches. The book not only targets students but also scholars pursuing higher research for further enhancement of the theoretical and practical applications of the subject.

Agroecology is the discipline that studies the ecological processes applied to agricultural production systems. It focuses on the study of a variety of agroecosystems and presents a specific manner of their observation. Agroecology is an interdisciplinary field that includes and uses natural science to understand the components of an agroecosystem such as soil properties, plant-insect interactions. It also uses social science to understand the effects of farming practices on rural communities. It does not entirely oppose the use of technology in agriculture but it evaluates how technology can be used in conjunction with social, human and natural assets. It studies and answers the questions concerning the four properties of agroecosystems such as productivity, sustainability, stability and equitability. It treats all of them as interconnected as well as necessary for a successful agroecosystem. This book explores all the important aspects of agroecology in the present day scenario. Its objective is to give a general view of the different areas of this field. It attempts to assist those with a goal of delving into the field of agroecology.

It was an honour to edit such a profound book and also a challenging task to compile and examine all the relevant data for accuracy and originality. I wish to acknowledge the efforts of the contributors for submitting such brilliant and diverse chapters in the field and for endlessly working for the completion of the book. Last, but not the least; I thank my family for being a constant source of support in all my research endeavours.

Editor

Reciprocal Relationships between Aggregate Stability and Organic Carbon Characteristics in a Forested Ecosystem of Northern Nigeria

H.M. Lawal*, J.O. Ogunwole and E.O. Uyovbisere

Department of soil science, Institute for Agricultural Research Ahmadu Bello University, Zaria -Nigeria
Email: leem8810@yahoo.com
*Corresponding Author

SUMMARY

Soil organic matter associated with different size aggregates differ in structure and function; therefore, play different roles in soil organic carbon (SOC) turnover. This study assessed the relationship between aggregate stability and soil organic carbon fractions in a forested soil. Aggregate stability characterized by mean weight diameter (MWD) was correlated with the various pools of SOC in a regression model. Mean weight diameter presented a 46% influence on total organic carbon (TOC) while, TOC accounts for 21.8% of aggregate stability. The unprotected and physically protected soil organic carbon did not significantly dictate stability of these soils. However, chemically protected and biochemically protected SOC influenced significantly aggregate stability of these forested soils.

Keywords: soil organic matter; mean weight diameter; aggregate stability; soil quality; savanna.

INTRODUCTION

The total organic matter in soil encompasses the various pool of soil organic matter (SOM) namely; active, slow and passive pools. Therefore, recognizing the susceptibility of the different pools of SOM to microbial degradation is a most useful approach to define SOM quality (Brady and Weil, 1999). The turnover time and carbon storage potential of these various pools affect the formation and stability of soil aggregates.

Soil organic matter associated with different soil size aggregates differ in structure and function (Christensen, 1992), and therefore, play different roles in soil organic carbon (SOC) turnover. This relationship is crucial to understand SOC dynamics, because SOM influences soil physical and chemical properties which are paramount to nutrient cycling and will consequently have effect on forest productivity. Forest vegetation plays a key role in atmospheric carbon sequestration, thereby mitigating greenhouse effect.

The formation distribution and turnover of SOM are largely affected and controlled by climate (chiefly, temperature and moisture) and modified by vegetation, parent materials and human activities. Overtime, SOC is allocated into various pools as a result of root growth, litter fall and subsequent microbial decomposition; these pools are defined on the basis of relative recalcitrance which governs their residence and turnover times. This study aimed to assess the relationship between aggregate stability and SOC fractions in a forested soil.

MATERIALS AND METHODS

Location and Soil sampling

Soil samples were collected from the Afaka Forest Reserve, Kaduna, Northern Guinea savanna ecological zone (Keay, 1959) of Nigeria, coordinates 10° 33' and 10° 40' north and 07° 15' East, annual rainfall of 1270 mm and 593 m altitude. Sampling was done in April, 2006 from seven (7) selected treatment plots planted to: *Acacia senegal, Azadirachta indica, Eucalyptus camadulensis, Khaya senegalensis, Prosopis africana, Tamarindus indica* and *Tectona grandis*. A stratified random sampling method was adopted to collect soil samples by hypothetically dividing each plot into three (3) subdivisions, representing three replications, since plots were not replicated and plot sizes were very large. In each replication, composite soil samples at four (4) different points were taken and then bulked. Soil samples from each location were taken from 0-25 cm depth and labeled for ease of identification.

Laboratory Analyses

Aggregate fractionation

A 200 g of air dried bulk soil (passed through a 5 mm sieve) was wetted by rapid immersion (slaked), then sieved in water for one minute using sieve sizes of 2 mm, 0.25 mm and 0.053 mm, with an average stroke per minute of 36, 14 and 3 for each of the sieve sizes respectively.

The sieving was carried out in the order of decreasing mesh size. After sieving with the initial sieve size (2 mm) the filtrate was transferred onto the proceeding sieve (0.25 mm), and then 0.053 mm sieve. The <0.053mm aggregate fraction was allowed to settle down and the water was decanted gently. The fractionated aggregates were oven dried at 60°C for 48 h and weighed.

The proportion of each aggregate fraction was corrected for sand and mean weight diameter determined. However large macroaggregate >2mm were not used in computing or calculating MWD, because the proportion of aggregates >2mm recovered after the wet sieving were too small in weight to be corrected for sand (sand free fraction).

The proportional weight of sand free aggregates is given as

Weight of aggregate fraction - % sand content
Weight of bulk soil - % sand content

(Masri and Ryan, 2006)

Mean weight diameter (MWD) was determined thus:

$$MWD = \sum_{1=1}^{n} xi \ wi$$

Where xi = mean diameter of sieve proceeding and preceeding

Wi = proportional weight of sand free aggregates

Soil organic matter fractionation

Soil organic carbon content was determined in each of the aggregate fractions (2-0.25, 0.25-0.053 and <0.053mm) by the dichromate oxidation method (Nelson and Sommers, 1982).

The following fractions were identified.

Sieve size	Measure organic matter	Conceptual soil organic matter
2-0.25 mm	Fine particulate organic matter	Unprotected
0.25-0.053 mm	Intra aggregate particulate organic matter	Physically protected
<0.053 mm	Silt and clay associated organic matter	Chemically protected

Sand free carbon concentration was calculated in aggregate fractions greater than 0.053mm as:

Sand free C fraction = ..Cfraction.........................
                                    1-[sand proportion] fraction

(Denef *et al*., 2001)

Biochemically protected soil organic matter (SOM)

The biochemically protected SOM was determined by acid hydrolysis (Tan *et al*., 2004). Non hydrolysable carbon was determined in the acid hydrolyzed samples by dichromate oxidation method as described by Nelson and Sommers (1982). The quantity of the non hydrolysable SOM was referred to as the biochemically protected SOM.

Data analysis

Soil aggregate stability characterized by mean weight diameter (MWD) was correlated with the various pools of SOC and macroaggregates in a regression model in accordance with the procedures of Steel and Torrie (1984).

RESULT AND DISCUSSION

Relationship between Soil Macroaggregates And Mean Weight Diameter

When the regression graph of large macroaggregate (>2mm)was plotted against mean weight diameter (Figure 1a) and vice versa (Figure1b), a quadratic equation gave the best fit with r^2 values of 0.5364 and 0.5408 respectively, implying that the proportion of large macroaggregate fraction in any soil dictates more than 50% of the MWD value.

Furthermore, both mean weight diameter and large macroaggregate fraction could be dependent on one another establishing a significant relationship (P>0.05) and indicating that MWD was directly proportional to large macroaggregate. Thus in a well aggregated soil (with high value of MWD), a large proportion of macroaggregate fraction is expected. The non-use of large macroaggregate in computing MWD in this study could be responsible for its low r^2 value (P≥0.05) in the regression graph (Figure 1a and b), compared to the r^2 values of the regression graph of small macroaggregate (2-0.25mm) versus MWD (Figure 1c and d).

The graph of small macroaggregate against mean weight diameter and vice versa revealed that the best fit equation was quadratic as indicated by r^2 values of 0.9936 and 0.972 (Figures. 1 c and d respectively), indicating that the MWD is dependent upon small macroaggregates and showing strong dependence of both variables (small macroaggregate and MWD) on each other(P≥0.01).

Relationship between Soil Organic Carbon Pools and Aggregation

The relationship between total organic carbon (TOC) and MWD was best presented with a cubic regression model. TOC vs MWD and MWD vs TOC had r^2 values of 0.464 and 0.218 (Figs. 2a and b respectively). In either case, the relationship was not significant (P≥0.05), indicating the independence of these factors.

Regression graph (Figure 2c and d) showed that there is a weak relationship between unprotected organic carbon (UPOC) and MWD. It was best presented by

a cubic equation and showed r^2 value of 0.1885 when UPOC was plotted against MWD, and a r^2 value of 0.0306 when MWD was plotted against UPOC (Figure 2c and d respectively). The non significance of this regression further confirms that UPOC have little influence on soil aggregate stability, since UPOC provides readily accessible nutrition/nourishment for soil microbes (Six *et al.,* 2000). Furthermore at the time of soil sampling in this study, (The month of April which marks end of dry season, shortly before the onset of the rainy seasons), the rate of litter decomposition may exceed litter input in the forest ecosystem, thus little of the UPOC will be available, therefore minimal contribution of this carbon pool to soil aggregate stability is expected at this time.

The relationship between MWD and physically protected organic carbon (PPOC) was also best shown by a cubic regression model, though the relationship was not significant. An r^2 value of 0.3172 was obtained when PPOC was plotted against MWD (Figure 2e) and an r^2 value of 0.2081 was presented by graph of MWD against PPOC (Figure 2f). The relationship between PPOC and MWD (Figure 2e and f) indicated a better influence of soil aggregation than that of the UPOC and MWD (Figure 2 c and d) hence explaining the longer turnover time of PPOC in soil (Brady and Weil, 1999) than the UPOC. The influence of PPOC on MWD and vice versa showed that these parameters explain between 21 and 32%. A high value MWD may suggest that there is a reasonable amount of PPOC in that soil.

The relationship between soil aggregate stability as characterized by MWD and chemically protected organic carbon (CPOC) was also best represented by a cubic equation regression model. An r^2 value of 0.3899 was obtained when CPOC was a dependent variable (Figure 2g) while an r^2 value of 0.8454 was obtained when CPOC was an independent variable (Figure 2h). The extent of soil aggregate stability is a function of its CPOC. Furthermore it could be deduced from the regression graphs that CPOC influences soil aggregate stability (Figure 2h) more than other aggregate associated carbon evaluated in this study. The value of MWD could explain less than 40% of CPOC however, CPOC explains over 80% of the MWD value. The formation of stable organomineral complexes is an important mechanism of soil carbon sequestration (Lal *et al.,* 2003) and its stabilization effect is principally observed at microaggregation level. However, it can also indirectly increase macroaggregation through a stimulation of microbial activity particularly, when Ca is present (Six *et al.,* 2004).

The reciprocal relationship between aggregate stability and CPOC is not well pronounced as the amount of CPOC present in soil is not a function of the soil aggregate stability rather; CPOC dictates the level of stability of soil aggregates (Figure 2g and h). These aggregates indicate that CPOC increases the stability of soil aggregates which in this case was represented by MWD.

Regression graph showed that a cubic equation gave the best fit, when Biochemically protected organic carbon (BPOC) was plotted against soil aggregate stability (Figure 2i), this was characterized by mean weight diameter (MWD), with an r^2 value of 0.5241. However, when MWD was plotted against BPOC an r^2 value of 0.744 was obtained (Figure 2j). The significance of this relationship is an indication that both BPOC and MWD are interdependent

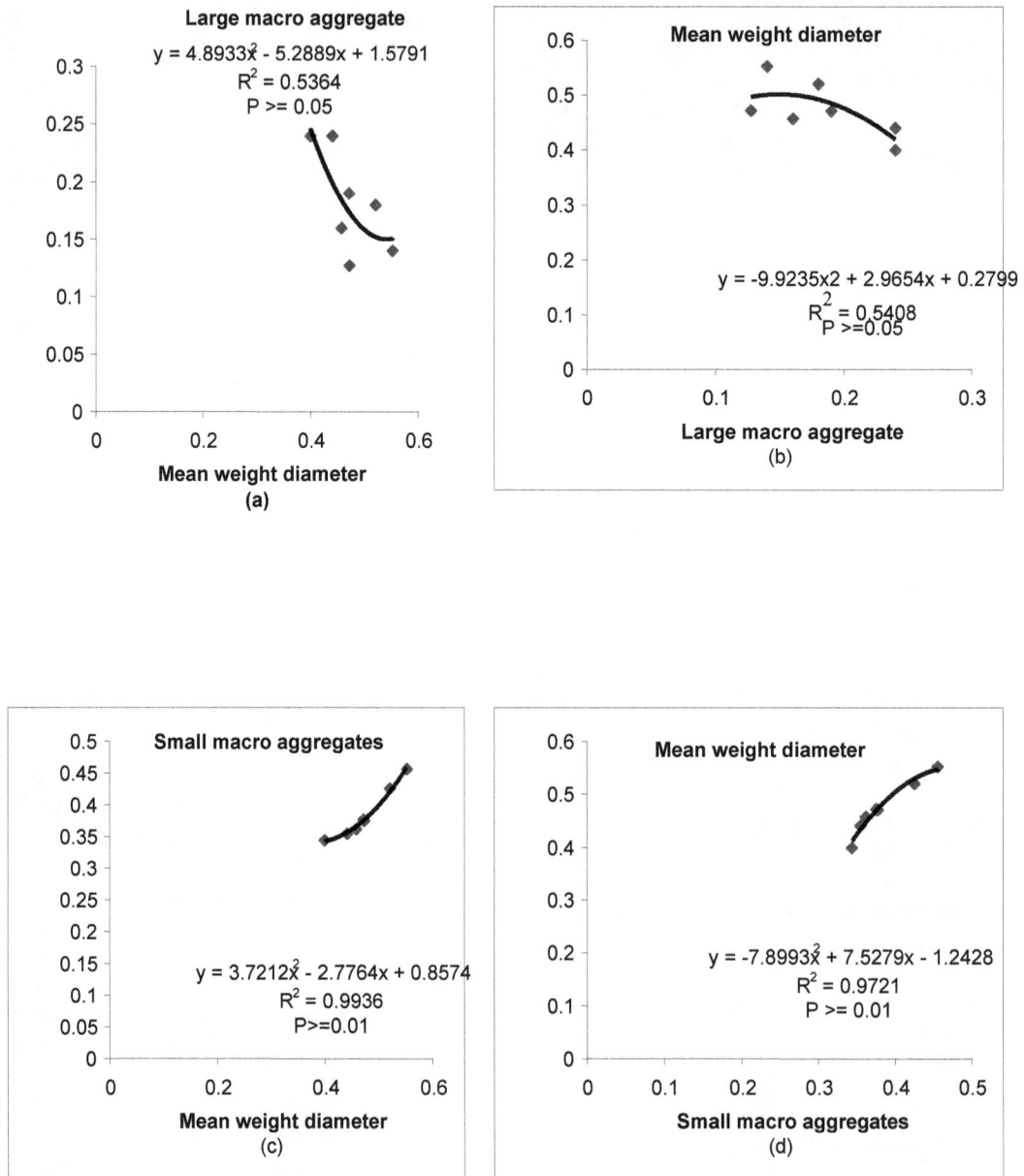

Figure 1: Relationship between Macro aggregate and Mean weight diameter

However, the extent of aggregate stability in a soil may be more dependent on BPOC content which is at a higher significant (P≥ 0.01) level (Figure 2j). Hence explaining that a higher BPOC content in soil would induce a stronger soil aggregate stability than a high MWD would induce increase in BPOC content in soil. The BPOC involves biomasses that have been converted into humic substances which are relatively resistant to microbial decomposition and have a long turnover time. Carbon pools are recalcitrant pools and therefore, vital in soil aggregation

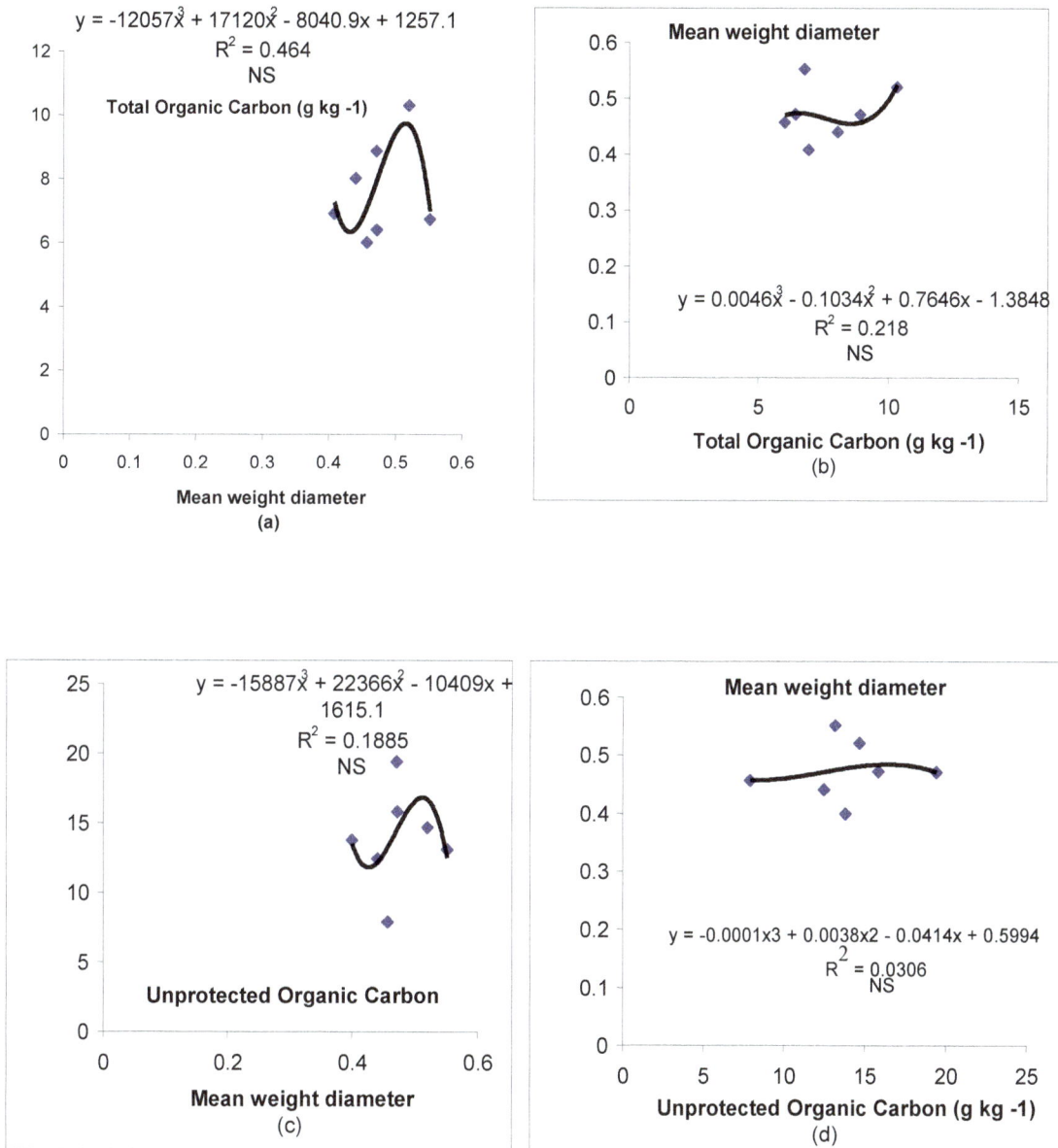

Figure 2: Relationship between Soil Organic Carbon and aggregation

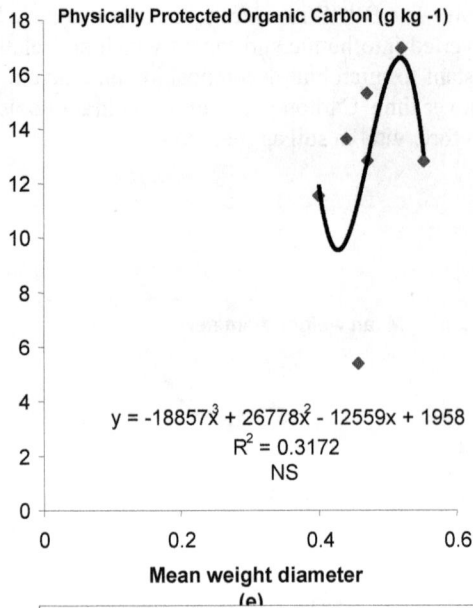

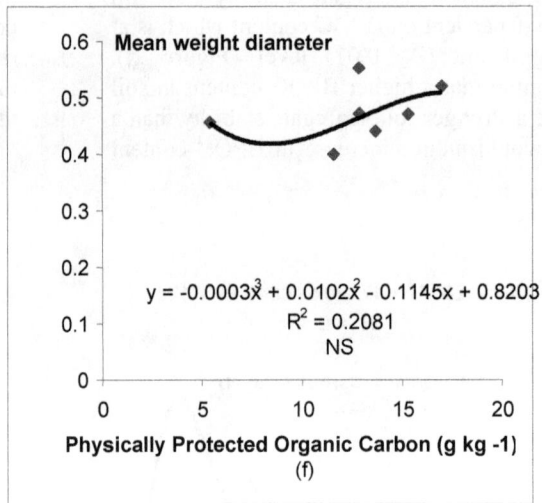

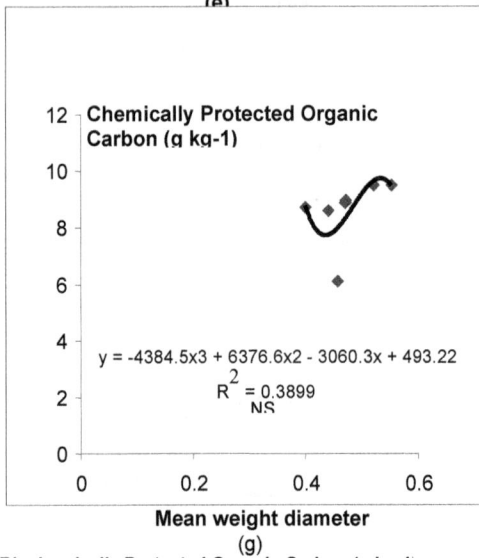

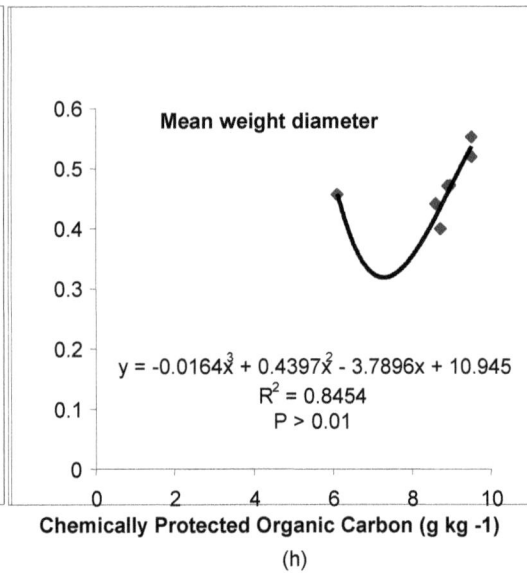

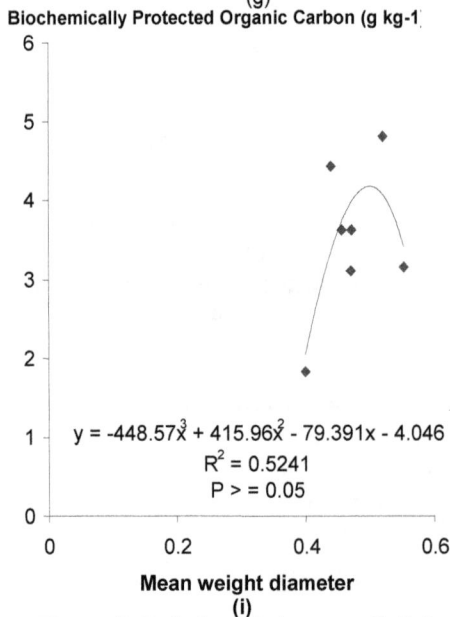

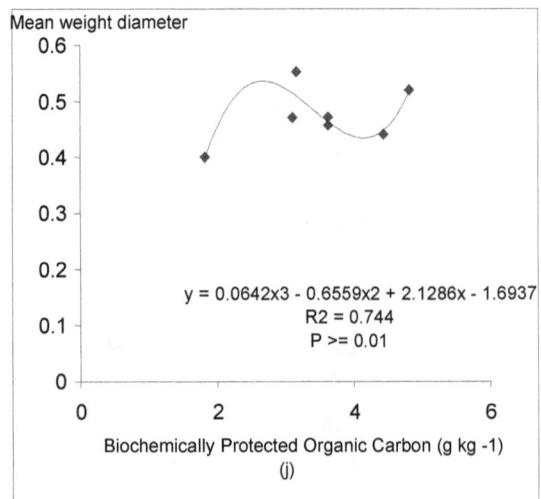

Figure 2: Relationship between Soil Organic Carbon and aggregation. (continues from previous page)

CONCLUSION

A strong relationship was observed between SOC pool and aggregation especially with BPOC. In a structurally stable soil, with high proportion of macroaggregate, Reciprocal relationship may not exist between TOC content and soil aggregate stability. However, whenever there is a significant relationship between soil aggregate stability and any of the SOC pools, specifically the BPOC (recalcitrant) and CPOC, then a good soil aggregate stability may be expected concluding that, in certain instances, TOC may not give precise prediction of the degree of aggregate stability in soil.

REFERENCES

Brady, N.C. and Weil R.R (eds) 1999. The nature and properties of soil. 12th edition prentice Hall Inc. London, 881pp.

Christensen, B.T. 1992. Physical fractionation o soil and organic matter in primary particle size and density separates. Soil Science. 20: 2 – 90.

Denef, K., Six, J., Bossuyt, H., Frey, S.D., Elliot, E.T., Merckx, R. and Paustian, K. 2001. Influence of dry-wet cycles on the interrelationship between. aggregate, particulate organic matter and microbial community dynamics. Soil Biology and Biochemistry. 33: 1599 – 1666.

Denef, K., Six, J., Merckx, D., and Paustian K. 2004. Carbon sequestration in micro aggregates of no tillage soils with different clay mineralogy Soil Science Society America Journal. 68.1935-1944.

Elliott, E.T. 1986. Aggregate structure and carbon, nitrogen and phosphorus in native and cultivated soils. Soil Science Society American Journal. 50:627 – 633.

Keay, R.W.J. 1959. An outline of Nigerian vegetation 3rd edition London pp 45.

Ladd, J.N., Amato, M. and Oades, J.M. 1985. Decomposition of plant materials in Australian soils. III. Residual organic and microbial biomass C and N from isotope labelled plant material and organic matter decomposing under field conditions. Australian Journal Soil Research. 23: 603-611.

Lal, R., Follet, R.F and Kimble J.M. 2003. Achieving soil carbon sequestration in the united states: A challenge to the policy makers. Soil Science. 168: 827 – 845.

Lal, R. 2004. Agricultural activities and the global carbon cycle. Nutrient cycling in Agroecosystems. 70:103-116.

Lal, R. and Okigbo B. 1990. Assessment of soil diction in the southern states of Nigeria. The World Bank Sector Policy and Research staff Environment working paper No. 39.

Masri, Z and J. Ryan 2006. Soil organic matter and related physical properties in a mediterranean wheat based rotation tried. Soil and Tillage Research 87: 146 – 154.

Metting, F.B., Smith, J.L. Amthor J.S. 1998. Science needs and new technology for soil carbon sequestration In: Rosenberg, N.J., R.C. Izaurralde, E.L. Malone (eds) Carbon sequestration in soils: Science, Monitoring and Beyond. A proceedings of the St. Micheals Workshop, December 1998. 199pp.

Nelson D.W and Sommers, L.E 1982: Total Carbon, Organic Carbon and Organic Matter. In page, A.: Miller, R.H and Keeney, D.R (eds) Methods of Soil Analysis II American Society of Agronomy Madison NY 539-579.

Six J., Elliott, E.T. and Paustian K. 2000. Soil structure and soil organic matter: II. A normalized stability index and the effect of mineralogy. Soil Science Society of America Journal. 64: 1042 – 1049.

Six, J. H. Bossuyt, S. Degryze and K. Denef. 2004. A history of research on the link between (micro) aggregate. Soil biota and soil organic matter dynamics. Soil Tillage Research. 79: 7 – 39

Steel, R.G. and Torrie, J.W. 1984. Principle and Procedures of Statistics. A Biometrical Approach. 2nd Ed. Mc. Graw Hill International Book Company, Aukland, 633pp.

Tan, Z.X., R. Lal, C. Izaurralde and W.M. Post. 2004. Biochemical Protected Soil organic carbon at the North Appalachian Experimental watershed. Soil Science. 169: 1-10.

Tomlison, P.R. 1957. A preliminary soil report on Afaka Forest reserve Experimental Area

Farmers' Perception of Soil Fertility Problems and their Attitude towards Integrated Soil Fertility Management for Coffee in Northern Tanzania

Godsteven P. Maro[1*], Jerome P. Mrema[2], Balthazar M. Msanya[2]
and James M. Teri[1]

[1]TaCRI, Lyamungu P.O. Box 3004, Moshi, Tanzania.
E-mail: godsteven.maro@tacri.org

[2] Department of Soil Science, Faculty of Agriculture, Sokoine University of
Agriculture. P.O, Box 3008, Morogoro, Tanzania
*Corresponding author

SUMMARY

A study was conducted in Hai and Lushoto districts, Northern Tanzania to establish the farmers' perception of soil fertility problem and their attitudes towards integrated soil fertility management (ISFM) for coffee, thereby identifying the appropriate intervention strategies. The study was based on a structured questionnaire involving 126 respondents. Both farmers' awareness of the problem and their attitude were highly significant ($P<0.01$). Age, household size, and adoption of improved coffee varieties affected farmers' awareness significantly ($P<0.05$). As for farmers' attitudes, six of the eight predictors were significant ($P<0.05$). Age, household size, adoption of new varieties and total farm income were highly significant ($P<0.01$). Age, total land under coffee and total off-farm income showed to negatively affect farmers' attitude. As farmers get older, they tend to refrain from innovation. Larger farms are likely to exert more pressure on the available organic resources. With multiple farms, distant farms are likely to receive less attention. As regards off-farm income, multiple ventures compete for the farmers' time, resources and attention. For the two districts, ISFM interventions will make a better impact to younger and more energetic farmers with enough land for commercial coffee production and who depend largely on this resource for their livelihood.

Key words: Soil fertility; farmers' perception; ISFM; coffee; Africa.

INTRODUCTION

Coffee is one of the major export crops in Tanzania, contributing to 23% of the agricultural GDP (URT, 2007). It contributes directly to the livelihoods of over 420,000 farm families and indirectly to over 2 million people employed in the coffee value-chain industry (Carr *et al.*, 2003). Arabica coffee contributes 65% of the Tanzanian total coffee export. The Tanzanian coffee, especially the washed Arabica, is one of the best in the world, ranked among the rare category of "Colombian Milds" used to blend other inferior coffees.

Coffee is also grown in many countries in East and Central Africa. Other important coffee producers are Ethiopia, Uganda, Kenya, Rwanda and Burundi. According to statistics from International Coffee Organization (ICO, 2011), total production for the six countries was 10.6, 11.4 and 12.9 million bags for 2008, 2009 and 2010 respectively. Tanzania's share was 11.14%, 6.2% and 7.08%, while Kenya's share was 5.08%, 5.51% and 6.56%. Ethiopia and Uganda together commanded over 70% of the share for all the three years.

The Tanzanian average smallholder coffee productivity per hectare ranges between 250 and 300 kg of parchment which is very low compared to the potential yield of over 1000 kg per tree (Baffes, 2003, Hella *et al.*, 2005). In Kenya, coffee yields were reported to have fallen from 892 kg.ha^{-1} in 1980 to 284 kg.ha^{-1} in 2006, much lower than average yields for Arabica coffee worldwide of 698 kg/ha and yields of 1160 kg/ha in Rwanda and 995 kg/ha in Ethiopia.

Soil fertility degradation is one of the major problems facing coffee productivity in Tanzania. It is defined by Stocking and Murnaghan (2000) and Maro *et al.* (2010) as the loss of soil physical and nutritional qualities. It has been an issue of concern throughout the Sub-Saharan Africa (SSA), and cuts across many different soils and crops (Okalebo *et al.*, 2007). In Tanzania, the problem covers all coffee growing zones and all types of coffee growers (Envirocare, 2004). Reports from Kenya indicate that decline in coffee yields were caused by farmers' reluctance to invest in fertilizers (Condliffe *et al.*, 2008), which translates to poor soil fertility.

Integrated soil fertility management (ISFM) has been cited by many authors, including Okalebo *et al.* (2007), Gumbo (2006) and Raab (2002), as the key approach in raising productivity levels in agricultural systems while maintaining the natural resource base. It is described by Vanlauwe and Zingore (2011) as a set of soil fertility management practices that necessarily include the use of fertilizer, organic inputs, and improved germplasm combined with the knowledge on how to adapt these practices to local conditions, aiming at maximizing agronomic use efficiency of the applied nutrients and improving crop productivity. Because of the pressing need for global food security, many articles have been published which relate ISFM to the production of annual food crops like maize (Ikerra *et al.*, 2007; Kimani *et al.*, 2007), and rice (Kaizzi *et al.*, 2007), giving lesser attention to perennial crops like coffee. It's no wonder then that the role of ISFM for coffee in Tanzania and the socio-economic perception of it have not been studied to any significant detail.

The coffee producing zone of Northern Tanzania comprises four regions, namely Arusha, Kilimanjaro, Manyara and Tanga (a total of 12 districts). Coffee production is both historical and traditional, especially in Kilimanjaro region which was the first to grow coffee as a commercial crop (Maro *et al.*, 2010). Annual coffee production trend for the zone indicates a decline over the years. A number of constraints have been suggested as the cause of this decline. Currently, as reflected during the coffee stakeholders' forum (Tanzania Coffee Board TCB, 2009), soil fertility degradation has emerged as the most limiting factor. This is, however, a very generic perception which needs to be studied in detail, by targeting specific locations and farming communities.

The current study was therefore conducted in Hai and Lushoto districts to establish the magnitude of soil fertility problem as perceived by farmers in the two districts, and to establish the baseline farmers' attitudes towards ISFM, thereby identifying the appropriate intervention strategies.

METHODOLOGY

A structured questionnaire was administered to farmers in Hai and Lushoto districts to solicit the farmers' opinion on soil fertility and coffee productivity. Generic questions included personal details (gender, age, level of education, position in the household, household size and sources of coffee management information) and farm details (size, number of trees and varieties). Additionally, respondents were requested to give an account of their knowledge of soil problems, source of ISFM knowledge if any, experience in industrial fertilizer use with coffee and negative effects if any, usage of organics (manure, coffee processing by-products, mulches, green manure plants), major and subsidiary income sources and income ranges last season. A total of 60 respondents were interviewed in Lushoto and 66 in Hai, making a total of 126 respondents. The data were processed and analyzed by using the Statistical Package for Social Sciences (SPSS version

16) (SPSS Inc, 2007). The analysis involved computations of mean and frequency, together with two linear regressions: one on farmers' appreciation of soil fertility problem and the other on farmers' attitude towards ISFM.

Defining the variables

The degree of appreciation of soil fertility deterioration as a problem (aP) was described as a mean of two ratings, one qualifying the farmers' knowledge of their soils (0, 1 and 2 for no, slight and basic knowledge respectively) and the other qualifying farmers' understanding of soil related problems (0 = no idea, 1 = could identify other problems, 2 = could identify crop-related problems and 3 = was able to identify nutritional disorders). The ratings were categorized as 0, 0.5, 1.0, 1.5 and 2.0 for unaware, slightly aware, moderately aware, sufficiently aware and fully aware respectively. The assumption was that, as noted by D'Emden *et al* (2005), awareness of a problem is a motivator in devising (or adopting) problem-solving techniques.

Attitude towards ISFM (α) was described as a mean of eight ratings including the two stated above (R_{soil} and R_{prob}) and six others. R_{ind}, R_f and R_b are dummy variables qualifying whether a farmer uses (1) or does not use (0) industrial fertilizers, farmyard manures or coffee by-products respectively. R_{fp} and R_{bp} at the scale of 0, 1, 2 and 3, are the ratings qualifying farmers who do not process because they do not use farmyard manure or pulp, those who use the organics raw without any processing, those who just heap the material to stabilize in the open, and those who compost the material in a pit. R_{train} is a rating that qualifies whether and how many times last year a farmer received training on ISFM (an aggregate of four topics – soils, ISFM, identification of nutritional problems and making of organic composts): 0 = no training, 1 = trained once, 2 = trained twice and 3 =

trained more than twice. The resultant ratings varied between 0 and 2, and were clustered at maximum values in terms of readiness to adopt ISFM interventions as shown in Table 1 below:

Table 1: Description of clustered ratings.

Cluster	Maximum value	Description
0	0	Minimum likelihood of adoption
0.1-0.5	0.5	Will need a lot of time and conviction to adopt
0.6-1.0	1.0	Will need some time and conviction to adopt
1.1-1.5	1.5	Will need little time and conviction to adopt
1.5-2.0	2.0	Will adopt readily.

Descriptive statistics

The two variables aP and α were exposed to descriptive statistics following the models of Nkamleu (2007) and Zhou *et al.* (2008), which involved physical counts and percentage frequency, and were compared per district.

Regression modelling

The defined variables aP and α were separately exposed to a linear regression model as functions of demographic predictors (age and level of education of the household head, the size of the household, farm and non-farm income) as defined by Doss (2003) and farm related predictors (such as land size and types of coffee trees). Both models used the same predictors as shown in the example below which represents aP:

$$aP = b_0 + b_1A + b_2ED + b_3HS + b_4FEX + b_5LS + b_6CV + b_7FI + b_8NFI + \ell$$

Where:
  b_0 represent the constant
  b_1A = coefficient related to age
  b_2ED = coefficient related to level of education
  b_3HS = coefficient related to household size
  b_4FEX = coefficient related to coffee farming experience in years
  b_5LS = coefficient related to total coffee land size
  b_6CV = coefficient related to coffee varieties (whether improved varieties are adopted)
  b_7FI = coefficient related to farm income last year
  b_8NFI = coefficient related to non-farm income last year
  e = random error of prediction.

Each of the eight predictors were then assessed in terms of the significance level at which it influences the farmers' awareness of soil fertility decline as a problem on one hand, and the farmers' readiness to adopt ISFM interventions on the other.

RESULTS AND DISCUSSION

The significance of predictors per district

The eight selected predictors were compared per district (t-test) and were all highly significant ($p < 0.001$). Means and their 95% confidence intervals are shown in Table 2. Average age of respondents is around 60 years, implying that coffee is still held by old people. This observation is in line with Morris and Venkatesh (2000), Baerenklau and Knapp (2007), Mateos-Planas (2003) and Tiamiyu et al. (2009). Education level was mainly primary, with fewer cases of post-primary education. The majority of households have 2-8 persons, which is normal for many Tanzanian households (ILFS, 2001, Kamuzora, 2001). With the mean coffee farming experience of 30 years, it implies that most of the coffee farmers in the study districts have immense experience in their business, and their perception of soil problems and best ways to manage soil fertility should be

considered in devising appropriate ISFM packages (Douthwaite et al., 2002).

Land size of mean 1.96 acres (CI 1.67-2.27) implies that the people we are dealing with are truly smallholders who are resource-poor, and therefore the ISFM packages should have that in mind. An average of 33% of the respondents have adopted the new improved varieties released by TaCRI. This implies that there is still an uphill task for TaCRI and other coffee stakeholders to promote these varieties among farmers.

The distribution of farm and off-farm incomes in 2009/10 is given in Table 3. Farm income appears to be fairly normally distributed with the majority ranging between 0.3m and 2m TZS. With off-farm income, 74.6% of the respondents reported to have none, thus depending entirely on the farm for their livelihood. Those who have subsidiary off-farm incomes (25.4%) may portray variable pictures as regards farm attention. For some it may be a deterrent factor, keeping the farmer busy with the off-farm ventures at the expense of the farm. For elite farmers however, a subsidiary off-farm income can act as a buffer against fluctuating coffee prices, and/or a stimulant in adopting good agricultural practices (Karki and Bauer, 2004).

Table 2: A comparison of the selected predictors per district

Predictor	Unit	Means	95% C.I lower	Upper	Notes
A	Years	60.83	58.37	63.29	Coffee is a crop for old people
ED	Rating	1.23	1.09	1.37	Majority primary, fewer ordinary
HS	Rating	2.37	2.21	2.54	2 to 8 persons per household
FEX	Years	30.08	27.3	32.86	People with immense coffee exp.
LS	Acres	1.96	1.67	2.27	Typical smallholders
CV	0=no, 1=yes	0.33	0.24	0.41	Adoption of 24-41%
FI	Rating	9.13	8.39	9.86	600,000 to 900,000 TZS
NFI	rating	1.83	1.05	2.6	Maximum of 200,000 TZS

Table 3: A summary of farm and off-farm incomes in 2009/10.

Category	Farm income (%)	Off-farm income (%)
None	0	74.6
<0.3m	14.4	11.0
0.3 – 2.0m	76.0	9.6
>2.0m	9.6	4.8

The distribution of variables per district

The frequency of farmers' awareness of soil fertility degradation as a problem is shown in Figure 1. The majority of respondents from Lushoto are either unaware (25%) or slightly aware (60%). On the other hand, 9% had sufficient awareness while none is fully aware. In Hai, the unaware and slightly aware groups were 13.6% and 45.4% respectively, while 3.0% are fully aware. An abnormal saddle was observed with the moderately aware group having 10.6% between 45.4% and 27.3%. As for attitude towards ISFM (Figure 2), the distribution of respondents in Hai was fairly normal, with a peak at 50% for moderate attitude group, tailing at very low (1.5%) and very high (7.6%). The Lushoto distribution was rather irregular, with only one interesting feature, that the percentages that have moderate and high attitudes are equal at 40% each, therefore constituting the bulk of the sample.

Analysis of regression models

A summary of the regression models for problem appreciation and attitude towards ISFM is given in Table 4 below.

Problem appreciation

The regression model for problem appreciation (aP) was highly significant (at $p<0.01$) even though there was a rather poor correlation (Adjusted R^2 of 0.133) among the parameters entered. Only household size and adoption of improved coffee varieties were highly significant ($p<0.01$), while age was significant at $p<0.05$. Age was seen to negatively affect the farmers' awareness of soil fertility problem as older people tend to become more passive about what happens in their farms (Truong and Yamada, 2002). The rest did not show any statistical significance; including level of education. The relationship between household size and problem appreciation is not very clear. However, if family members are trained in diagnosing unusual characteristics in the field, the bigger the hosehold size, the more likely it is for problems to be identified.

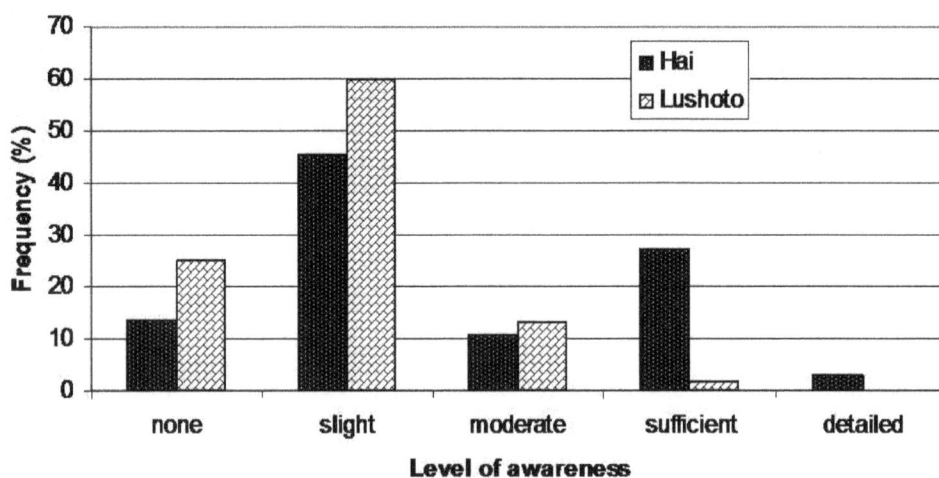

Figure 1: Distribution of awareness of soil fertility decline as a problem

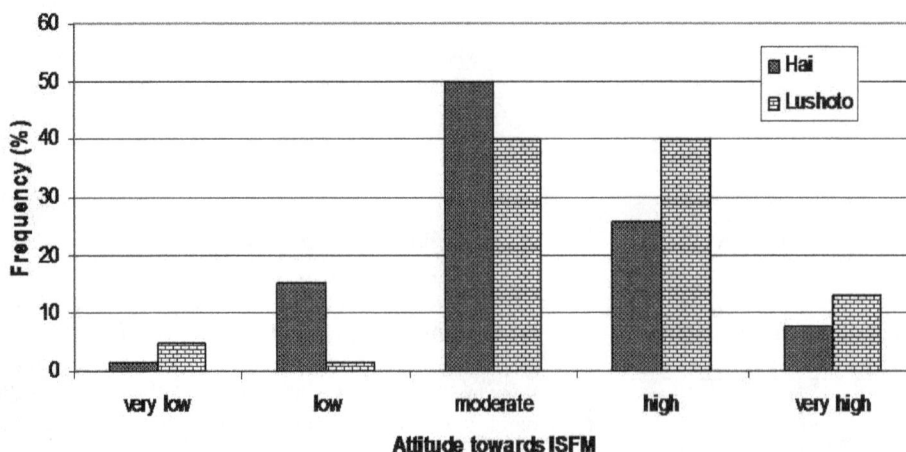

Figure 2: Distribution of attitudes of farmers towards soil fertility management

During the survey, and especially in Lushoto, respondents showed considerable awareness about soil fertility degradation, as also noted in Kenya by Kimiti *et al.* (2007). Their indigenous technical knowledge (ITK) showed that mishai trees (*Albizzia maranguensis)* contribute in restoration/maintenance of soil fertility as noted by Maro *et al.* (2010). Awareness of a problem has been cited as a motivation to the adoption of problem-solving technologies (D'Emden *at al.*, 2005). Other ITKs learnt during the survey include the "tugutu" bush (*Adhatoda engleriana* Lindau, family Acanthaceae) which is also medicinal (Moshi *et al.*, 2005). It has been tested with other crops and found to be highly nutritive. A formulation for making liquid fertilizer from their leaf extract was described. This opens an avenue for further research on the nutrient content of the "tugutu" leaves and ways in which this, where present, can be integrated in the local ISFM packages for coffee.

Attitude towards ISFM

The regression model was also highly significant (at $p<0.01$). Of the 8 parameters used in predicting α (attitude towards ISFM), 4 were highly significant (Age, household size, adoption of new varieties and total farm income) and 2 were significant at $p<0.05$ (land size and coffee farming experience). These observations are partly in line with those of Jamala *et al.* (2011). Level of education showed positive but insignificant influence on farmers' attitudes. The significance of education level in affecting adoption was reported by Barungi and Maonga (2011), Tiamiyu *et al.* (2009), Ono (2006) and Ani *et al.* (2004), which does not appear to be true in the study areas.

Age, total land under coffee and total off-farm income had negative B, β and t values. Age showed to negatively influence the capacity and willingness to adopt new approaches including ISFM. This is in line with the observations by Nzomoi *et al.* (2007). The fact that total coffee land showed negative relationship with attitude towards ISFM (contrary to the observation by Karki and Bauer, 2004) can only be explained in two scenarios: larger farms exerting more pressure on the limited amounts of available organic sources of nutrients like FYM; and farmers having multiple farms, some a distance away from their households, thus precluding efforts to use organic sources in those distant farms (Vanlauwe and Giller, 2006; Nkamleu, 2007). Off-farm income showed negative influence on farmers' attitudes, observations that are in line with those of Adolwa et al (2010). If this source of income contributes substantially to the total family income, the farmers' attention gets skewed from coffee towards the other ventures.

Table 4: Model summaries for problem appreciation and attitude towards ISFM.

Predictors	Problem appreciation			Attitude towards ISFM		
	β	t	sign	β	T	sign
Age	-0.163	-1.597	0.113	-0.350	-3.103	0.002
Level of education	0.041	0.447	0.656	0.113	1.319	0.190
Household size	0.251	2.761	0.007	0.235	2.785	0.006
Years growing coffee	0.079	0.763	0.447	0.288	2.530	0.013
Coffee land size	0.165	1.743	0.084	-0.185	-2.083	0.039
New varieties adoption	0.228	2.553	0.012	0.422	5.022	0.000
Farm income last year	0.087	0.956	0.341	0.227	2.659	0.009
Off-farm income last year	-0.110	-1.159	0.249	-0.145	-1.654	0.101
(Constant)		1.747	0.083		3.953	0.000

CONCLUSIONS

The perception of soil fertility degradation as a problem in the study areas is influenced by several household and farm variables. Among the eight predictors, only the age of household head, the size of the household and adoption of new improved varieties showed to be responsible for variation in perception, with the former having a negative impact.

Attitudes towards ISFM showed to be highly influenced by age, household size, adoption of new varieties and total farm income; and moderately influenced by total land under coffee and number of years spent by the household head in coffee business. Again here, age showed a negative relationship to attitude towards ISFM, implying that older people are usually skeptical in adopting new approaches.

In the two districts, ISFM interventions will make a better impact to younger and more energetic farmers with enough land for commercial coffee production and who depend largely on this source for their livelihood. These are the ones who can easily adopt improved varieties and good agricultural practices, including ISFM practices like mulching, composting of farmyard manure, coffee pulp and other field residues. . Long-term plans should be to encourage younger people to take up the coffee farming business, build the capacity to monitor the soil fertility regularly and give quick, site-specific recommendations. Also, promotion of the improved coffee varieties among farmers should continue.

Way forward

This is the first in a series of studies aimed at developing an effective and spatial ISFM decision support system for coffee in Northern Tanzania. It has effectively opened up our knowledge of soil fertility problems as perceived by farmers. The next steps will be to explore the extent of the problem through soil fertility surveys, and then decide on the right ISFM packages that will make impact in the study areas. The findings will be useful for TaCRI in planning for ISFM intervention in the Northern Zone.

REFERENCES

Adolwa, I.S., Esilaba, A.O., Okoth, P.O and Mulwa, M.R. 2010. Factors influencing uptake of integrated soil fertility management knowledge among smallholder farmers in Western Kenya. Proc. KARI Biennial Scientific Conference, 8-12 November, 2010: 9 pp.

Ani A.O., Ogunnika, O and Ifah, S.S. 2004. Relationship between socio-economic characteristics of rural women farmers and their adoption of farm technologies in Southern Ebonyi State, Nigeria. International Journal of Agriculture and Biology. 6: 802-805.

Baffes, J. 2003. Tanzania's Coffee Sector: Constraints and Challenges in a Global Environment.http://www.aec.msu.edu/fs2/inputuseworkshop/Baffes_Tanzania_Coffee.pdf. Site visited on 30th June 2007.

Baerenklau, K.A and Knapp, K.C. 2007. Dynamics of agricultural technology adoption: Age structure, reversibility and uncertainty. American Journal of Agricultural Economics. 89: 190-201.

Barungi, M. and Maonga, B.B. 2011. Adoption of soil management technologies by smallholder farmers in Central and Southern Malawi. Journal of Sustainable Development in Africa. 13(3): 28-38.

Carr, M.K.V., Stephens, W., Van der Vossen, H.A.M and Nyanga, A. 2003. Tanzania Coffee Research Institute Strategic Action Plan 2003-2008: Contributing towards a profitable and sustainable coffee industry in Tanzania. Report to TaCRI, ICPS, Cranfield University, Silsoe, UK.   161pp. http://www.dev.tacri.org/uploads/media/TaCRI_Strategic_Action_Plan_2003-2008_01.pdf visited 21 February 2012

Condliffe, K., Kabuchi, W., Love C and Ruparell, R. 2008. Kenya coffee: A cluster analysis. Harvard Business School, May, 2008: 34 pp. http://www.isc.hbs.edu/pdf/Student_Projects/Kenya_Coffee_2008.pdf visited 22 February, 2012.

D'Emden, F.H., Llewellyn, R.S. and Burton, M.P. 2005. Adoption of conservation tillage in Australian cropping regions: An application of duration analysis. Technological Forecasting and Social Change 73: 630-647.

Doss, C.R. 2003. Understanding Farm Level Technology Adoption: Lessons Learned from CIMMYT's Micro Surveys in Eastern Africa. CIMMYT Economics Working Paper 03-07. Mexico, D.F.: CIMMYT.

Douthwaite, B., Manyong, V.M., Keatinge, J.D.H and Chianu, J. 2002. The adoption of alley farming and mucuna: Lessons for research, development and extension. Kluwer Academic Publishers, Agroforestry Systems 56: 193-202.

Envirocare. 2004. A study on the importance of coffee industry in Kilimanjaro region. A NORAD Project, Moshi, Tanzania. October 2004. 33pp.

Gumbo, D. 2006. Integrated soil fertility management. Technical Brief, Practical Action Southern Africa, Harare, Zimbabwe; 06 September, 2006. 5pp.

Hella, J.P., Mdoe N.S and Lugole, J.S. 2005. Coffee baseline report for Tanzania Coffee Research Institute. Bureau for Agricultural Consultancy and Advisory Service, Sokoine University of Agriculture, Morogoro, Tanzania. 40pp.

ICO (International Coffee Organization), 2011. Coffee export statistics by country of origin. http://www.ico.org/historical/2000+/PDF/EXPORTS.pdf (visited 30 January, 2012)

Ikerra, S.T., Semu, E. and Mrema, J.P. 2007. Combining Tithonia diversifolia and Minjingu Phosphate Rock for improvement of P availability and maize grain yields on a Chromic Acrisol in Morogoro, Tanzania. In: A. Batiano et al (eds). Advances in ISFM in Sub-Saharan Africa: Challenges and Opportunities. Springer, Pp. 333-344
.

ILFS (Integrated Labour Force Survey), 2001. Household analysis in Tanzania. Chapter 10: 123-128. http://www.tanzania.go.tz/ilfs%5CChapter%2010.doc visited 21 February, 2012.

Jamala, G.Y., Shehu, H.E and Garba, A.T. 2011. Evaluation of factors influencing farmers' adoption of irrigated rice production in Fadama soil of North Eastern Nigeria. Journal of Development and Agricultural Economics. 3(2): 75-79.

Kaizzi, C.K., Ssali, H., Nansamba, A. and Vlek, P.L.G. 2007. The potential benefits of Azolla, velvet bean and N fertilizers in rice production under contrasting systems in Eastern Uganda. In: A. Batiano et al (eds). Advances in ISFM in Sub-Saharan Africa: Challenges and Opportunities. Springer, pp. 423-433.

Kamuzora, C.L. 2001. Poverty and family size patterns: Comparison across African countries. Research Report No. 01.3, REPOA: 44 pp. http://www.repoa.or.tz/documents/01.3_Poverty_and_Family_Kamuzora_fin_17-04.pdf visited 21 February, 2012

Karki, L.B and Bauer, S. 2004. Technology adoption and household food security. Analyzing factors determining technology adoption and impact of project intervention: A case of smallholder peasants in Nepal. Proceedings of Deutscher Tropentag Workshop, 5-7 October, 2004, Humboldt University, Berlin: 1-8.

Kimani, S.K., Esilaba, A.O., Odera, M.M., Kimenye, L., Vanlauwe, B. and Bationo, A. 2007. Effects of organic and mineral sources of nutrients on maize yields in three districts of Central Kenya. In: A. Batiano et al (eds). Advances in ISFM in Sub-Saharan Africa: Challenges and Opportunities. Springer, pp. 353-357.

Kimiti, J.M., Esilaba, A.O., Vanlauwe, B. and Batiano, A. 2007. Participatory diagnosis in the Eastern Drylands of Kenya: Are farmers aware of their soil fertility status? In: A. Batiano et al (eds). Advances in ISFM in Sub-Saharan Africa: Challenges and Opportunities. Springer, pp. 961-967.

Maro, G.P., Kitalyi, A., Nyabenge, M. and Teri, J.M. 2010. Assessing the impact of land degradation on coffee sustainability in Kilimanjaro region, Tanzania. In: Proceedings of the 23rd ASIC Conference, 3 -8 October, 2010, Nusa Dua, Bali, Indonesia. 8 pp

Mateos-Planas, X. 2003. Technology adoption with finite horizons. Elsevier Science-Direct, Journal of Economic Dynamics & Control 28: 2129-2154

Morris, M.G and Venkatesh, V. 2000. Age differences in technology adoption decisions: Implications for a changing work force. Personnel Psychology. 53: 375-403.

Moshi, M.J., Kagashe, G.A.B and Mbwambo, Z.H. 2005. Plants used to treat epilepsy by Tanzanian traditional healers. Journal of Ethnopharmacology. 97: 327-336.

Nkamleu, G.B. 2007. Modelling farmers' decisions in ISFM in Sub-Saharan Africa: A multinomial logit analysis in Cameroon. In: Batiano, A. et al (eds). Advances in ISFM in Sub-Saharan Africa: Challenges and Opportunities. TSBF-CIAT, Nairobi, Kenya. 891-903.

Nzomoi J.N., J.K. Byaruhanga, H.K. Maritim and P.I. Omboto 2007. Determinants of technology adoption in the production of horticultural export produce in Kenya. Journal of Business Management. 1: 129-135.

Okalebo, J.R., Othieno, C.O., Woomer, P.L., Karanja, N.K, Semoka, J.M.R., Bekunda, M.A., Mugendi, D.N., Muasya, R.M., Bationo, A. and Mukhwana, E.J. 2007. Available technologies to replenish soil fertility in East Africa. In: Batiano, A. et al (eds). Advances in ISFM in Sub-Saharan Africa: Challenges and Opportunities. TSBF-CIAT, Nairobi, Kenya. 45-62.

Ono, Y. 2006. Technology adoption in a community of heterogeneous education level: Who are your good neighbours? Economic Bulletin 15(8): 1-11.

Raab, R.T. 2002. Fundamentals of Integrated Soil Fertility Management. IFDC Training materials for the "Training Program on Integrated Soil Fertility Management (ISFM) in the Tropics", Lome, Togo, October 7-12, 2002. 10 pp.

Sanginga, N. and Woomer, P.L. 2009. Integrated soil fertility management in Africa: Principles, practices and developmental processes. http://webapp.ciat.cgiar.org/tsbf_institute/pdf/t sbf_isfm_book09_contents.pdf visited 30 January, 2012.

Stocking, M. and Murnaghan, N. 2000. Land degradation: Guidelines for field assessment. Overseas Development Group, University of East Anglia, Norwich, UK. 120 pp.

Tanzania Coffee Board (TCB) 2009. Way forward in the Tanzanian coffee sector. Proceedings of the first Coffee Stakeholders' Conference, Arusha, Tanzania, 30 November – 1 December, 2009. 39 pp

Tiamiyu, S.A., Akintola, J.O and Rahji, M.A.Y 2009. Technology adoption and productivity difference among growers of new rice for Africa in savanna zone of Nigeria. Tropicultura. 27: 193-197.

Truong, T.N.C. and Yamada, R. 2002. Factors affecting farmers' adoption of technologies in farming system: A case study in O'Mon district, Can Tho province, Mekong Delta. Omorice 10: 94-100.

United Republic of Tanzania (URT) 2007. The National Economic Situation. Ministry of Planning, Economy and Empowerment. Topcom Printers, Dar es Salaam, Tanzania. 217pp.

Vanlauwe, B. and Giller, K.E. 2006. Popular myths around soil fertility management in Sub-Saharan Africa. Agriculture, Ecosystems and Environment 116: 34-46.

Vanlauwe, B. and Zingore, S. 2011. Integrated soil fertility management: An operational definition and consequences for implementation and dissemination. Better Crops, 95(3): 4-7.

Zhou, S., Herzfeld, T., Glauben, T., Zhang, Y. And Hu, B. 2008. Factors affecting Chinese farmers' decisions to adopt a water saving technology. Canadian Journal of Agricultural Economics 56: 51-61.

3

RAPD, Microsatellites Markers in the Genetic Diversity Characterization of *Beauveria bassiana* (Bals.) Vuill. Isolates

M. da C. Mendonça[1]*, M. da F. Santos[2], R. Silva-Mann[3], J. M. Ferreira[4]

[1]*Emdagro/Embrapa Tabuleiros Costeiros Av. Beira Mar, n. 3250, 13 de Julho, CEP 49025-040, Aracaju, Sergipe, Brazil. Email: marcelom@cpatc.embrapa.br*
[2]*Doutoranda do Departamento de Genética da Escola de Superior de Agricultura Luiz de Queiroz, Av. Pádua Dias, 11 CP 9, CEP 13418-900, Piracicaba, São Paulo, Brazil*
[3]*Departamento de Engenharia Agronômica da Universidade Federal de Sergipe, Av. Marechal Rondon s/n, Jardim Rosa Elze, CEP 49100-000, Aracaju, Sergipe, Brazil*
[4]*Embrapa Tabuleiros Costeiros Av. Beira Mar, n. 3250, 13 de Julho, CEP 49025-040, Aracaju, Sergipe, Brazil*
*Corresponding Author

SUMMARY

Beauveria bassiana has been used for the control of agricultural pests. However, the genetic diversity of these fungi in the State of Sergipe, Brazil has been poorly studied. The objective of this study was to quantify the genetic diversity of *B. bassiana* isolates by means of RAPD markers, microsatellites, and rDNA-ITS regions, relating their origins and different geographic hosts. Nineteen primers were used in the RAPD analysis, and four primer pairs were used as microsatellite markers. The rDNA-ITS of 18S rDNA region were evaluated using primers ITS 1 and ITS 4. The RAPD analysis showed genetic similarities from 2% to 77%. Isolates 057.99 and 053.96 had the highest similarity (77%), followed by 064.99 and 053.96, with 73% similarity. With regard to microsatellites, isolates BC.05 and 032.91 showed the highest genetic similarity (82%), followed by 064.99 and 057.99 isolates, with 80% genetic similarity. The analysis of the rDNA-ITS region showed an intense fragment with 550pb for all isolates. PCR products were digested with *EcoRI* and *HaeIII*, but no restrictions were found for such products.

Key words: *Beauveria bassiana*; biological control; entomopathogenic fungi; molecular markers.

INTRODUCTION

Beauveria is a globally distributed genus of soil-borne entomopathogenic hyphomycetes of interest as a model system for the study of entomopathogenesis and the biological control of pest insects. Species recognition in *Beauveria* is difficult due to a lack of taxonomically informative morphology. This has impeded assessment of species diversity in this genus and investigation of their natural history (Rehner & Buckley, 2005). This globally distribution associated to difficulties for recognition species support the objective of this current study in Sergipe State, due its important agriculture and the necessity to reduce the use of agrochemical products to control pests aiming the sustainability of agroecossistems.

The fungus *Beauveria bassiana* (Bals.) Vuill. has been used to control various insect species because of its great adaptation and persistence in the population of specific hosts and in the environment, producing effects, in the long run, on pest suppression (Alves, 1986).

Nowadays, many methods are available to analyze genetic diversity and to monitor entomopathogenic fungi. The methodologies can be separated based on the technique used, and vary with regard to their ability to detect differences between individuals, cost, ease of use, consistency, and reproducibility (Milach, 1998). Among these methods, RAPD and isozymes markers were used to discriminate 24 *B. bassiana* isolates collected from the lesser mealworm, *Alphitobius diaperinus* (Coleoptera: Tenebrionidae) in North Carolina and West Virginia. The RAPD marker showed better resolution in the discrimination of isolates when compared with the isozymes, producing 141 bands for the 24 isolates and separating each as a unique class. Variation was detected not only within and between isolates from different regions, but also between isolates collected from the host insect (Castrilho & Brooks, 1998).

The detailed discrimination of the source of isolates, population structure, and genetic relations has turned the microsatellite technique into a frequently used method to differentiate entomopathogenic fungi. Rehner & Buckley (2003) described the isolation and characterization of eight microsatellite loci that successfully amplify some isolates representative of the phylogenetic diversity in the *B. bassiana* complex.

In addition, another genomic portion that can be of help in genetic diversity studies is ribosomal DNA (rDNA). The rDNA unit has components in its sequence which involve variations and can be used in systematic studies for different taxonomic levels PCR-restriction fragment length polymorphism (RFLP) assay of internal transcribed spacer (ITS) rDNA sequences (Fouly et. al, 1997).

Traditionally the identification of microorganisms has been done by assessing the phenotype by morphological characteristics; however, this can lead to controversy regarding the identification of several species, and are not safe to separate individuals into species. Traditional methods for identification of studies involving fungi require infection, morphology and culture media. These methods are usually time consuming and can be inconclusive and lead to misinterpretation. The use of molecular techniques in

study of enthomopathogenic fungi has been employed. These tools have also been used in the analysis of genetic diversity and the study of intra-and interspecific relationships of different species, populations and individuals. Such studies have been performed by analysis of macromolecules through the use of molecular markers, which are based on the variation found in protein or in the sequence of nucleotides that compose the nucleic acids DNA and RNA, which has been used to facilitate understanding of the taxonomy, phylogeny and ecology of various fungi (Kao *et al.*, 2002; Gauthier *et al.*, 2007).

Therefore, the objective of this study was to evaluate the genetic variability among isolates of the fungus *B. bassiana*, collected from different hosts and regions in the State of Sergipe and other localities, by means of RAPD markers, microsatellites, and amplification of the rDNA ITS region.

MATERIAL AND METHODS

Beauveria bassiana isolates and culture conditions

We used 11 isolates of the entomopathogenic fungus *B. bassiana*, collected from different locations and naturally parasitized insect species in the field (Table 1).

The isolates were deposited in Embrapa *Tabuleiros Costeiros*' Biological Agents Bank and then transferred to liquid medium (40 mL) consisting of: sucrose - 10g L^{-1}, L-asparagine – 2g L^{-1}, yeast extract – 2g L^{-1}, KH_2PO_4 – 1g L^{-1}, $MgSO_4.7H_2O$ – 0.1g L^{-1}, $ZnSO_4.7H_2O$ – 0.44mg L^{-1}, $FeCl_3.6H_2O$ – 0.48mg L^{-1}, $MnCl_2.H_2O$ – 0.36mg L^{-1} (Alfenas, 1998). The fungi were maintained in this medium at 25°C±1°C for 7 days, on a stirring table at 80 *g* in the dark. After that period, the mycelia were vacuum-filtered through a Büchner funnel, stored at -21°C in freezer and then freeze-dried (Benchtop device, Virtis, NY, USA).

DNA extraction

For genomic DNA extraction, 700 µL of a solution consisting of 549 µL sterile water, 300 µL 10% SDS, 100 µL 0.5M EDTA pH 8, and 50 µL 1M Tris-HCl pH 8, and 2-mercaptoethanol/sample were added to the mycelium macerated in liquid nitrogen, which was maintained at 65°C for 60 min.

The mixture was centrifuged at 12,000 *g* for 10 min and the supernatant was collected. Six-hundred µL of a saturated phenol and chloroform mixture (1:1) were added, followed by centrifugation at the same rotation and for the same time. The supernatant was transferred to a microtube, and the same volume of chloroform and isoamyl alcohol (1:1) was added. The mixture was centrifuged and the supernatant was removed; 1/10 of its volume of 3M sodium acetate and 2/3 of ice-cold isopropanol (-20°C) were added.

After resting for 60 min at -20°C, the mixture was centrifuged and the precipitate thus obtained was washed twice with 200 µL 70% ethanol. The precipitate was resuspended with 100 µL TE (Tris-EDTA buffer) containing RNAase (10 ng µL^{-1}) and incubated at 65°C for 20 min. Two new precipitations and washes were performed, and the precipitate was resuspended with 50 µL TE. All amplification reactions were performed in a thermocycler (Biômetra, complete description).

Amplification of RAPD markers

The amplification reactions had a 25 µL volume [14.3 µL sterile nuclease-free water, 3 µL 10X buffer, 0.6 µL d-NTP's (10mM), 0.9 µL MgCl$_2$ (50mM), 0.2 µL taq polymerase (2U), 3 µL primer, and 3 µL DNA]. Twenty arbitrary sequence decamer primers (IDT 1 to 10) were tested.

Amplification consisted of an initial temperature of 94°C/5 min followed by 45 amplification cycles at 94°C/1 min; 36°C/2 min; and 72°C/1 min.

The amplification products were separated on 0.8% agarose gel under electrophoresis in TBE buffer 0.5X (90mM Tris-borate, 1mM EDTA, pH 8.0) at 100V for 90 minutes, stained with ethidium bromide (0.5 µg mL) and visualized under UV light.

Microsatellite markers in *Beauveria bassiana* **diversity analysis**

The DNA extracted from *B. bassiana* isolates was submitted to PCR reactions using four pairs of microsatellite marker-based primers (Rehner & Buckley, 2003).

The amplification reactions were conducted in a 25µL volume [3µL buffer 10X, 0.9µL MgCl$_2$ (50mM), 0.6µL d-NTP's (10mM), 0.4µL taq polymerase (2U), 18.1µL ultrapure water, 0.5µL of each primer (sense and antisense at 0.2 µM), and 1 µL (30ng) DNA] in a thermocycler programmed for 94°C/5 min, followed by 30 cycles at 94°C/1 min, 50°C/1 min, 72°C for 2 min, and finally 72°C/7 min. The products were separated in 1.5% agarose gel under electrophoresis in TBE buffer 1X, and then visualized under UV light.

Amplification of rDNA ITS regions

The primers used were synthesized by GIBCO (Life Technologies) (ITS1- TCC GTA GGT GAA CCT GCG G and ITS4- TCC TCC GCT TAT TGA TAT GC) (White *et al.*, 1990), and suspended in sterile ultrapure water to a concentration of 200 µM. The working solutions were used at concentrations from 20 to 30 µM.

The reactions were conducted in 25 µL volumes [16.4 µL sterile ultrapure water; 2.5 µL PCR buffer 10x; 2.5 µL MgCl$_2$; 0.5 µL d-NTP's; 0.5 µL of each primer; and 0.1 µL Taq DNA polymerase per reaction, and 3 µL DNA]. The amplification program consisted of initial denaturation at 95°C/3 min; followed by 32 cycles at 94°C/1 min, 57°C/1 min, and 72°C/1 min; and one cycle at 72°C/3 min.

The PCR products were separate on a 1.5% agarose gel under electrophoresis at 1.5%/TBE 1X, stained with ethidium bromide, photographed using UV light source.

Table 1 - Monospore *Beauveria bassiana* (Bals.) Vuill. cultures from the Biological Agents Bank at Embrapa *Tabuleiros Costeiros*.

Isolates	Isolation location	Host species/Family
BC.05	Boquim/SE – C	*Cratosomus flavofasciatus*/Curculionidae
027.94	São Cristóvão/SE – C	*Rhinostomus barbirostris*/Curculionidae
032.91	Aracaju/SE – C	*Rhynchophorus palmarum*/Curculionidae
052.96	Betume1/SE – C	*Homalinotus coriaceus*/Curculionidae
053.96	Moju/PA – I	*Brassolis sophorae*/Nymphalidae
057.99	Betume2/SE – C	*Homalinotus coriaceus*/Curculionidae
058.99	Saquarema/RJ – I	*Homalinotus coriaceus*/Curculionidae
059.99	Malhador/SE – C	*Rhynchophorus palmarum*/Curculionidae
062.00	Neópolis/SE – C	*Coraliomela brunnea*/Chrysomelidae
064.99	Egypt – I	*Rhynchophorus ferruginus*/Curculionidae
065.03	Umbaúba/SE – C	*Rhinostomus barbirostris*/Curculionidae

C – Collected; I – Introduced
ITS-RFLP

The ITS region product amplified by PCR was digested with *HaeIII* and *EcoRI* endonucleases (5μL), for 4 hours at 37°C. Electrophoresis was run from the digested product at 80 V and 80mA for 3 hours, in 1.5% agarose gel, using TBE buffer 1X, and then stained with ethidium bromide (0.5 μg mL^{-1}) and visualized under UV light.

Data analysis

When evaluating the gels, the presence (1) or absence (0) of bands were used to construct matrices. The genetic similarity estimates (Sgij) between each pair of genotypes were obtained using Jaccard's coefficient by means of the expression: Sgij = a/a+b+c.

The variables for the expressions were obtained according to the following scheme:

		Isolate i	
		1	0
Isolate j	1	a (1, 1)	b (1, 0)
	0	c (0, 1)	d (0,0)

Similarities were obtained using the statistical package NTSYS-pc version 2.1, while clusters were obtained by the UPGMA method (Rohlf, 2000). The errors associated with the similarities were estimated according to Skroch *et al.* (1992), by the following expressions:

V= ns (1-s) (n-1), estimated standard error (V/n)$^{1/2}$

where:
V= variance for the genetic similarity between each pair of isolates;
s= genetic similarity between each pair of isolates;
n= total number of bands used to estimate similarities.

RESULTS AND DISCUSSION

In the RAPD analysis, 38 polymorphic bands and 14 monomorphic bands were obtained, some unique for certain isolates. These were used to calculate genetic similarity values, which ranged from 2% to 77%.

The isolates showed a mean similarity of 38.6%, forming distinct groups on the dendrogram, even among those isolates that came from common hosts; isolate 65.03 was the most divergent as compared to the others (Figure 1).

Based on the genetic similarities between the analyzed genotypes (Table 2), isolates 57.99 (Betume2-SE) and 53.96 (Moju-PA) showed the highest similarity (77%),

followed by 64.99 (Egypt) and 53.96 (Moju-PA), with 73% similarity.

Ferreira *et al.*, (2001) conducted bioassays comparing the pathogenicity of isolates 57.99 and 53.96 on *B. sophorae*, the coconut leaf caterpillar. The virulence shown during control of this insect was similar, around 92.5%, corroborating the RAPD test results in which the highest genetic similarity value was observed (77%). These isolates also showed potential for the biological control of other coconut pests, such as: *H. coriaceus* - black coconut bunch weevil (87%), *R. palmarum* - American palm weevil (91.5%), and *R. barbirostris* - bottle brush weevil (95.3%).

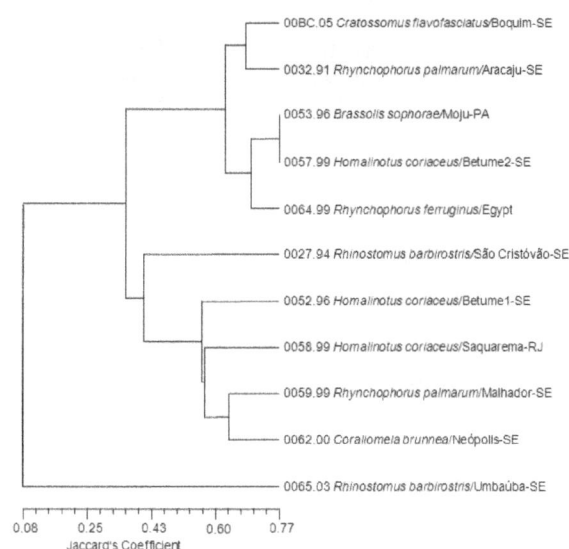

Figure 1. Dendrogram of similarities between 11 *Beauveria bassiana* isolates for RAPD analysis, based on Jaccard's similarity coefficient (UPGMA method).

Isolates 57.99 and 58.99, obtained from the same host (*H. coriaceus*), showed 30% genetic similarity. A similar result was observed between isolate pairs 32.91 and 58.99, and 32.91 and 59.99, obtained from *R. palmarum*, with a mean genetic similarity of 36.5% (Table 3). Among these isolates, 32.91 has caused up to 100% mortality in *R. palmarum* adults in laboratory tests, followed by isolates 59.99, 57.99, and 58.99 (Ferreira *et al*, 2001).

The isolates from Sergipe with the highest genetic similarity were BC.05 (Boquim-SE) and 32.91 (Aracaju-SE) (68%), which were obtained from insect pests on different crops (*Citrus* and *Cocus nucifera*), but belonged to the same family (Curculionidae). Conversely, those with the lowest similarity were 65.03 (Umbaúba-SE) and 59.99 (Malhador-SE), both isolated from a coconut pest.

Isolate BC.05 was collected in 2005. So far, no evaluation tests on its pathogenicity have been carried out. Nevertheless, due to its close genetic proximity with isolate 32.91, it has a potential for the control of *R. palmarum*, since it was obtained from an insect - *C. flavofasciatus* - in the same family. Similarly, it can be suggested that isolate 32.91 has a potential to control *C. flavofasciatus*.

The isolates with the smallest similarity values were 65.03 and 59.99 (2%). These isolates were obtained from distinct edaphic-climatic regions: one was from Umbaúba (latitude 11°23'00" south/longitude 37°39'28" west), at an elevation of 130 meters, and the other was from Malhador (latitude 10°39'28" south/longitude 37°18'17" west), at an elevation of 251 meters, with variable rainfall regimes and mean temperatures in those regions (SEPLANTEC, 2006). These factors may act as selection pressure for the occurrence of mutations between isolates.

The microsatellite analysis for *B. bassiana* generated 41 bands, all of which were polymorphic; the electrophoretic migration distances between

amplification products were different from the populations described by McGuire *et al.* (2005), possibly due to their various origins and hosts of the isolates used in the analysis.

A total of 20 amplified bands were observed for primer sequence Ba01; the highest polymorphism was observed for primer Ba05, with only two amplified bands. The length of the amplified fragments varied from 100 to 2,000 bp.

Rehner and Buckley (2003) evaluated a wide geographic population of *B. bassiana* and observed that the size of the fragments amplified by the Ba05 sequence varied from 110 to 175 bp.

The highest genetic similarity values observed among the isolates analyzed (Table 3) were 82% for pairs BC.05 (Boquim - SE) and 32.91 (Aracaju - SE), and 80% for 64.99 (Egypt) and 57.99 (Betume2-SE). For the first pair, the microsatellites confirmed the RAPD analysis results, where a higher genetic similarity value was observed for these isolates, both from Sergipe (Figure 2).

Table 2. Genetic similarities (%) (located below the diagonal line) and standard error associated with similarity (above the diagonal line) between *Beauveria bassiana* isolates based on Jaccard's coefficient for RAPD markers.

Isolates	BC05	27.94	32.91	52.96	53.96	57.99	58.99	59.99	62.00	64.99	65.03
BC05		0.003	0.003	0.003	0.003	0.003	0.003	0.003	0.003	0.003	0.001
27.94	44		0.003	0.003	0.003	0.003	0.003	0.003	0.003	0.002	0.001
32.91	68	37		0.003	0.003	0.002	0.003	0.003	0.003	0.003	0.001
52.96	43	47	46		0.003	0.003	0.003	0.003	0.003	0.003	0.001
53.96	55	33	68	44		0.002	0.003	0.003	0.003	0.002	0.001
57.99	64	34	69	37	77		0.002	0.003	0.003	0.003	0.001
58.99	33	43	35	55	35	30		0.003	0.003	0.002	0.0004
59.99	32	35	38	56	38	32	58		0.003	0.003	0.0002
62.00	36	37	36	57	36	32	55	64		0.002	0.0006
64.99	51	30	67	35	73	66	29	32	30		0.001
65.03	11	9	8	9	10	10	4	2	5	8	

Table 3. Genetic similarities (%) (below the diagonal line) and standard error associated with similarity (above the diagonal line) between *Beauveria bassiana* isolates using microsatellites, based on Jaccard's coefficient.

Isolates	BC05	27.94	32.91	52.96	53.96	57.99	58.99	59.99	62.00	64.99	65.03
BC05		0.003	0.002	0.003	0.003	0.003	0.003	0.003	0.003	0.003	0.000
27.94	54		0.003	0.003	0.002	0.003	0.003	0.003	0.003	0.003	0.001
32.91	82	48		0.003	0.003	0.002	0.002	0.003	0.003	0.002	0.000
52.96	59	59	45		0.002	0.003	0.003	0.003	0.003	0.003	0.0004
53.96	32	26	37	20		0.003	0.001	0.002	0.002	0.003	0.000
57.99	64	44	74	42	63		0.002	0.003	0.003	0.002	0.000
58.99	37	46	31	50	12	26		0.003	0.003	0.003	0.001
59.99	56	56	50	61	19	41	53		0.003	0.003	0.001
62.00	50	50	43	62	17	35	54	67		0.003	0.0004
64.99	62	36	72	40	45	80	32	44	37		0.000
65.03	0	9	0	4	0	0	14	8	4	0	

Enkerli *et al.* (2004) used microsatellite markers for the genetic characterization of *B. brongniartii* isolates applied during 14 years for the biological control of *Melolontha melolontha* (Coleoptera: Scarabaeidae), and demonstrated that after that period some isolates persisted in the environment and new populations were derived from them, coexisting in the same habitat. This fact has probably contributed for other factors to act as selection pressure over this species.

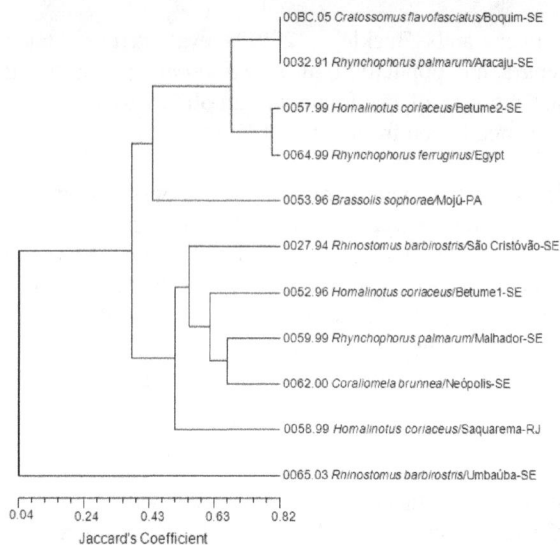

Figure 2. Dendrogram of similarities between 11 *Beauveria bassiana* isolates for microsatellites, based on Jaccard's similarity coefficient (UPGMA method).

A similar result as obtained in the RAPD analysis was observed in the microsatellite analysis for isolate 65.03, with no genetic similarity with isolates BC.05, 32.91, 53.96, 57.99, and 64.99. This fact validates the information obtained with multiple molecular markers in studies on populations of entomopathogenic fungi.

Figure 3 confronts the genetic diversity data for *Beauveria* utilizing isolates using RAPD molecular markers and microsatellites in the consensus analysis. It can be seen that two great groups were formed, with approximately 38% genetic divergence. One group includes isolates BC.05, 32.91, 53.96, 57.99, and 64.99, while the other group includes isolates 27.94, 52.96, 58.99, 59.99, and 62.00. Isolate 65.03 formed a third independent branch, with 80% divergence, when compared with the others.

In order to elucidate the taxonomic position of isolate 65.03 in relation to the others with regard to the genetic proximities as evaluated by RAPD and microsatellites, a rDNA ITS region analysis was conducted, since this region is conserved within a species but is variable between species. A band of approximately 550 bp was observed for all isolates, in addition to the occurrence of a few unspecific bands with poor resolution.

Since no variations were observed in the amplification of the ITS regions, the amplification products were restricted for the various isolates. After restriction and visualization of the products, it was verified that no results occurred that would allow their differentiation.

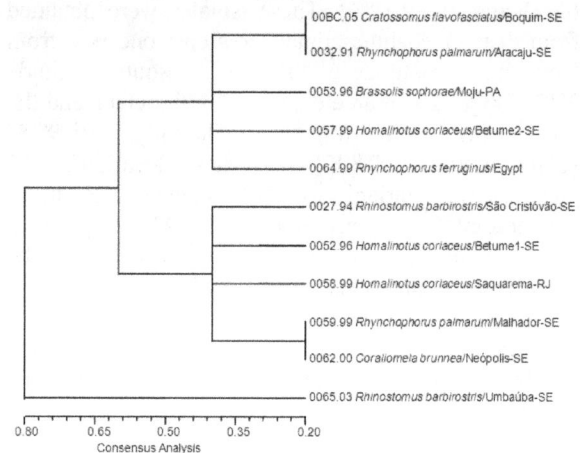

Figure 3. Dendrogram of consensus cluster analysis between 11 *Beauveria bassiana* isolates for RAPD and microsatellites, based on Jaccard's coefficient (Strict method).

On the other hand in a study of *Metarhizium anisopliae var. anisopliae* isolates the ITS regions resulted in a single PCR-amplification product for all isolates, while digestion of the ITS1 and ITS2 PCR-amplification products with several restriction enzymes detected several molecular differences. The conservative nature of our ITS fingerprinting analysis meant that it was very difficult to discriminate between isolates because many of the isolates analyzed shared the same genotype (Becerra Velásquez *et al.*, 2007).

The formation of two great groups was observed in the dendrogram. Group 1 consisted of isolates BC.05, 32.91, 64.99, 27.94, 58.99, 62.00, and 65.03, while group 2 consisted of isolates 52.96, 53.96, 57.99, and 59.99.

In evaluating isolates from the same genus, *Rhynchophorus*, isolates 32.91 (*R. palmarum*) and 64.99 (*R. ferruginus*) showed 50% genetic similarity. However, considering the intraspecific values for this genus, a low similarity (17%) was observed in isolates 32.91 and 59.99 (*R. palmarum*). A similar fact was found in the species *H. coriaceus*, for which a low intraspecific genetic similarity was observed in pair 52.96 and 58.99 (25%) (Figure 3).

The populations of isolates from the State of Sergipe showed genetic diversity. Knowledge about the distribution and magnitude of intra- and interpopulation genetic variation is of great importance to understand the fungus biology, which could be used to make inferences about the potential impact of different ecological factors that influence the evolution process of the species (Wang, *et al.*, 2003).

Wang *et al.* (2003) worked with microsatellite markers and observed that the genetic relation between *B. bassiana* isolates was more associated with geographic location than with insect host species, suggesting that the location factor constitutes selection pressure. In this work, no direct relation was observed between similarity data, geographic location, and host, as previously suggested.

Wang *et al.* (2004) used polymerase chain reaction-restriction fragment length polymorphism PCR-RFLP of the *pr1* gene, microsatellite markers and 28S rDNA haplotyping detecting the presence or absence of group I introns in the population study of the entomopathogenic fungus, *Beauveria bassiana*. The findings showed that the average genetic diversity index of geographical populations was significantly smaller than that of populations derived from insect host orders, indicating that the genetic relatedness of *B. bassiana* strains was highly associated with geographical locality rather than insect host species.

In the Sergipe conditions, successive applications of *B. bassiana* have been used for the biological control of coconut and citrus pests. Under favorable conditions, the fungus persists in the environment through the generations of insects present, and may undergo the action of different management practices and environmental conditions which, throughout the years, would imply selection pressures that might generate new genetic types.

Thirty-eight strains of the entomopathogenic *B. bassiana*, isolated from diverse species of Lepidoptera (Pyralidae) or Coleoptera (Curculionidae, Chrysomelidae, and Scolytidae) from various geographical sites, were examined by RFLP and RAPD analysis, showing clustering genus host specifity (Mauer *et al.*, 1997), which differs from those results reported in this study.

Using *Paecilomyces farinosus* genus a study was determined by comparison of the products of polymerase chain reaction amplification of anonymous regions of genomic DNA with single arbitrary sequence oligonucleotide primers (RAPD analysis). Isolates were collected from seven insect species in eastern Canada and they differed greatly in cultural and morphological phenotype. All *P. farinosus* isolates were clearly distinguished from three other entomopathogenic fungi, including *P. fumosoroseus*. However, RAPD banding patterns did not, correlate with ecological backgrounds or morphological phenotypes of *P. farinosus* isolates. These observations support the conclusion that *P. farinosus* from eastern Canada is not composed of strains which can be separated on the basis of the ecological or morphological criteria selected (Chew *et al.*, 1998).

New perspectives do come up and may lead to a necessity for researches focused on the specific identification of the most divergent isolates of this pathogen for various crops. Such identification may suggest a new taxonomic position, and the identification of strains of these *B. bassiana* isolates which, after characterization, would allow the identification of markers associated with pathogenicity, contributing toward the selection of more virulent types in a more rapid and precise manner, associated with the evaluation via inoculation and later determination of pest mortality.

CONCLUSIONS

Beauveria bassiana isolates showed a mean similarity of 38.6% for RAPD analysis. The isolates 57.99 (Betume2-SE) and 53.96 (Moju-PA) showed the highest similarity (77%) by RAPD analysis. By microsatellite analysis, the isolates BC.05 (Boquim-SE) and 32.91 (Aracaju-SE) showed the highest genetic similarity (82%). The ITS products, obtained for the different isolates of *Beauveria bassiana*, confirmed higher homology of the sequences for the analyzed isolates.

REFERENCES

Alfenas, A. C. 1998. Eletroforese de isoenzimas e proteínas afins. Viçosa, MG: Universidade Federal de Viçosa.

Alves, S. B. 1986. Controle microbiano de insetos. São Paulo: Manole.

Becerra Velásquez, V.; Paredes Cárcamo, M.; Rojo Meriño, C.; France Iglesias, A.; Franco Durán, J. 2007. Intraspecific differentiation of Chilean isolates of the entomopathogenic fungi *Metarhizium anisopliae* var. *anisopliae* as revealed by RAPD, SSR and ITS markers *Genetics and Molecular Biology*. 30 (1):89-99.

Castrillo, L. A.; Brooks, W. M. 1998. Differentiation of *Beauveria bassiana* isolates from the Darkling Beetle, *Alphitobius diaperinus*, using isozyme and RAPD analysis. Journal of Invertebrate Pathology, North Carolina. 72:190-196.

Chew, J. S. K.; Strongman, D. B.; Mackay, R. M. 1998. Comparisons of twenty isolates of the entomopathogen *Paecilomyces farinosus* by analysis of RAPD markers. Mycological Research. 102:1254-1258.

Enkerli, J.; Widmer, F.; Keller, S. 2004. Long-term field persistence of *Beauveria brongniartti* strains applied as biocontrol agents against European cockchafer larvae in Switzerland. Biological Control, San Diego, CA. 29:115-123.

Ferreira, J. M. S.; Araújo, R. P. C.; Saro, F. B. 2001. Perspectivas para o uso de fungos entomopatogênicos no controle microbiano das pragas do coqueiro. Aracaju: Embrapa Tabuleiros Costeiros. (Embrapa Tabuleiros Costeiros. Circular Técnica. 26).

Fouly, H.; Wilkinson, H. T.; Chen, W. 1997. Restriction analysis of internal transcribed spacers and the small subunit gene of ribosomal DNA among four *Gaeumannomyces* species. Mycologia, New York. 89:590-597.

Gauthier, N.; Dalleau-Clouet, C.; Fargues, J.; Bom, M.-C. Microsatellite variability in the entomopathogenic fungus *Paecilomyces fumosoroseus*: genetic diversity and population structure. Mycologia, 99(5), 2007, pp. 693–704.

Kao, S.-S.; Tasai, Y.-S.; Yang, P.-S.; Hung, T.-H. 2002. Use of Randon Amplified Polymorphic DNA to characterize Entomopathogenic Fungi, *Nomuraea* spp., *Beauveria* spp., and *Metarhizium anisopliae* var. *anisopliae* from Taiwan and China. Formasan Entomology, 22:125-134.

Maurer , P.; Couteaudier, Y.; Girard, P. A.; Bridge, P. D.; Riba G. 1997. Genetic diversity of *Beauveria bassiana* and relatedness to host insect range. Mycological Research. 101:159-164.

Mcguire, M. R.; Ulloa, M.; Park, Y. H.; Hudson, N. 2005. Biological and molecular characteristics of *Beauveria bassiana* isolates from California *Lygus Hesperus* (Hemiptera: Miridae) populations. Biological Control, San Diego. 33:307-314.

Milach, S. C. K. 1998. Marcadores moleculares em plantas. Porto Alegre: UFRGS.

Pu, S.-C., Chen, M.-J., Ma, Z.-Y., Xie, L., LI, Z.-Z., Huang, B. 2010. Genotyping isolates of the entomopathogenic fungus *Beauveria bassiana* *sensu lato* by multi-locus polymerase chain reaction-denaturing gradient gel electrophoresis (PCR-DGGE) analysis. African Journal of Biotechnology. 9(27): 4290-4294.

Rehner, S. A.; Buckley, E. P. 2003. Isolation and characterization of microsatellite loci from the entomopathogenic fungus *Beauveria bassiana* (Ascomycota: Hypocreales). Molecular Ecology Notes, Oxford. 409-411.

Rehner, S. A.; Buckley, E. P. 2005. A *Beauveria* phylogeny inferred from nuclear ITS and EF1-a sequences: evidence for cryptic diversification and links to *Cordyceps* teleomorphs. Mycologia. 97(1): 84–98.

Rohlf, F. J. 2000. Numeral taxonomy and multivariate analysis system – version 2.10. New York: [s.n.].

Seplantec - SUPERINTENDENCIA DE RECURSOS HIDRICOS DO ESTADO DE SERGIPE. 2006. Análise sazonal. Disponível em: <http://www.seplantec-srh.se.gov.br> Access: 12 maio.

Skroch, P. W.; Tivang, J.; Nienhuis, J. 1992. Analysis of genetic relationship using RAPD marker data. In: INTERNATIONAL UNION OF FORESTRY RESEARCH ORGANIZATIONS. Proceedings of IUFRO International Conference, Cali: Breeding tropical trees.26-30 (Section 202-08).

Wang, C., Fan, M., LI, Z., Butt, T.M. 2004a. Molecular monitoring and evaluation of the application of the insect-pathogenic fungus *Beauveria bassiana* in southeast China. Journal of Applied Microbiology. 96(4):861–870.

Wang, C.; Shah, F. A.; Patel, N.; LI, Z.; Butt, T. M. 2003b. Molecular investigation on strain genetic relatedness and population structure of *Beauveria bassiana*. Environmental Microbiology, Oxford. 5 (10):908-915.

White, T.J.;Bruns, T.;Lee, S.; Taylor, J. 1990. Amplification and direct sequencing of fungal ribosomal RNA genes for phylogenetics. In: INNIS, M.A., GELFAND, D.H., SNINSKY, J.J. AND WHITE T.J. (Eds.) PCR Protocols - A Guide to Methods and Applications. Academic Press, London. 315 – 322.

4

Neuroendocrine Effects of Insulin, IGF-I and Leptin on the Secretion of the Gonadotropin-Releasing Hormone (GnRH)

Adrian Guzmán Sánchez[1], Ana María Rosales-Torres[1*]
and Carlos G. Gutiérrez Aguilar[2]

[1] *Departamento de producción Agrícola y Animal Universidad Autónoma Metropolitana Xochimilco. Calzada del Hueso 1100, 04960. México Distrito Federal*

[2] *Departamento de Reproducción Facultad de Medicina Veterinaria y Zootecnia. Universidad Nacional Autónoma de México. Ciudad Universitaria 04510 México Distrito Federal*

E-mails: alexanderguz_san@yahoo.com.mx; anamaria@correo.xoc.uam.mx; ggcarlos@servidor.unam.mx
* Corresponding author

SUMMARY

Animal energy balance greatly determines his reproductive success. In the majority of mammals, under a negative energy balance, there is a decrease in the synthesis of the gonadotropin-releasing hormone (GnRH) which reduces the activity of the hypothalamic-pituitary-gonadal axis. When the energy balance is improved, the hypothalamus reacts to this change and reestablishes the secretion of GnRH. Insulin, the insulin growth factor I (IGF-I) and leptin seem to be the main messengers that signal the hypothalamus on the animal energy balance. It has been seen that the peripheral concentrations of these hormones under positive or negative energy states are associated with changes in GnRH secretion. This review shows how IGF-I acts directly on GnRH neurons affecting its synthesis and secretion, whereas that insulin and leptin act on neurons in the arcuate nucleus (ARC), which synapses with GnRH neurons in the medial preoptic zone. Both insulin and leptin decrease the expression of neuropeptide Y (NPY) and therefore the negative effect of NPY on GnRH secretion. On the other hand insulin and leptin stimulating the synthesis of galanin-like peptide (GALP) and proopiomelanocortin (POMC). Both GALP, as well as the POMC metabolites (mainly the melanocyte stimulating hormone) increase the synthesis of GnRH. Finally, leptin increases the expression of kisspeptin in ARC neurons. Kisspeptin has a positive effect on the synthesis and secretion of GnRH.

Key words: GnRH secretion, insulin, IGF-I, leptin, neurotransmitters, nutritional stress

INTRODUCTION

Energy and/or energy reserves control the reproductive function in several species. In humans, anorexia, cachexia and excessive exercise block the reproductive cycle (De Souza et al., 1998). In rodents, overnight fasting, show a decrease in the secretion of the luteinizing hormone (LH; Gamba and Pralong, 2006). In beef and dairy cows, low body condition score (BCS) at calving prolong postpartum anestrus compared to cows in good BCS (Crowe, 2008). Heifers in nutritional anestrus fed a high energy diet (16.2 Mcal EM) resume ovulatory cycles in a shorter time span (57 days) than heifers on a moderate energy diet (10.2 Mcal EM) (80 days) (Bossis et al., 2000).

The loss of the reproductive function under negative energy balance is associated to a decrease in the synthesis, concentration and secretion of GnRH, and therefore reduction of LH which blocks follicular maturation, ovulation and the reproductive cycle itself (Gamba and Pralong, 2006; Crowe, 2008; Hill et al., 2008). The concentrations of LH during anestrus induced by food restriction in cows and heifers were low in comparison to animals that were fed a maintenance diet during anestrus. In both cases, re-feeding increased animal body weight, BCS and reestablished LH concentrations (Richards et al., 1989a; Bossis et al., 2000). It has been reported that feed restrictions, and extreme exercise in rodents, sheep and monkeys causes a shutdown of GnRH pulses within minutes to hours while reverting (usually between an hour or two) when an adequate amount of energy is available (Hill et al., 2008).

Although it is clear that the availability of energy reserves control the reproductive function regulating the secretion of GnRH, the mechanisms by which the hypothalamus monitors the animal energy status and the regulation of the neurotransmitters involved in signaling have not been well established.

The main molecules involved in transmitting the message to the hypothalamus on the animal energy status are insulin, IGF-I and leptin. The objective of this review is to gather information on the regulation of GnRH synthesis and secretion by insulin, IGF-I abd leptin under nutritional stress. In the first part of the review, a brief description is made on the hypothalamic nuclei and the neurotransmitters involved in the secretion of GnRH. Next an analysis of the mechanisms of action of insulin, IGF-I and leptin, as well as the role these molecules on the regulation of the function of GnRH neurons under conditions of nutritional stress.

Hypothalamic nuclei that control GnRH secretion

The hypothalamus is located at the level of the third ventricle under the thalamus and above the hypophysis. Frontal and limited by the optic chiasma and caudal by the mammillary bodies (Daniel, 1976; Saleem et al., 2007). Anatomically and functionally, the hypothalamus is divided into the anterior, intermediate and posterior lobes, as well as a medial and lateral area where different hypothalamic nuclei are found (Table 1; Loes et al., 1991; Sahar et al., 2007). The best characterized nuclei are: the supraoptic nucleus (SON) and the paraventricular nucleus (PVN) almost totally formed by long neurons that mainly produce oxytocin and vasopressin (Daniel, 1976). The medial preoptic nucleus (MPO) and the ARC are important hypothalamic structures involved in the control of GnRH secretion. The MPO is located in the preoptic area and within are the cellular bodies of GnRH neurons (Schneider, 2004). In the ARC, there are three main neuronal populations, those POMC, NPY (Coll et al., 2007) and GALP (Crown et al., 2007). A fourth population of neurons in this nucleus are those producing kisspeptin (Goodman et al., 2007; Gottsch et al., 2004a), although in the anteroventral periventricular nucleus, there is a large amount of kiss-1 neurons (Gottsch et al., 2004a). It has been shown that 50-75% of the axonic projections of the GnRH neurons in the MPO in rodents are directed towards the median eminence (ME; Smith and Jennes, 2001) where GnRH is released into the perivascular space of the fenestrated capillaries that form part of the primary plexus of the hypophysiary portal system (Jennes and Conn, 1994). The axons of the neuronal populations in ARC are directly or indirectly connected to the GnRH neurons. This may serve as a path for regulating the metabolism of this neuropeptide (Crown et al., 2007; Hill et al., 2008; Xu et al., 2009a).

Neurotransmitters involved in the secretion of GnRH

Although there is a large amount of neurotransmitters involved in the control of GnRH secretion only four neurotransmitters relate with the energetic state of the animal and with the concentrations of the metabolic hormones, insulin, IGF-I and leptin.

Neuropeptide Y (NPY)

The NPY is a 36 amino acid peptide widely distributed in the central nervous system, both in humans and rodents. This peptide is expressed in the hypothalamus (mainly ARC), amygdala, hippocampus, nucleus of the solitary tract and cerebral cortex. The NPY interacts with at least six receptors linked to protein G (Y1, Y2, Y3, Y4, Y5 and Y6). NPY´s functions are very diverse (Eva *et al.*, 2006). Among other functions NPY acts as the link between animal energy status and its reproductive function (Crown *et al.*, 2007). The axons of NYP neuron are in contact with the cell bodies of GnRH neuron in MPO, as well as with the axons of GnRH neurons in the median eminence (Xu *et al.*, 2009a), for which the NPY neurons could regulate both the synthesis, as well as the release of GnRH. Although NPY seems to stimulate LH in cycling animals (Eva *et al.*, 2006), in general this peptide blocks the gonadotrophic axis in situations of negative energy balance (Gamba and Pralong, 2006; Hill *et al.*, 2008). Fasting for 48 h in ovariectomized rats decreases LH concentrations and increases NPY expression even in the presence of estradiol (Kalamatianos *et al.*, 2008). It has recently been shown that the binding of NPY with its Y5 receptor, hyperpolarizes the membrane potential of GnRH neurons inhibiting the secretion of GnRH (Xu *et al.*, 2009a). In addition, the inhibition of the Y5 receptor by the use of an antagonist depolarizes the membrane of GnRH neurons (Xu *et al.*, 2009a) and eliminates the inhibitory effect that NPY exerted on LH (Raposinho *et al.*, 1999). In ob/ob infertile rats with Y4 receptor knockouts, total male fertility and partial female estral cycles are reestablished, moreover there is an increase of GnRH expression (Sainsbury *et al.*, 2008).

Propiomelancortin (POMC)

The POMC gene is actively transcribed in several tissues, including ARC neurons. In the central nervous system, POMC is enzymatically processed and gives rise to at least four small peptides: β-endorphins, and α-, β-, g-melanocyte-stimulating hormone (α-MSH, β-MSH and g-MSH, respectively; Coll *et al.*, 2004). The nerve endings of POMC in ARC project towards MPO where they release β-endorphin and α-MSH, both involved in the control of GnRH (Hill *et al.*, 2008;

Ward *et al.*, 2009). In fasting rats, LH concentrations decrease in comparison to that in well-fed rats and this decrease in LH is associated to a decrease in the release of GnRH by the hypothalamus. Under these conditions, a lesser baseline concentration of α-MSH in MPO and in the ARC-ME region in fasting rats has been seen in comparison to those well-fed animals (Watanobe, 2002). *In vitro* studies with immortalized GnRH neurons show that both α-MSH, and g-MSH increase AMPc concentration and the GnRH secretion in dose-dependent way (Stanley *et al.*, 2003). In addition, the infusion of g-MSH directly into MPO increases the plasmatic concentrations of LH in rats (Stanley *et al.*, 2003). Recent reviews by Crown *et al.* (2007) and Hill *et al.* (2008) suggest that β-endorphin inhibits the secretion of GnRH/LH. In sheep, it has been reported that β-endorphin is involved in the inhibition of the pulsatile release or surges of GnRH in an endocrine medium dominated by P_4 (Taylor *et al.*, 2007).

Galanin-like Peptide (GALP)

The GALP is a 60 amino acid neuropeptide originally isolated from the hypothalamus of pigs that is partially homologous to the galanin orexigenic neurotransmitter. Although GALP is widely distributed in the central nervous system, GALP is mainly found in ARC and the median eminence of rats (Man and Lawrence, 2008). The role of GALP as a link between the status of energy and the animal´s reproductive function has been widely reviewed (Gottsch *et al.*, 2004b; Kageyama *et al.*, 2005; Crown *et al.*, 2007). These reports have shown that GALP neurons in ARC project their axons towards the GnRH neurons in MPO and that GALP stimulates the secretion of GnRH and sexual behavior both in male, and in female rats.

Table 1. Location of hypothalamic nuclei with respect to the medial lateral and rostral caudal axes and nuclei involved in the control of GnRH*

Region	Medial Area	Lateral Area
Anterior	Medial preoptic nucleus (MPO)*	Lateral preoptic nucleus (LPN)
	Supraoptic nucleus (SON)	Lateral nucleus
	Anteroventral periventricular nucleus (AVPV)*	
	Paraventricular nucleus (PVN)	
	Anterior nucleus (AN)	
	Suprachiasmatic nucleus (SC)	
Intermediate	Dorsal-medial nucleus (DMN)	Lateral nucleus
	Ventromedial nucleus (VMN)	Tuberolateral nucleus (TLN)
	Arcuate nucleus (ARC)*	
Posterior	Mammillary Nucleus (MN)	Lateral nucleus
	Posterior Nucleus (PN)	

Modified by Saleem *et al.*, 2007, Hill *et al.*, 2008

The intracerebroventricular infusion of GALP in ovariectomized rats increases the mean concentration and pulsing frequency of LH in the presence of estrogens, suggesting that the effect of GALP on LH is estrogen-dependent (Uenoyama *et al.*, 2008). In contrast, knockout male and female rats for the GALP gene did not show any differences in LH concentrations with respect to the wild-type rats, both while fasting as well as well-fed (Dungan-Lemko *et al.*, 2008).

Kisspeptin

Kisspeptin is a ligand codified by the kiss-1 gene that acts through receptors coupled to the G protein (GPR54). Kisspeptin is the main regulator of the GnRH neurons (Kadokawa *et al.*, 2008). The kiss-1 gene is expressed by the AVPV nucleus and ARC, while the GPR54 receptor is expressed by GnRH neurons and its mutation causes hypogonadotrophism and hypogonadism in humans and rats (Hill *et al.*, 2008; Catellano *et al.*, 2009; Clarkson and Herbison, 2009). Treatment with kiss-10, in prepubertal heifers increases mean LH blood concentrations (Kadokawa *et al.*, 2008). The effect of kisspeptin is proposed as the stimulator of GnRH secretion (Roseweir and Millar, 2009) or acting directly on the gonadotrope in the hypophysis (Suzuki *et al.*, 2008).

IGF-I

IGF System and Mechanism of Action

The IGF-I is a basic polypeptide consisting of 70 amino acids (Trojan *et al.*, 2007) that together with IGF-II, two receptors of IGF, IGF binding proteins (IGFBP) and IGFBP proteases form the IGF system which regulates somatic growth, cell proliferation and apoptosis (Trojan *et al.*, 2007).

The IGF-I has important effects on the metabolism of glucose together with insulin (Sandhu *et al.*, 2002). Growth hormone (GH) is the main stimulator for the production of IGF-I in the liver, but this polypeptide can be produced by the majority of tissues in response to GH and other factors such as insulin (Le Roith, 2003; Werner *et al.*, 2008).

The biological effects of IGF-I are mediated by its interaction with its receptor I (IGFR-I), although it may also bind to the insulin receptor and the receptor related to insulin. The IGFR-I is a member of the super family of receptors linked to kinase proteins, is a tetramere formed by two α-subunits located towards the extracellular space where the IGF-I is linked and two β subunits within the cell where the kinase domain is found (Adams *et al.*, 2000). The union of the ligand with the receptor causes a conformational change of the receptor allowing the linking of ATP and the phosphorylation of the tyrosine dominion of the receptor. This phosphorylation increases the receptor kinase activity in order to phosphorylate a series of cytoplasmatic substrates that together are known as downstream signaling transductions mediators (Werner *et al.*, 2008). Among the mediators that are activated in response to the IGF-I are, the mitogen activated protein kinases (MAPK), phosphatidylinositol-3-kinase/protein kinase C (IP3K/PKC) and the blockage of the glycogen synthetase enzyme. The activation of these mediators triggers cell mitosis (via MAPK), inhibits apoptosis (since IP3K inhibits caspases, BAD and increases the expression of bcl-2). Due to the increase of the kinase activity in the cell, IGF-I can activate glucose and amino acids transporters, stimulate protein synthesis by activating transcription factors and inhibit gluconeogenesis (Le Roith, 2003; Trojan *et al.*, 2007; Werner *et al.*, 2008). The bioavailability of IGF-I is regulated by at least 6 IGF binding proteins (IGFBPs). The IGFBP3 is the protein of highest molecular weight and found predominantly in serum. The binding of this protein with IGF-I blocks proteolysis of the factor increasing its half life (Baxter, 2000). In general, the IGFBPs inhibit the metabolic and proliferative effects of IGF-I (Werner *et al.*, 2008).

IGF-I and GnRH

The onset of lactation after calving in dairy and beef cows causes negative energy balance in the animal. Schillo (1992) suggests that IGF-I concentrations may be inversely related with the duration of the postpartum anestrus. In cycling beef cows, IGF-I concentrations are greater than in cows in anestrus, while the relative abundance of IGFBP-2 in serum was less in cycling cows (Roberts *et al.*, 1997). In cattle, IGF-I serum concentrations are associated with the amount of energy reserves (Roberts *et al.*, 1997; Guzmán *et al.*, 2008). Similar results are seen in dairy cows (Kawashima *et al.*, 2007). In cows and heifers, food restriction decreases IGF-I concentrations which in turn has been associated to a decrease in LH serum concentrations (Richards *et al.*, 1991). It is known that in heifers, re-feeding increases IGF-I concentrations (León *et al.*, 2004) while shortening the time between the start of re-feeding and the first ovulation after anestrus (Bossis *et al.*, 2002). In post-partum beef cattle, the use of beta-adrenergic agonists reduces the response to an estrous induction program associated with a reduction of IGF-I serum concentration (Guzmán *et al.*, 2009b).

The presence of IGFR-I in the hypothalamus (median eminence and MPO) and hypophysis (Bach and Bondy, 1992; Daftary and Gore, 2005) suggests a direct effect of IGF-I on the secretion of GnRH and

LH. In rodents, IGF-I is required for triggering the effects of positive feedback of estradiol on LH (Etgen et al., 2006), while in the culture of neurons expressing GnRH, the addition of 10 ng/mL of IGF-I to the culture medium increases the mRNA of GnRH (Daftary and Gore, 2005), which shows that this growth factor has an important effect on the production of GnRH. On the other hand, it has been shown that IGF-I can increase the release of GnRH when stimulating the axons of GnRH neurons in the median eminence of the hypothalamus (Ojeda et al., 2008). Although IGF-I is expressed in astrocytes and GnRH neurons, the larger part of this factor in the median eminence during puberty and the estral cycles comes from the general circulation (Ojeda et al., 2008). The inactivation of IGFR-I in adult rats result in the abolition of the synaptic plasticity in the hypothalamus (Fernández- Galaz et al., 1999). In the rat, IGF-I acts on the axons of GnRH neurons in the medial eminence to stimulate a dose-dependent release of GnRH (Hiney et al., 1991). In rats close to puberty, small doses of IGF-I administered intraventricularly increase the secretion of LH. In addition, in late proestrus, the IGF-I serum levels and the expression of IGFR-I in the median eminence increase (Hiney et al., 1996). Although it is clear that IGF-I directly regulates GnRH neurons, it has recently been shown that IGF-I is capable of increasing the expression of kiss-1 in female rats before puberty (Hiney et al., 2009).

Insulin

Structure and Mechanism of Action

Insulin is a 51 amino acids peptide hormone produced as a pre-hormone by the beta cells of the pancreatic Langerhans islets. It is produced as a pre-prohormone. The pre-proinsulin is composed by an acid peptide chain (A) and a basic peptide chain (B) bound by a peptide called peptide C. Proinsulin is formed by the elimination of peptide C and the binding of two chains through a disulphide bridge. In this way, insulin is stored in secretory granules until a stimulus causes its release (Hayirli, 2006).

The insulin receptor is a tyrosine kinase receptor that is homologous to the IGF-I receptor in 85% (Corcoran et al., 2007). Binding of insulin to the receptor's α-subunit causes autophosphorylation of the kinase tyrosine domain present in the β subunits. This coincides with the internalization of the receptor which is one of the mechanisms by which insulin's action is regulated (Hayirli, 2006). The activation of the kinase tyrosine domain in the insulin receptor phosphorilates the insulin receptor substrates (mainly IRS-I and IRS-2), and the Sch adaptor proteins. After the phosphorylation of the receptor, the Grb2 protein binds to the IRS-I and to Shc activating the ras complex and the MAPK cascade. These events stimulate the cell growth and the expression of genes. The IRS-I, and in a lesser proportion the IRS-2, recruit Src (homology 2 domain-containing proteins) and the p85 regulating subunit of PI3K for activation. The PI3K phosphorilates phosphatidylinositol-diphosphate to convert it to phosphatidylinositol-triphosphate (IP3) that activates IP3-dependent kinase proteins such as PKC and protein kinase B (PKB). The PKC stimulates the translocation towards the plasmatic membrane of the glucose-4 transporter (GLUT4) to stimulate the uptake of glucose, while PKB enters the nucleus for stimulating the transcription of genes protein synthesis and phosphorylation of glycogen synthetase to promote the synthesis of glycogen. Finally, the activation of PI3K is associated to lipogenesis in response to insulin (Hayirli, 2006; Corcoran et al., 2007; Gerozissis, 2008). An increase in nutrient intake, glucagon and parasympathetic stimulation triggers the synthesis and secretion of insulin whereas fasting, hunger, exercise, galanin, somatostatin, sympathetic stimulus, IL-6 and PGF2-α can inhibit its synthesis and secretion (Hayirli, 2006).

Insulin and GnRH

Insulin levels vary throughout the day. However, the amount of circulating insulin is in direct proportion to the amount of adipose tissue (León et al., 2004; Crown et al., 2007). The central infusion of insulin in sheep stimulates the secretion of LH (Miller et al., 1995), while in sheep with an increase in food intake there is a positive correlation (0.73) between LH concentrations and insulin, however, this relationship is lost in feed restricted sheep (Miller et al., 2007). The supply of nutrients to nutritionally anestrus beef cows, increases insulin levels and the animals recover ovarian activity (Richards et al., 1989b). Similar results were reported in heifers (León et al., 2004). Sinclair et al. (2002) showed that animals with more than 8 U/L of insulin tended to have a shorter interval between calving to first estrus than animals with a lesser concentration. In dairy cows, there seem to be a similar effect (Gong et al. 2002; Gutiérrez et al., 2006).

Although evidence shows that insulin regulates reproductive function possibly by affecting the GnRH secretion, the presence of the insulin receptor in GnRH neurons has not been demonstrated in vivo (Hill et al., 2008). In vitro studies with GnRH immortalized neurons show that insulin can activate these neurons through the MAPK pathway (Kim et al., 2005; Salvi et al., 2006). However, although these results suggest a direct effect of insulin on the GnRH neurons, it is possible that this hormone regulates GnRH indirectly by acting on other hypothalamic nuclei.

As previously mentioned, ARC is an important regulator of GnRH neurons since the axons of some neurons in this nucleus synapse directly to the bodies of the GnRH neurons in the MPO. The NPY and POMC neurons (Hill et al., 2007) in ARC have receptors to insulin. In lactating rats, there is an increase in the expression of NPY mRNA and a decrease in the mRNA of POMC in ARC. However, the treatment with insulin inverts the expression pattern of these two neurotransmitters (Xu et al., 2009b). Yang et al (2010) reports that insulin regulates Ca^{+2} conductance channels activated by K+ channels in NPY neurons in ARC. The effect of insulin on these channels is via PI3K and can be the mechanism through which this hormone regulates the synthesis of NPY. In rats that did not express an insulin receptor in POMC exclusively, the metabolic and reproductive phenotype was not affected (Könner et al., 2007) suggesting that POMC does not participate in the regulation of GnRH mediated by insulin. Fasting for 48 hours in rats decreases the expression of mRNA of GALP. However, the intracerebroventricular infusion of insulin increases GALP (Fraley et al., 2004). Although insulin seems to regulate GALP, its receptor has not been reported in GALP neurons. However, GALP neurons have NPY receptors (Gottsch et al., 2004b). By which NPY reduces production of GALP the same way as NPY reduces GnRH. Therefore it is possible that that as insulin reduce NPY synthesis, it reduction removes the inhibitory effect of NPY on GALP and in this way insulin may increase GALP expression. Finally, in male diabetic rats with hypogonadism, insulin does not affect kiss-1 mRNA concentrations (Castellano et al., 2006).

Leptin

Mechanism of Action

Leptin is a hormone produced by the ob gene in adipose tissue (Zhang et al., 1994). It is a polypeptide hormone of 167 amino acids and secreted by adipose tissue in direct proportion to fat tissue content, nutritional status and tissue location (Chilliard et al., 2005). The leptin receptor (Ob-R) is a glycoprotein with a single transmembrane domain that belongs to the family of receptors linking to cytokines. There are six isoforms of the leptin receptor derived by the splicing alternative of immature mRNA. However, it is only the long isoform (Ob-Rb = 1162 amino acids) that translates the cell signal in response to leptin. The six isoforms of the leptin receptor possess the same binding domain to the ligand differing in the length of the intracellular dominion. The intracellular domain of Ob-Rb contains approximately 300 amino acids with several coupling sites for proteins essential in signal transduction. On the other hand, the intracellular domain of the short isoforms only have 30 to 40 amino

acids and do not have these coupling sites (Robertson et al., 2008).

The Ob-Rb does not have intrinsic enzymatic activity and should associate with janus kinase 2 (JAK2) in order for the signal to be translated. The Ob-Rb has two sites rich in proline (Box1 and Box2) that mediate the binding with JAK2. The binding of leptin to the receptor stimulates the autophosphorylation of JAK2 in two tyrosine residues. The phosphorylated JAK2 binds phosphate groups to Ob-Rb in three highly conserved tyrosine residues. Phosphorylated tyrosine residues 1077 and 1138 linked to signal transducers and activators of transcription (STAT3 and STAT5) to activate it by phosphorylation. The phosphorylation of the 985 tyrosine residue recruits SHP2 (SH2 domain with phosphatase 2 activity) and SOCS-3 (suppressor of cytokine signaling-3). The SHP2 activates the extracellular kinase regulators (ERK) or MAPK, while SOCS-3 blocks the translation of the signal by Ob-Rb. The Ob-Rb activates IRS 1 and 2 to phosphorylate IP3K (Zieba et al., 2008; Robertson et al., 2008).

Leptin and GnRH

Mice mutanr for the ob gene (ob/ob) or the receptor Ob-Rb (db/db) are obese and infertile but the administration of exogenous leptin in ob/ob rats reverts the phenotype but not so in db/db rats (Chilliard et al., 2005). Leptin increases the releases of GnRH by the hypothalamus (Amstalden et al., 2002). In beef heifers, the nutritional restriction decreases leptin concentrations and when the animals are re-fed, the hormone levels increase (León et al., 2004). Two-day fasted cows had decrease concentrations of insulin and leptin, and the infusion of recombinant leptin normalizes insulin concentrations and causes hyperstimulation of LH secretion (Amstalden et al., 2002). In heifers, fasting decreases concentrations of leptin, insulin and IGF-I, and LH pulsing frequency (Amstalden et al., 2000). Multiparous Brahman cows with a short anestrous postpartum (≤ 37 days) have greater leptin concentrations during the first 6 weeks postpartum than cows with long anestrus postpartum (≥ 78 days). Similarly, there is a negative correlation between anestrous postpartum and leptin concentrations before calving, at calving and postpartum (Strauch et al., 2003). The reduction of back fat by the use of a beta adrenergic receptor agonist, decreases leptin concentrations and the response to an estrous induction program in lactating beef cows (Guzmán et al., 2009a).

Although leptin is involved in GnRH regulation, the presence of its receptor in GnRH neurons has not been shown. However, Ob-Rb is expressed in ARC where leptin may reduce NPY (orexigenic) expression and increase the expression of POMC (anorexigenic) to decrease food intake, increasing energy expenditure (Friedman and Halaas, 1998; Coll *et al.,* 2007) and regulating GnRH secretion (Schneider, 2004). Just as insulin, the infusion of leptin in lactating rats decreases NPY expression and increases POMC (Xu *et al.,* 2009b). The infusion of leptin after 48 hours of fasting in rats increases the expression of mRNA for GALP (Juréus *et al.,* 2000). Similarly, in ob/ob or diabetic rats, the expression of GALP is decreased in comparison to the control and the phenomenon can be reverted by the infusion of leptin (Cunninham, 2004; Gottsch *et al.,* 2004b). Leptin seems to be the main regulator of kisspeptin under conditions of nutritional stress. It has been recently reported that leptin increases the expression of kiss-1 in ARC neurons (Hill et al, 2007). Castellano *et al.* (2009) reports that there at least three lines of evidence that show that leptin regulates kiss-1. First, thre expression of kiss-1 mRNA decreases significantly in both intact and gonadectomized leptin deficient mice (ob/ob). In both cases, the infusion of leptin reverts the reduction of kiss-1. Secondly, in diabetic rats where there is hypoleptinemia or hypoinsulinemia, the chronic infusion of leptin, but not insulin normalizes the levels of kiss-1 in the hypothalamus. Thirdly, *in vitro* studies with murine kiss-1 neurons, leptin is capable of increasing the expression of kiss-1 mRNA.

CONCLUSION

The information presented here shows that under nutritional stress, the decrease in circulating concentrations of IGF-I, insulin and leptin is associated with reduced release of GnRH. Under low energy intake or reserves, IGF-I concentrations are low and has a low impact on kiss-1 and GnRH secretion. Similarly, the reduction in insulin and leptin concentrations allows an increase in NPY and a reduction in POMC and GALP, which will inhibit GnRH secretion. In addition, the reduction in leptin decreases kiss-1 and the positive effect of kiss-1 on GnRH neurons. In contrast, when energy balance is positive, IGF-1 increases and directly stimulates the synthesis and secretion of GnRH and kiss-1. Whereas the increase in leptin increases kiss-1 and leptin and insulin decrease NPY and increase POMC, GALP to reestablish the function of the hypothalamic-pituitary-gonadal axis (Figure 1).

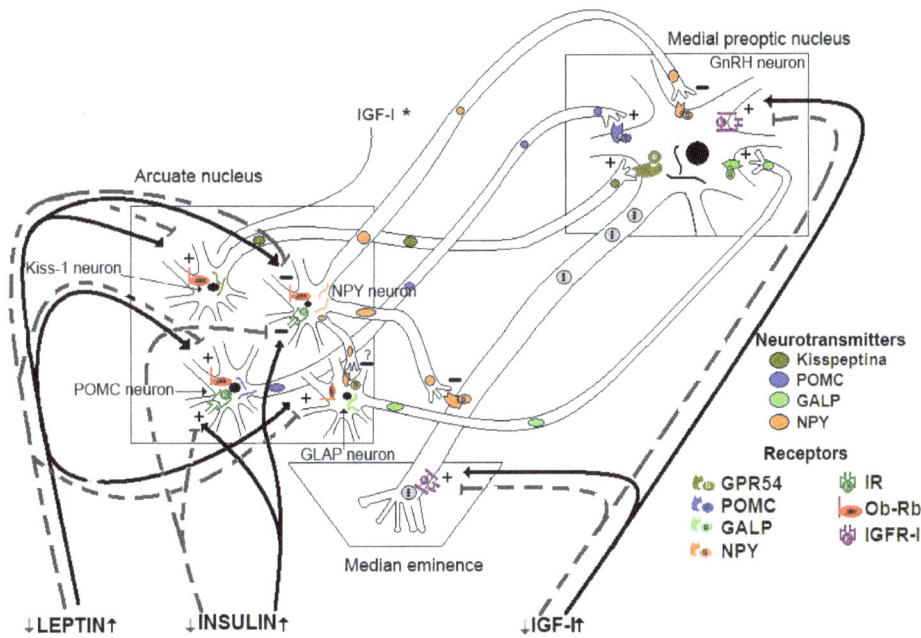

Figure 1. Neuronal integration of the pathway by which insulin, IGF-I and leptin regulate the functioning of the GnRH neurons.
+,- = Stimulating or inhibitory effect of the hormone on the synthesis or release of the neurotransmitter; ↓↑ = low or high peripheral concentrations of insulin, IGF-I and leptin;* = the IGF-I receptor has not been reported in kiss-1 neurons but IGF-I increases the synthesis of kiss-1. Dotted line = low concentrations of insulin, IGF-1 and leptin when they cannot exert their effect on neurons that have their receptor; Continuous line = when the balance of energy is positive, the concentrations of insulin, IGF-I and leptin increase and may then affect the neurons that have their receptors; POMC = proopiomelanocortin; GALP = galanin-like peptide; NPY = neuropeptide Y, GPR54= kisspeptin receptor, IR = insulin receptor, Ob-Rb = Leptin receptor, IGFR-I = insulin like growth factor receptor type I (IGF-I)

REFERENCES

Adams, T.E., Epa, V.C., Garrett, T.P., Ward, C.W. 2000. Structure and function of the type 1 insulin-like growth factor receptor. Cellular and Molecular Life Sciences. 57, 1050-1093.

Amstalden, M., Garcia ,M.R., Stanko, R.L., Nizielski, S.E., Morrison, C.D., Keisler, D.H., Williams, G.L. 2002. Central infusion of recombinant ovine leptin normalizes plasma insulin and stimulates a novel hypersecretion of luteinizing hormone after short-term fasting in mature beef cows. Biology of Reproduction. 66, 1555-1561.

Amstalden, M., Garcia, M.R., Williams, S.W., Stanko, R.L., Nizielski, S.E., Morrison, C.D., Keisler, D.H., Williams, G.L. 2000. Leptin gene expression, circulating leptin, and luteinizing hormone pulsatility are acutely responsive to short-term fasting in prepubertal heifers: relationships to circulating insulin and insulin-like growth factor I. Biology of Reproduction. 63, 127-133.

Bach, M.A., Bondy, C.A. 1992. Anatomy of the pituitary insulin-like growth factor system. Endocrinology. 131, 2588-2594.

Baxter RC. 2000. Insulin-like growth factor (IGF)-binding proteins: interactions with IGFs and intrinsic bioactivities. American Journal of Physiology Endocrinology and Metabolism. 278, 967-976.

Bossis, I., Wettemann, R.P., Welty, S.D., Vizcarra, J., Spicer, L.J. 2000. Nutritionally induced anovulation in beef Heifers: ovarian and endocrine function during realimentation and resumption of ovulation. Biology of Reproduction. 62, 1436-1444.

Castellano, J.M., Navarro, V.M., Fernández-Fernández, R., Roa, J., Vigo, E., Pineda, R., Dieguez, C., Aguilar, E., Pinilla, L., Tena-Sempere, M. 2006. Expression of hypothalamic KiSS-1 system and rescue of defective gonadotropic responses by kisspeptin in streptozotocin-induced diabetic male rats. Diabetes. 55, 2602-2610.

Castellano, J.M., Roa, J., Luque, R.M., Dieguez, C., Aguilar, E., Pinilla, L., Tena-Sempere, M. 2009. KiSS-1/kisspeptins and the metabolic control of reproduction: physiologic roles and putative physiopathological implications. Peptides. 30, 139-145.

Chilliard, Y., Delavaud, C., Bonnet, M. 2005. Leptin expression in ruminants: nutritional and physiological regulations in relation with energy metabolism. Domestic Animal Endocrinology. 29, 3-22.

Clarkson, J., Herbison, A.E. 2009. Oestrogen, kisspeptin, GPR54 and the pre-ovulatory luteinizing hormone surge. Journal of Neuroendocrinology. 21, 305-311.

Coll, A.P., Farooqi, I.S., Challis, B.G., Yeo, G.S., O'Rahill, S. 2004. Proopiomelanocortin and energy balance insights from human and murine genetics. The Journal of Clinical Endocrinology and Metabolism. 89, 2557-2562.

Coll, AP., Farooqi, I.S., O'Rahilly, S. 2007. The hormonal control of food intake. Cell. 20;129, 251-262.

Corcoran, M.P., Lamon-Fava, S., Fielding, R.A. 2007. Skeletal muscle lipid deposition and insulin resistance: effect of dietary fatty acids and exercise. The American Journal of Clinical Nutrition. 85, 662-677.

Crowe, M.A. 2008. Resumption of ovarian cyclicity in post-partum beef and dairy cows. Reproduction in Domestic Animal. 43 Supplement 5, 20-28.

Crown, A., Clifton, D.K., Steiner, R.A. 2007. Neuropeptide signaling in the integration of metabolism and reproduction. Neuroendocrinology. 86, 175-182.

Cunninham, M.J. 2004. Galanin-like peptide as a link between metabolism and reproduction. Journal of Neuroendocrinology. 16, 717-723.

Daftary, S.S., Gore, A.C. 2005. IGF-1 in the brain as a regulator of reproductive neuroendocrine function. Experimental Biology and Medicine (Maywood). 230, 292-306.

Daniel, P.M. 1976. Anatomy of the hypothalamus and pituitary gland. Journal of Clinical Pathology Supplement (Association Clinical Pathology). 7, 1-7.

De Souza, M.J., Miller, B.E., Loucks, A.B., Luciano, A.A., Pescatello, L.S, Campbell, C.G, Lasley,

B.L. 1998. High frequency of luteal phase deficiency and anovulation in recreational women runners: blunted elevation in follicle-stimulating hormone observed during luteal-follicular transition. The Journal of Clinical Endocrinology and Metabolism. 12, 4220-4232.

Dungan-Lemko, H.M., Clifton, D.K., Steiner, R.A., Fraley, G.S. 2008. Altered response to metabolic challenges in mice with genetically targeted deletions of galanin-like peptide. American Journal of Physiology Endocrinology and Metabolism. 295, E605-612.

Etgen, A.M., González-Flores, O., Todd, B.J. 2006. The role of insulin-like growth factor-I and growth factor-associated signal transduction pathways in estradiol and progesterone facilitation of female reproductive behaviors. Frontiers in Neuroendocrinology . 27, 363-375.

Eva, C., Serra, M., Mele, P., Panzica, G., Oberto, A. 2006. Physiology and gene regulation of the brain NPY Y1 receptor. Frontiers in Neuroendocrinology. 27, 308-339.

Fernandez-Galaz, M.C., Naftolin, F., Garcia-Segura, L.M. 1999. Phasic synaptic remodeling of the rat arcuate nucleus during the estrous cycle depends on insulin-like growth factor-I receptor activation. The Journal of Neuroscience. 1;55(3), 286-292.

Fraley, G.S., Scarlett, J.M., Shimada, I., Teklemichael, D.N., Acohido, B.V., Clifton, D.K., 2004. Steiner RA. Effects of diabetes and insulin on the expression of galanin-like peptide in the hypothalamus of the rat. Diabetes. 53, 1237-1242.

Friedman, J.M., Halaas, J.L. 1998. Leptin and the regulation of body weight in mammals. Nature. 22; 395, 763-770.

Gamba, M., Pralong, F.P. 2006. Control of GnRH neuronal activity by metabolic factors: the role of leptin and insulin. Molecular and Cellular Endocrinology. 25, 254-255:133-139.

Gerozissis, K. 2008. Brain insulin, energy and glucose homeostasis; genes, environment and metabolic pathologies. European Journal of Pharmacology. 6;585(1), 38-49.

Gong, J.G., Lee, W.J., Garnsworthy, P.C., Webb, R. 2002. Effect of dietary-induced increases in circulating insulin concentrations during the early postpartum period on reproductive function in dairy cows. Reproduction. 123, 419-427.

Goodman, R.L., Lehman, M.N., Smith, J.T., Coolen, L.M., de Oliveira, C.V., Jafarzadehshirazi, M.R., Pereira, A., Iqbal, J., Caraty, A., Ciofi, P., Clarke, I.J. 2007. Kisspeptin neurons in the arcuate nucleus of the ewe express both dynorphin A and neurokinin B. Endocrinology. 148, 5752-5760.

Gottsch, M., Cunningham, M.J., Smith, J.T., Popa, S.M., Acohido, B.V., Crowley, W.F., Seminara, S., Clifton, D.K., Steiner, R.A. 2004a. A role for kisspeptins in the regulation of gonadotropin secretion in the mouse. Endocrinology. 145, 4073–4077.

Gottsch, M.L., Clifton, D.K., Steiner, R.A. 2004b. Galanin-like peptide as a link in the integration of metabolism and reproduction. Trends in Endocrinology and Metabolism. 15, 215-221.

Gutiérrez, C.G., Gong, J.G., Bramley, T.A., Webb R. 2006. Selection on predicted breeding value for milk production delays ovulation independently of changes in follicular development, milk production and body weight. Animal Reproduction Science. 95, 193-205.

Guzmán A., Gonzalez-Padilla, E., Garces-Yepez, P., y Gutiérrez, C.G. 2009a. Decreased body fat reduces the response to an estrus induction program in beef suckling cows, associated to a reduction in leptin concentration. Biology of Reproduction, Journal. 81, suppl 1. 494.

Guzmán, S.A., Garcés, Y.P, González, P.E., Rosete, F.J., Calderón, C.R., Gutiérrez, C.G. 2008. La Respuesta a un programa de inducción de la ciclicidad de vacas productoras de carne con cría es afectada por la condición corporal. XXI Congreso panamericano de Ciencias Veterinarias. Guadalajara México.

Guzmán, S.A., González-Padilla, E., Garcés-Yepez, P., Rosete, F.J., Calderón, C.R, Murcia, C., Gutiérrez, C.G. 2009b. Reducción en las concentraciones séricas de insulina e IGF-I pero no leptina, se asocia a una reducción en la respuesta a un programa de inducción de estros en vacas de carne amamantando. XLV

Reunión Nacional de Investigación Pecuaria Saltillo, Coahuila México.

Hayirli, A. 2006. The role of exogenous insulin in the complex of hepatic lipidosis and ketosis associated with insulin resistance phenomenon in postpartum dairy cattle. Veterinary Research Communications. 30, 749-774.

Hill, J.W., Elmquist, J.K., Elias, C.F. 2008. Hypothalamic pathways linking energy balance and reproduction. American Journal of Physiology Endocrinology and Metabolism. 294, E827-832.

Hiney, J.K., Ojeda, S.R., Dees, W.L. 1991. Insulin-like growth factor I: a possible metabolic signal involved in the regulation of female puberty. Neuroendocrinology. 54, 420-423.

Hiney, J.K., Srivastava, V., Nyberg, C.L., Ojeda, S.R., Dees, W.L. 1996. Insulin-like growth factor I of peripheral origin acts centrally to accelerate the initiation of female puberty. Endocrinology. 137, 3717-3728.

Hiney, J.K., Srivastava, V.K., Pine, M.D., Les Dees. W. 2009. Insulin-like growth factor-I activates KiSS-1 gene expression in the brain of the prepubertal female rat. Endocrinology. 150, 376-384.

Jennes, L., Conn, P.M. 1994. Gonadotropin-releasing hormone and its receptors in rat brain. Frontiers in Neuroendocrinology. 15, 51-77.

Juréus, A., Cunningham, M.J., McClain, M.E., Clifton, D.K., Steiner, R.A. 2000. Galanin-like peptide (GALP) is a target for regulation by leptin in the hypothalamus of the rat. Endocrinology. 141, 2703-2706.

Kadokawa, H., Matsui, M., Hayashi, K., Matsunaga, N., Kawashima, C., Shimizu, T., Kida, K., Miyamoto, A. 2008. Peripheral administration of kisspeptin-10 increases plasma concentrations of GH as well as LH in prepubertal Holstein heifers. Endocrinology. 196, 331-334.

Kageyama, H., Takenoya, F., Kita, T., Hori, T., Guan, J.L., Shioda, S. 2005. Galanin-like peptide in the brain: effects on feeding, energy metabolism and reproduction. Regulatory Peptides. 15,126:21-26.

Kalamatianos, T., Grimshaw, S.E., Poorun, R., Hahn, J.D., Coen, C.W. 2008. Fasting reduces KiSS-1 expression in the anteroventral periventricular nucleus (AVPV): effects of fasting on the expression of KiSS-1 and neuropeptide Y in the AVPV or arcuate nucleus of female rats. Journal of Neuroendocrinology. 20, 1089-1097.

Kawashima, C., Fukihara, S., Maeda, M., Kaneko, E., Montoya, C.A., Matsui, M., Shimizu, T., Matsunaga, N., Kida, K., Miyake, Y., Schams, D., Miyamoto, A. 2007. Relationship between metabolic hormones and ovulation of dominant follicle during the first follicular wave post-partum in high-producing dairy cows. Reproduction. 133, 155-163.

Kim, H.H., DiVall, S.A., Deneau, R.M., Wolfe, A. 2005. Insulin regulation of GnRH gene expression through MAP kinase signaling pathways. Molecular and Cellular Endocrinology. 20, 242, 42-49.

Könner, A.C., Janoschek, R., Plum, L., Jordan, S.D., Rother, E., Ma, X., Xu, C., Enriori, P., Hampel, B., Barsh, G.S., Kahn, C.R., Cowley, M.A., Ashcroft, F.M., Brüning, J.C. 2007. Insulin action in AgRP-expressing neurons is required for suppression of hepatic glucose production. Cell Metabolism. 5, 438-449.

Le Roith, D.. 2003. The insulin-like growth factor system. Experimental Diabesity Research. 4, 205-212.

León, H.V., Hernández-Cerón, J., Keislert, D.H., Gutierrez, C.G. 2004. Plasma concentrations of leptin, insulin-like growth factor-I, and insulin in relation to changes in body condition score in heifers. Journal of Animal Science. 82, 445-451.

Loes, D.J., Barloon, T.J., Yuh, W.T., DeLaPaz, R.L., Sato, Y. 1991. MR anatomy and pathology of the hypothalamus. American Journal of Roentgenology. 156, 579-585.

Man, P.S., Lawrence, C.B. 2008. Galanin-like peptide: a role in the homeostatic regulation of energy balance? Neuropharmacology. 55, 1-7.

Miller, D.W., Blache, D., Martin, G.B. 1995. The role of intracerebral insulin in the effect of nutrition on gonadotrophin secretion in mature male sheep. Journal of Endocrinology. 147, 321-329.

Miller, D.W., Harrison, J.L., Bennett, E.J., Findlay, P.A., Adam, C.L. 2007. Nutritional influences on reproductive neuroendocrine output: insulin, leptin, and orexigenic neuropeptide signaling in the ovine hypothalamus. Endocrinology. 148, 5313-5322.

Ojeda, S.R., Lomniczi, A., Sandau, U.S. 2008. Glial-gonadotrophin hormone (GnRH) neurone interactions in the median eminence and the control of GnRH secretion. Journal of Neuroendocrinology. 20, 732-742.

Raposinho, P.D., Broqua, P., Pierroz, D.D., Hayward, A., Dumont, Y., Quirion, R., Junien, J.L., Aubert, M.L. 1999. Evidence that the inhibition of luteinizing hormone secretion exerted by central administration of neuropeptide Y (NPY) in the rat is predominantly mediated by the NPY-Y5 receptor subtype. Endocrinology. 140, 4046-55.

Richards, M.W., Wettemann, R.P., Schoenemann, H.M. 1989a. Nutritional anestrus in beef cows: body weight change, body condition, luteinizing hormone in serum and ovarian activity. Journal of Animal Science. 67, 1520-1526.

Richards, M.W., Wettemann, R.P., Schoenemann, H.M. 1989b. Nutritional anestrus in beef cows: concentrations of glucose and nonesterified fatty acids in plasma and insulin in serum. Journal of Animal Science. 67, 2354-2362.

Richards, M.W., Wettemann, R.P., Spicer, L.J., Morgan, G.L. 1991. Nutritional anestrus in beef cows: effects of body condition and ovariectomy on serum luteinizing hormone and insulin-like growth factor-I. Biology of Reproduction. 44, 961-6.

Roberts, A.J., Nugent, R.A. 3rd, Klindt, J., Jenkins, T.G. 1997. Circulating insulin-like growth factor I, insulin-like growth factor binding proteins, growth hormone, and resumption of estrus in postpartum cows subjected to dietary energy restriction. Journal of Animal Science. 75, 1909-17.

Robertson, S.A., Leinninger, G.M., Myers, M.G. Jr. 2008. Molecular and neural mediators of leptin action. Physiology and Behavior. 6;94, 637-642

Roseweir, A.K., Millar, R.P. 2009. The role of kisspeptin in the control of gonadotrophin secretion. Human Reproduction Update. 15, 203-212.

Sainsbury, A., Schwarzer, C., Couzens, M., Jenkins, A., Oakes, S.R., Ormandy, C.J., Herzog, H. 2008. Y4 receptor knockout rescues fertility in ob/ob mice. Genes and Development. 1;16, 1077-1088.

Saleem, S.N., Said, A.H., Lee, D.H. 2007. Lesions of the hypothalamus: MR imaging diagnostic features. Radiographics. 27, 1087-1108.

Salvi, R., Castillo, E., Voirol, M.J., Glauser, M., Rey, J.P., Gaillard, R.C., Vollenweider, P., Pralong, F.P. 2006. Gonadotropin-releasing hormone-expressing neurons immortalized conditionally are activated by insulin: implication of the mitogen-activated protein kinase pathway. Endocrinology. 147, 816-826.

Sandhu, M.S., Heald, A.H., Gibson, J.M., Cruickshank, J.K., Dunger, D.B., Wareham, N.J. 2002. Circulating concentrations of insulin-like growth factor-I and development of glucose intolerance: a prospective observational study. Lancet. 18;359, 1740-1745.

Schillo, K.K. 1992. Effects of dietary energy on control of luteinizing hormone secretion in cattle and sheep. Journal of Animal Science. 70, 1271-1282.

Schneider, J.E. 2004. Energy balance and reproduction. Physiology and Behavior. 81, 289-317.

Sinclair, K.D., Molle, G., Revilla, R., Roche, J.F., Quintans, G., Marongiu, L., Sanz, A., Mackey, D.R., Diskin, M.G. 2002. Ovulation of the first dominant follicle arising after day 21 post partum in suckling beef cows. Animal Science. 75, 115-126.

Smith, M.J., Jennes, L. 2001. Neural signals that regulate GnRH neurones directly during the oestrous cycle. Reproduction. 122, 1-10.

Spicer, L.J. 2001. Leptin: a possible metabolic signal affecting reproduction. Domestic Animal Endocrinology. 21, 251-70.

Stanley, S.A., Davies, S., Small, C.J., Gardiner, J.V., Ghatei, M.A., Smith, D.M., Bloom, S.R.

2003. gamma-MSH increases intracellular cAMP accumulation and GnRH release in vitro and LH release in vivo. FEBS Lett. 22;543, 66-70.

Strauch, T.A., Neuendorff, D.A., Brown, C.G., Wade, M.L., Lewis, A.W., Keisler, D.H., Randel, R.D. 2003. Effects of lasalocid on circulating concentrations of leptin and insulin-like growth factor-I and reproductive performance of postpartum Brahman cows. Journal of Animal Science. 81, 1363-70.

Suzuki, S., Kadokawa, H., Hashizume, T. 2008. Direct kisspeptin-10 stimulation on luteinizing hormone secretion from bovine and porcine anterior pituitary cells. Animal Reproduction Science. 30;103, 360-365.

Taylor, J.A., Goubillon, M.L., Broad, K.D., Robinson, J.E. 2007. Steroid control of gonadotropin-releasing hormone secretion: associated changes in pro-opiomelanocortin and preproenkephalin messenger RNA expression in the ovine hypothalamus. Biology of Reproduction. 76, 524-531.

Trojan, J., Cloix, J.F., Ardourel, M.Y., Chatel, M., Anthony, D.D. 2007. Insulin-like growth factor type I biology and targeting in malignant gliomas. Neuroscience. 30;145, 795-811.

Uenoyama, Y., Tsukamura, H., Kinoshita, M., Yamada, S., Iwata, K., Pheng, V., Sajapitak, S., Sakakibara, M., Ohtaki, T., Matsumoto, H., Maeda, K.I. 2008. Oestrogen-dependent stimulation of luteinising hormone release by galanin-like peptide in female rats. Journal of Neuroendocrinology. 20, 626-631.

Ward, D.R., Dear, F.M., Ward, I.A., Anderson, S.I., Spergel, D.J., Smith, P.A., Ebling, F.J. 2009. Innervation of gonadotropin-releasing hormone neurons by peptidergic neurons conveying circadian or energy balance information in the mouse. PLoS One. 4, e5322.

Watanobe, H. 2002. Leptin directly acts within the hypothalamus to stimulate gonadotropin-releasing hormone secretion in vivo in rats. The Journal of Physiology. 15;545, 255-268.

Werner, H., Weinstein, D., Bentov, I. 2008. Similarities and differences between insulin and IGF-I: structures, receptors, and signalling pathways. Archives of Physiology and Biochemistry. 114, 17-22.

Xu, J., Kirigiti, M.A., Cowley, M.A., Grove, K.L., Smith, M.S. 2009a. Suppression of basal spontaneous gonadotropin-releasing hormone neuronal activity during lactation: role of inhibitory effects of neuropeptide Y. Endocrinology. 150, 333-340.

Xu, J., Kirigiti, M.A., Grove K.L., Smith, M.S. 2009b. Regulation of food intake and gonadotropin-releasing hormone/luteinizing hormone during lactation: role of insulin and leptin. Endocrinology. 150, 4231-4240.

Yang, M.J., Wang, F., Wang, J.H., Wu, W.N., Hu, Z.L., Cheng, J., Yu, D.F., Long, L.H., Fu, H., Xie, N., Chen, J.G. 2010. PI3-k integrates the effects of insulin and leptin on large-conductance Ca2+-activated K+ channels in neuropeptide Y neurons of the hypothalamic arcuate nucleus. American Journal of Physiology Endocrinology and Metabolism. 298, E193–E201.

Zhang, Y., Proenca, R., Maffei, M., Barone, M., Leopold, L., Friedman, J.M. 1994 Positional cloning of the mouse obese gene and its human homologue. Nature. 372, 425–432.

Zieba, D.A., Szczesna, M., Klocek-Gorka, B., Williams, G.L. 2008. Leptin as a nutritional signal regulating appetite and reproductive processes in seasonally-breeding ruminants. Journal of Physiology and Pharmacology. 59;9, 7-18.

5

Chemical Composition and *In Situ* Evaluation of Fresh and Ensiled Sugarcane (*Saccharum officinarum*)

José Andrés Reyes-Gutiérrez[1], Cándido Enrique Guerra-Medina[2]
and Oziel Dante Montañez-Valdez[1*],

[1] *Centro Universitario del Sur de la Universidad de Guadalajara. Departamento de Desarrollo Regional. Prolongación Colón S/N. Ciudad Guzmán, Jalisco. CP 49000. México. Email montanez77@hotmail.com*

[2] *División de Desarrollo Regional, Centro Universitario de la Costa Sur, Universidad de Guadalajara, Autlán de Navarro. Jalisco, México*

.*Corresponding author

SUMMARY

This study evaluated chemical composition and *in situ* degradability of dry matter (DM), organic matter (OM) and ruminal pH of fresh (FSC) and ensiled (SCS) sugarcane (*Saccharum officinarum*) forage diets. *In situ* digestibility was determined using the nylon bag technique with four cows fitted with a rumen cannula. Cows were fed with fresh or ensiled sugar cane and supplemented with 1 kg of commercial dairy concentrate (18% CP). Ground sample (5g) for each sugar cane (FSC, and SCS) were incubated in rumen for 0, 8, 12, 24, 36, 48, 72 and 96 h Treatments were distributed in a completely randomized design with six replicates. ESC showed significant changes ($P<0.05$) in DM, ADF, NDF and ash. In situ digestibility of dry matter (ISDDM, %) was higher ($P<0.05$) for FCS in most incubation periods with respect to SCS, except at 24 h of incubation in which no difference ($P>0.05$) was noted. In situ digestibility of organic matter (ISDOM, %) was higher ($P<0.05$) for FCS at incubation periods of 8, 36, 48 and 96 h; however at 24 h of incubation was higher ($P<0.05$) in SCS. The ISDOM was similar ($P>0.05$) at 12 and 76 h of incubation. The ruminal pH showed no differences ($P>0.05$) between treatments. It is concluded that the silage of sugar cane is an alternative to provide forage in the season of low growth and quality of the grass.

Key words: Sugarcane silage; digestibility; forage; rumen; chemical composition; *in situ*.

INTRODUCTION

In the world, the ruminants that contribute to food security for humans are estimated at two billion. These animals provide 70% of the total animal protein consumed, 80% of the milk consumed and 10% of the natural fiber used by humans. In the next 25 years, it will be necessary to double the production of animal protein derived from ruminants to ensure the protein intake of a growing world population (Barahona and Sánchez, 2005). Forage resources play a fundamental role in ruminant nutrition and provide over 90% of the energy consumed by them worldwide (Fitzhugh *et al.,* 1978; Wilkins, 2000).

Ruminants have the ability to convert low-quality feed into high quality protein, and use feed produced on land unsuitable for growing crops for human consumption (Varga and Kolver, 1997). This is possible because the rumen microorganisms synthesize and secrete an enzyme complex of β-1-4 cellulases that allow the hydrolysis of forage cell walls. Sugarcane is found amongst these important resources (Espinoza *et al.,* 2006; Aranda *et al.,* 2010). Conventionally, it is harvested every day, chopped and served to the animals; however, the daily cut has some disadvantages, such as the demand for labor-intensive daily cuts, husked and chopped (Rocha *et al.,* 2009). In this scenario, cane of sugar as silage can be an option due to its persistence, wide distribution in tropical and subtropical areas, and a high biomass production (Molina *et al.,* 1997).

The appropriate supplementation with sugarcane is necessary to improve its use (Martin, 1997). Therefore, the best evaluation of feed quality is the animal response, in addition, nutritional value of feed is the combined effect of digestibility, consumption and feed efficiency (Van Soest, 1982). Digestibility and intake are the main parameters that define feed quality; however these are not routinely measured because of high costs and strong demand for labor and time required for *in vivo* experiments (Rodriguez *et al.,* 2007). The lack of information on chemical composition, digestibility and ruminal variables of sugar cane fresh or as silage induce this research, so that the objective of this study was to provide useful information about ruminal digestibility and chemical composition of fresh or ensiled sugarcane.

MATERIALS AND METHODS

The experimental work was done at the Nutrition Laboratory of the Centro Universitario del Sur de la Universidad de Guadalajara and at "Dos Pivotes" ranch located in the Municipality of Zapotlán El Grande, Jalisco, Mexico. The materials tested were: 1) fresh sugar cane, variety CP 72-2086, with 13 months of age of second cut and 2) sugar cane silage (same variety). Samples of ingredients were dried in a circulating air oven at 60 °C for 24 hours and then milled in hammer mill with 2 mm sieve for further analysis. Total dry matter (DM) was estimated a circulating air oven (100 °C for 24); crude protein (CP) was determined by the Kjeldahl method; ash (A) and organic matter (OM) was calculated by difference, using the technique described by the

AOAC (2007). The pH of the silage was determined as described by Tejada de Hernández (1985).

The determination of the fiber fractions (NDF and ADF) was performed using alpha amylase without ash correction as specified by Van Soest *et al.* (1991). Digestibility was determined *in situ* using four Holstein cows (625 ± 63 kg) of 4-years old, and fitted with permanent rumen cannula of 10 cm core diameter (Bar Diamond Lane, Parma, ID, USA). Cows were distributed at random in an experimental design in simple sequences of treatments. The experiment lasted 30 days, divided into two periods of 15 days each (10 for adaptation and 5 for collecting samples). The diets consisted of: the ingredient under study (FSC, and SCS) *ad libitum* plus 1.0 kg of commercial dairy concentrate (APILECHE ULTRA®, 18% PC, México) divided into two meals (AM - PM) to ensure greater cellulolytic activity of the microflora of the rumen. Fresh clean water was available *ad libitum.*

For *in situ* digestibility of DM and OM the procedure proposed by Vanzant *et al.* (1998) was followed. Nylon bags were used (10 x 15 cm, pore size 40 to 60 μm) with 5 g of sample. Each sample of the proposed treatments (FSC and SCS) were incubated in rumen for 0, 8, 12, 24, 36, 48, 72 and 96 h in triplicate, in addition at each time blanks secured with nylon thread to a piece of string (30 cm long, weight 150 g) were added and left suspended in the rumen. Subsequently, the bags were removed from the rumen according to the incubation times along with the zero hour, and then bags were washed with circulating water at low pressure, until the water came out just as clear as it had entered. Subsequently, the bags of waste were dried in a circulating air oven (48 h at 60 °C). Fluid ruminal samples were taken from the ruminal cannula at two hour intervals; one was taken 1 h before daytime feeding and the other 12 hours later. Ruminal pH of the fluid in the rumen was measured using a portable potentiometer (Model PC18) immediately after rumen fluid was collected.

Statistical analysis

Data from chemical composition, *in situ* digestibility of DM and OM were analyzed using PROC GLM SAS and (SAS, 1999); and ruminal pH was analyzed with PROC MIXED SAS (1999).

RESULTS AND DISCUSSION

The chemical analysis of the FSC and SCS (Table 1) showed changes in DM, CP, ADF, NDF and ash, due to fermentation. The OM value was similar between fresh or ensiled sugarcane. The DM content in FSC was 31.36%; this value was higher than the results found by different authors. Rocha *et al.* (2009)

reported 30.5% of DM with the RB72454 variety at 12 months old; Alli and Baker (1982) and Ferreira et al. (2007) reported 28.2% of DM in different varieties of sugarcane and harvested at seven months old. However, the DM content of sugarcane in present research was less than the reported by Peláez et al. (2008), they found 35.4% DM in sugar cane for 12 months old.

The DM in SCS was 36.0%, this value was higher than the reported by Rocha et al. (2009) and Ferreira et al. (2007), 28.6% and 21.58% respectively, Peláez et al. (2008), found a value of 38.0%, which was higher than what was found in this study. In present study, the CP for SCS was 14.6% higher than the FSC, this increase occurred as a result of the use of soluble carbohydrates during silage fermentation, that increased the percentage of CP. According to Rötz and Muck (1994), the CP content can increase from 1 to 2 percentage units in the DM with this process.

Table 1. Chemical composition (%) of fresh or ensiled sugarcane (DB)

	FSC	SCS
Components	%	
Dry Matter	31.36b	36.00a
Organic Matter	25.25	25.76
Crude Protein	4.37	5.01
ADF	20.89b	27.14[a]
NDF	49.54b	54.38[a]
Ash	6.11b	10.24[a]
pH	6.90a	3.58b

[a,b] Different superscripts following means in the same row indicate differences at P<0.05; FSC = Fresh sugarcane; SCS= sugarcane silage.

The structural components of cell wall, NDF and ADF (Table 1), the values were different between treatments; similar results were reported by several authors (Alli and Baker, 1982; Kung Jr. and Stanley, 1982; Pedroso et al., 2006; Bravo-Martins et al., 2006; Ferreira et al., 2007; Pelaez et al., 2008; Rocha et al., 2009). The increase in the proportion of fiber components in silage in relation to original material is due to the loss of water-soluble constituents, together with the tributaries produced during fermentation and loss of gas (Kung Jr and Stanley, 1982; Bolsen, 1995).

The ash content in the sugarcane is generally low. The concentrations of ash obtained in this study are considered high compared to values obtained by Rocha et al. (2009). These differences may be related to the varieties, plant age and fertilization. In the silages, the variations in the levels of ash can be used

to estimate the losses in DM during fermentation, which does not change during the fermentation process. Pedroso et al. (2005) noted that the ash content in sugar cane silages increased with the fermentation, due to loss of nutrients in the form of gas and effluent during the ensiling.

The pH value of the SCS (3.58) is within the limits reported for sugarcane silages (Pedroso et al., 2007). Regarding the ISDDM, Table 2 shows that at 24 h of incubation there were no differences between treatments (P>0.05), but in other periods of incubation were higher (P <0.05) in FSC. From 8 to 96 h of incubation reached the highest values of digestibility (39.92 to 61.93%, respectively). Other authors (Aranda et al., 2004; Lopez et al., 2003; Peláez et al., 2008) reported similar results as the ones in this study at 72 h of incubation (above 60%) exploring different varieties of sugarcane. Molina et al. (1999) in a study of 74 sugarcane varieties found digestibility values between 54.1 to 81.0% of the total DM, pointed out that sugar cane varieties for forage use must have at least 50% of DM digestibility. The reduction coefficient of ISDDM in SCS is reflected by the concentration of DM, NDF and ADF during the fermentation process. Pedroso (2003) observed a significant reduction in IVDDM of silage from sugarcane in relation to forage (47.1% vs 62.9%).

Table 2. Percentage of in situ digestibility of fresh and ensiled sugarcane.

Fraction	Incubation time (h)	FSC	SCS	SEM
	96	61.93a	56.60b	1.15
	72	60.75a	52.29b	0.89
	48	56.80a	51.45b	1.08
DM	36	47.21a	44.08b	0.66
	24	44.11	43.6	0.86
	12	38.61a	34.29b	0.92
	8	39.92a	32.06b	0.82
	96	57.88a	47.43b	2.35
	72	56.66	56.66	0.90
	48	50.70a	45.87b	1.50
OM	36	52.29a	47.50b	1.06
	24	50.13b	56.30a	0.85
	12	46.01	44.12	1.45
	8	54.20[a]	45.70b	2.96

[a,b] Different superscripts following means in the same row indicate differences at P<0.05; FSC = Fresh sugarcane; SCS= sugarcane silage; SEM= Standard error of the mean.

The in situ digestibility values of OM (Table 2) showed significance (P <0.05) between treatments in most incubation times. The FSC treatment had the highest values except at 24 h of incubation, which was favorable (P <0.05) for the SCS. At 12 and 72 h of incubation no difference (P <0.05) between treatments was observed. In this sense, the decline in overall OM digestibility observed in this study is similar to other reports, and it is attributable to the increase in the proportion of cell walls (López *et al.*, 2000). There were no differences (P> 0.05) in ruminal pH between treatments, mean rumen pH for treatments (FSC, SCS) of 7.04 and 7.12 respectively (Table 3). Similar results to this study were found by Garcia *et al.* (2008) with average values of 6.62 and 7.20. Görtler (1975) who suggests that the rumen pH is an indicator that may change the celullosis, and mention that the optimum value for celullosis is in a range of 6.70 to 7.00, that was found in this study.

Table 3. Effect of FSC and SCS on ruminal pH over time.

Time	TREATMENTS[1]		
	FSC	SCS	SEM[2]
-1	7.30	7.20	0.14
0	6.93	7.03	0.14
2	6.90	7.27	0.14
4	7.46	7.62	0.14
6	7.19	7.57	0.14
8	7.19	7.04	0.14
10	6.72	6.64	0.14
12	6.68	6.60	0.14
Average	7.04	7.12	0.14
SET[3]	0.21	0.21	

[1] Treatments: FSC= fresh sugar cane, SCS= sugarcane silage;
[2] SEM = standard error of the mean.
[3] SET = standard error of the treatments.

CONCLUSION

It is concluded that the conservation technique (silage) of sugarcane is a good alternative, because it preserve the nutrient content in the season when cutting fodder reach low nutrient levels. Moreover, the advantage of the FSC is by increases in the percentages of DM, CP and improving rumen pH conditions.

REFERENCES

Alli I, Baker B E. 1982. Studies on the fermentation of chopped sugarcane. Animal Feed Science Technolgy 7:411-417.

Aranda E M, Ruiz P, Mendoza G D, Marcoff C F, Ramos J A, y Elías A. 2004. Cambios en la digestión de tres variedades de caña de azúcar y sus fracciones de fibra. Revista Cubana de Ciencia Agrícola 38:137-144.

Aranda E M, Mendoza G D, Ramos J A, Da Silva I C and Vitti A C. 2010. Effect of fibrolitic enzymes on rumen microbial degradation of sugarcane fiber. Ciencia Animal Brasileña 11:448-495.

AOAC. 2007. Official Methods of Analysis of the Association of Official Agricultural Chemists. 18th Edition. Published by the Official Agricultural Chemists. Washington, D.C.

Barahona R, y Sánchez P. 2005. Limitaciones físicas y químicas de la digestibilidad de pastos tropicales y estrategias para aumentarla. Revista Corpoica 6:69-82.

Bolsen K. 1995. Silage: basic principles. In: Barnes, R.F.; Miller, D.A.; Nelson, C.J. (Eds.) Forages. 5. ed. Ames: Iowa State University. p.163-176.

Bravo-Martins C E, Carneiro L, and Castro-Gómez R J. 2006. Chemical and microbiological evaluation of ensiled sugarcane with different additives. Brazilian Journal of Microbiology 37:499-504.

Espinoza F, Argenti P, Carrillo C, Araque C, Torres A, y Valle A. 2006. Uso estratégico de la caña de azúcar (Saccharum officinarum) en novillas mestizas gestantes. Zootecnia Tropical 24: 95-107.

Ferreira D, Gonçalves L, Molina L R, Castro-Neto A, and Tomich T R. 2007. Fermentation of sugarcane silage treated with urea, zeolita, bacteria inoculant and bacteria/enzymatic inoculants. Arquivo Brasileiro de Medina Veterinaria e Zootecnia 59:423-433.

Fitzhugh H, Hodgson H, Scoville O J, Nguyen T. and Byerley T. 1978. The Role of Ruminants in Support of man: Winrock Report. Winrock Foundation, Morrilton, Arkansas.

García H, Abreu M, y Soto J M. 2008. Digestión de residuos de la cosecha cañera tratados con hidróxido de sodio. 1. Determinación de la digestibilidad *in situ*. REDVET Revista Electrónica de Veterinaria IX:(11). http://www.veterinaria.org/revistas/redvet/n1 11108/111106.pdf

Görtler H, 1975. Fisiología de la digestión y de la absorción. Fisiología Veterinaria. 2da edición española de la 3ra edición alemana: Edit. Acribia (España) 1:274-311.

Kung Jr L, Stanley R W. 1982. Effect of stage of maturity on the nutritive value of whole-plant sugarcane preserved as silage. Journal Animal Science 54:689-696.

Molina A S, Febles I y Sierra J F. 1997. Ensilaje de caña de azúcar con síntesis proteica. Formulación de los aditivos. Revista Cubana de Ciencia Agrícola 31:271-274.

Molina A S, Sierra J F and Febles I. 1999. Sugarcane silage with protein synthesis: combined effect of additives and density. Cuban Journal Agricultura Science 33:205-208.

Martín P C y Brito M. 1997. Cantidad y tipo de proteína en dietas de forraje de caña de azúcar para toros. Revista Cubana de Ciencia Agrícola 31:265-269.

López I, Aranda E M, Ramos J A y Mendoza G D. 2003. Evaluación nutricional de ocho variedades de caña de azúcar con potencial forrajero. Revista Cubana de Ciencia Agrícola 37:381-386.

Pedroso A F. 2003. Chemical additive and microbial inoculants effects on the fermentation and on the control of the alcohol production in sugarcane silages. 2003. 120f. Tese (Doutorado) - Escola Superior de Agricultura Luiz de Queiróz, Universidade de São Paulo, Piracicaba.

Pedroso A F, Nussio L and Paziani S F. 2005. Fermentation and epiphytic microflora dynamics in sugar cane silage. Scientia Agricola 62:427-432.

Pedroso A F, Nussio L. and Barioni J R. 2006. Performance of holstein heifers fed sugarcane silages treated with urea, sodium benzoate or *Lactobacillus buchneri*. Pesquisa Agropecuária Brasileira 41:649-654.

Pedroso A F, Nussio L, and Lourdes D R. 2007. Effect of treatment with chemical additives and bacterial losses and quality of silage from sugarcane. Revista Brasileira de Zootecnia 36:558-564.

Peláez A, Meneses M, Miranda R L. Mejías M R, Barcena G R y Loera O. 2008. Ventajas de la fermentación sólida con Pleurotus sapidus en ensilajes de caña de azúcar. Archivos Zootécnicos 57:25-33.

Rodriguez M, Simoes S C y Guimaraes J R. 2007. Uso de indicadores para estimar consumo y digestibilidad de pasto. LIPE, lignina purificada enriquecida. Revista Colombiana de Ciencias Pecuarias 20:518-525.

Rocha V A, Cardoso P J, Da Silva A C, Ricardo E A, Botego T V and Freitas S R. 2009. Effect of the addition of Lactobacillus sp. *In*: sugarcane silages. Revista Brasileira de Zootecnia 38:1009-1017.

Rotz CA, Muck RE (1994). Changes in forage quality during harvest and storage. *In*: Forage quality, evaluation and utilization. Fahey, G. C. J., Collins, M., Mertens, D. R. and Moser, L. E. (Eds.). American Society of Agronomy, Madison, WI. pp. 828-868

SAS. 1999. User's Guide: Statistics, version 8.0. Ed. SAS Institute, Inc., Cary N.C.

Tejada de Hernández I. 1985. Manual de laboratorio para análisis de ingredientes utilizados en la alimentacion animal. Patronato de apoyo a la investigación y experimentación pecuaria en México. México pp. 387.

Van Soest P J. 1982. Nutrition ecology ruminant. O. B. Books. Inc. Corvallis O. R., USA.

Van Soest P J, Robertson J B and Lewis B A. 1991. Methods for dietary fiber, neutral detergent fiber and non starch polysaccharides in relation to animal nutrition. Journal Dairy Science 74:3583-3597.

Vanzant E S, Cochran R C and Titgemeyer E C. 1998. Standardization of *in situ* techniques for ruminant feedstuff evaluation. Journal Animal Science 76: 2717-2729.

Varga G A and Kolver E S. 1997. Microbial and animal limitations to fiber digestion and utilization. Conference: New Developments in Forage Science Contributing to Enhanced Fiber Utilization by Ruminants. Journal of Nutrition 127: 819-823.

Wilkins R J. 2000. Forages and their Role in Animal Systems. *In:* D.I. Givens, E. Owen, R.F.E. Axford and H.M. Omed (editors) Forage Evaluation in Ruminant Nutrition, CAB International, U.K. pp 1-14.

6

Isolation of Native Strains of *Trichoderma Spp*, from Horticultural Soils of the Valley of Toluca, for Potential Biocontrol of *Sclerotinia*

Hilda G. García-Núñez[1], Sergio de J. Romero-Gómez[2],
Carlos. E. González-Esquivel[3], E. Gabino Nava-Bernal[1],
A. Roberto Martínez-Campos[1*].

[1] *Univ. Autónoma del Estado de México. Instituto de Ciencias Agropecuarias y Rurales. Km. 14.5 Autopista Toluca-Atlacomulco. San Cayetano de Morelos. Toluca, Estado de México. C.P. 50295.*

[2] *Univ. Autónoma de Querétaro. Fac. Química. Av. Hidalgo S/N, Col. Niños Héroes. Querétaro, Qro. C.P. 76010*

[3] *UNAM. Centro de Investigaciones en Ecosistemas. Antigua Carretera a Pátzcuaro 8701. Col. Ex-Hacienda de San José de la Huerta. CP 58190,Morelia, Michoacán. E-mail: armartinezc@uaemex.mx*

*Corresponding author

SUMMARY

The presence of *Trichoderma* mold strains was evaluated in seven localities in the southern part of the Valley of Toluca in the State of Mexico. This area has a high potential for growing vegetables. In the study, native strains of *Trichoderma* were isolated from soil samples, physiographic factors were identified, as well as the physicochemical properties of the soil which may affect *Trichoderma* occurrence. The potential of *Trichoderma* strains for control of *Sclerotinia spp.*, a pathogenic fungus which causes soft rot in lettuce, was evaluated. Eleven strains were isolated, most of them associated with the type of soil found in the San Francisco Putla and San Francisco Tetetla localities. Logistic regression analysis showed no relationship between the soil properties (organic matter content and pH) and the presence of *Trichoderma*. Tukey test (p<0.05) showed significant differences between the percentage of inhibition of *Sclerotinia* by the eleven native strains of *Trichoderma*. The TF10, TL4 y TX8 strains had a high biocontrol potential, with inhibition percentages of 80%, 86% y 88%, respectively. These strains are an ecological alternative for the control of S*clerotinia* spp.

Key words: *Trichoderma, Sclerotinia,* biological control

INTRODUCTION

Lettuce (*Lactuca sativa* L.*)* is one of the crops with more planted surface in the horticultural zone of the Valley of Toluca, in the State of Mexico. The production of this vegetable satisfies the country´s internal demand. Soft rot is the main diseases that affects this crop and can cause losses of up to 60% of the production (Hao and Subbarao, 2005; Wu and Subbarao, 2006).

Soft rot disease is caused by two species of fungus of the *Sclerotinia* genus, *S. minor* Jagger and *S. sclerotiorum* (Lib.) de Bary. Both species may be found in the same fields, being one of them normally predominant. Control of *Sclerotinia* infections is complicated since the sclerotia that are resistance structures and the primary inoculum for new infections may remain for long time in the soil. (Davis *et al.*, 2002). *Sclerotinia* species use different modes of infection in the plant, *S. minor* infects by eruptive germination of sclerotia while *S. sclerotiorum* infects by carpogenic germination (Hao and Subbarao, 2005). Soft rot disease is characterized by abundant growth of white and cottonlike mycelia, as well as by aqueous rotting of the plant's crown and root, and the pathogen can attack the crop in any phase of its development (Subbarao, 1998; Davis *et al.*, 2002; Rabeendran *et al.*, 2006).

Cultural, chemical and biological methods have been used to deal with soft rot disease with variable success. Application of chemical products, such as fludioxonil, fluazinam and iprodion, has been effective to reduce *Sclerotinia* infections (Hubbard *et al.*, 1997: Matheron and Porchas, 2004). However some of those products can produce damage the crops by phytotoxic effects (Hao *et al.*, 2003) eliminate beneficial organisms along with the pathogens as they are non-specific (Rey *et al.*, 2000), may cause health problems in the individuals who apply them and have a negative impact on the environment by accumulation due to persistence. All theses reasons have generated doubts about the convenience of their use.

.

Biological control (BC) is an ecological alternative for the management of plant diseases that are important in agriculture (Heredia and Delgadillo, 2000). It is known that some fungi are antagonistic to pathogenic organisms and thanks to this effect can decrease the damage caused by diseases in agroecosystems (Infante *et al.*, 2009) and it has been stated that BC must be done taking advantage of the diversity of native microorganisms in soil (Altieri, 1999).

Trichoderma genus has been widely studied and used for biological control. *Trichoderma* shows ecological plasticity, a high enzymatic ability to degrade substrates, are easily isolated, are rapidly cultivated and are very efficient to control a broad range of phytopathogens such as *Fusarium, Pythium, Rhizoctonia* and *Sclerotinia* (Quiroz-Sarmiento *et al.,* 2008). Products including *Trichoderma harzianum* have been used successfully in the suppression of damping-off in carrot caused by *Rhizoctonia solani* (Adams*,* 1990*)*. However, in other cases, formulas have not been effective, since they contain species that are not compatible with the environment or with the characteristics of the region where they are applied (Rabeendran *et al.*, 2006) for this reason, it is very important to use native strains in biocontrol. The objectives of this study were to identify the physical and chemical properties of the soil and the physiographic factors determining the establishment of *Trichoderma* in seven localities of the horticultural zone of the Valley of Toluca, as well as the isolation of native strains with the potential to act as biological control of *Sclerotinia*.

MATERIALS AND METHODS

Study site localization

The study zone is located at the south of the City of Toluca, between 19° 05´ and 19° 10´N and between 99° 30´ and 99° 40´W. In this area 50 *Sclerotinia* spp. infected lettuce plots were located (Table 1). Infected plots were distributed across 7 towns in the municipalities of Tenango del Valle, Rayón, Joquicingo and Texcalyacac.

Table 1. Plots found in each locality, for isolation of native strains of *Trichoderma* spp.

Municipality	Town	Plots
Tenango del Valle	Santa María Jajalpa	9
	Tenango de Arista	7
	San Francisco Putla	10
	San Francisco Tetetla	7
Texcalyacac	San Mateo Texcalyacac	5
Joquicingo	San Pedro Techuchulco	5
Santa María Rayón	San Juan la Isla	7

In order to draw a map including the edaphic and physiographic factors of the study zone and to relate these factors to the presence or absence of *Trichoderma* all plots were georeferenced using the Global Positioning System (GPS),

Soil Sampling

Soil sampling included winter-spring (irrigation) and summer (rainy season) culture cycles. Four soil samples of 250 g were obtained from each analyzed plot, at a depth of 15 cm. Two of these samples where obtained from nearby healthy plants and the other two from nearby plants with soft rot symptoms. Collected samples were kept at -72°C, until tests were performed. Sclerotia found on lettuce plants were isolated and propagated in the laboratory for further use in *Trichoderma – Sclerotinia* antagonism tests.

Determination of Physicochemical properties of soil

Soil samples were removed from the ultra-low temperature freezer, dried at room temperature (RT) and processed according to the applicable protocol. pH determination was performed to 1:2 soil-water suspensions according to AS-02 method of the NOM-021-RECNAT-2000. Organic matter quantification (OM) was performed using Walkley and Black (AS-07) method, texture was estimated by the Bouyoucos (AS-09) method.

Isolation of *Trichoderma* strains

Isolation of *Trichoderma* strains from soil samples was carried out by the serial dilution method of Guigón-López and González-González, (2004). Each soil sample was homogenized and a 10 g sub-sample was taken, placed in a test tube containing 90 mL of a saline isotonic solution (0.85% sterile sodium chloride) and shaken for 20 minutes. 1 mL of that mixture was diluted with 9 ml of isotonic solution in a test tube, and mixed by 2 minutes; this procedure was repeated until 5 dilutions were obtained. 1 mL from the last three dilutions was plated in phytone yeast extract agar plates and incubated at 25°C for 7 days. *Trichoderma* resembling colonies were selected and isolated to pure strains by consecutive culture in order to be further analyzed.

Microscopic identification of *Trichoderma* strains.

In order to confirm the identity of those strains suspected to be *Trichoderma* a sample of mycelium was taken from each one of the pure strains, placed on a slide, and stained with methylene blue was added. Microscopic structures as size and shape of conidia and phyalides were observed and determined with the dichotomous keys (Samuels *et al.,* 2008) and identified using an optical microscope coupled to the Motic Images Plus 2.0 program, just those strains confirmed by this method were kept and labeled with a code, for its handling in further analyses.

Sclerotinia strains isolation.

The sclerotia obtained from the lettuce plants were disinfected by immersion in 5% sodium hypochlorite for three minutes, washed three times with sterile distilled water, placed on sterile filter paper and allowed to dry. Later on, each sclerotium was placed in the center of a Petri dish, with Phytone Yeast Extract Agar and incubated at 25°C for 7 days (Mónaco *et al*, 1998). *Sclerotinia* pure strains were isolated from initial cultures by consecutive mycelium transfers.

In vitro confrontation tests

In vitro confrontation tests were carried out by the dual cultivation method using pure strains of *Trichoderma* (antagonist) and *Sclerotinia* (pathogen) (Martínez and Solano, 1994). A sclerotium was placed at one end of a Phytone Yeast Extract Agar plate and one cm^2 of active mycelium of *Trichoderma* was placed in the other end; plates were incubated at 25° C for seven days. This assay was done in triplicate for each isolated *Trichoderma* strain. Percentage of inhibition of radial growth (PIRG) was recorded every 24 h, using the following formula:

$$PIRG=[(R1-R2)/R1] \times 100$$

Where, R1 is the radial growth of the control non confronted *Sclerotinia* strain and R2 is the radial growth of the *Trichoderma* confronted *Sclerotinia* strain (Samaniego *et al*., 1998, cited by Martínez *et al*., 2008).

Experimental design.

The confrontation tests were performed according to a completely random design, using three repetitions for each one of the eleven isolated native strains of *Trichoderma* that were confronted against *Sclerotinia spp.*

Statistical analysis.

Statistical differences in pH and OM values between the localities were determined by variance analysis. The possible relationship between physical properties of the soil (pH and organic matter content) and the presence of *Trichoderma* strains was determined using logistic regression. The potential of each

Trichoderma strain as bio control agent for *Sclerotinia* infection was assessed by variance analysis and Tukey means comparison test (p<0.05), using the Statgraphics plus statistical package, version 4.1.

RESULTS AND DISCUSSION

Isolated native strains of *Trichoderma*

Eleven native strains belonging to the *Trichoderma* genus were found and identified by macroscopic and microscopic characteristics of the colonies, according to Infante (2009), strains were distributed along five of the seven sampled localities (Table 2). *Trichoderma* strains were expected to be found in more than 50% of the sampled plots because of its cosmopolitan character and the fact that it is a natural inhabitant of soils, but *Trichoderma* strains were found in just 22% of plots; these results differ to those reported by Michel-Aceves *et al.*, 2001, in that work native strains of *Trichoderma* were found in 88% of the sampled sites. Michel-Aceves *et al.* mention the collection season is the main factor that may affect the presence of the fungus in soils and reported spring and summer as the best seasons to find a high number of isolates; however, even when the present study cover both seasons the number of isolates was lower.

Edaphic and physiographic factors in the *Trichoderma* distribution zone

Figure 1 illustrates the presence or absence of *Trichoderma* (red and green points respectively). It can be seen that even when the sampling zone is near to water bodies in 78% of the plots *Trichoderma* was not found; this make it clear that at least in this zone the proximity to rivers is not a determining factor for the establishment of *Trichoderma*. On the other hand, the profile of the sampling area is slightly wavy and the mold was found at different altitudes, which confirms its ecological plasticity, as pointed out by Samuels, 2006 and Infante *et al.*, 2009. Although *Trichoderma* as genus can be found in all latitudes and in all types of soils, geographic distribution of species of *Trichoderma* are quite different, while some species are broadly spread, as is the case of *T. pseudokoningii*, others have a limited geographic distribution, as is the case of *T. viride* that is not commonly found in the colder northern regions and in even more so for *T. aureoviride,* whose distribution is limited to the United Kingdom and northern Europe (Samuels, 2006).

Table 2. Isolated *Trichoderma* strains, by locality

Town	Strains	Code
San Francisco Tetetla	4	TL2,TL4.TL5,TL6
San Francisco Putla	3	TF6,TF8,TF10
Santa María Jajalpa	2	TX7,TX8
San Juan la Isla	1	TJ6
Tenango de Arista	1	TT6
San Pedro Techuchulco	0	-
San Mateo Texcalyacac	0	-

With respect to the type of soil where *Trichoderma* was found, the greater number of *Trichoderma* isolates (8) was obtained from phaeozem, and only two isolates were found in leptosol soil, although 58% of the sampling sites are found in leptosol (Table 3). This opens the possibility that the type of soil may be an important factor in the establishment of this *Trichoderma*. However, it has been reported that the presence or absence of *Trichoderma* depends on many factors, mainly the species involved, environmental conditions and pathosystem (Duffy *et al.*, 1997).

Determining physicochemical properties of soil

pH values of soil at sampling sites went from 4.1 to 8, that is, from strongly acidic to slightly alkaline (Table 3). pH value was found to be significantly different among sampling localities (Tukey p<0.05). Although this property depends on many factors, it is worth mentioning that for this study, those localities that have a pH values similar are placed near each other while when geographic distance increase so does the difference in pH values.

Native *Trichoderma* strains were found in acid soils (Table 3). This can be explained by the fact that acidity is a factor which affects presence, density and longevity of this fungal genus (Michel-Aceves *et al.*, 2001); pH also has an influence on the production of enzymes involved in the degradation process of fungi attacks (Kredics *et al.*, 2003; Samaniego, 2008) and some species of this genus produce organic acids (gluconic, citric or fumaric acid) that decrease the pH of soils, allowing phosphates, micronutrients and minerals to become soluble, and which are necessary for plant metabolism (Vinale *et al.*, 2008). The results of the present study coincide with what was reported by Okoth *et al.*, (2007), who reported that *Trichoderma* is abundant in acid soils.

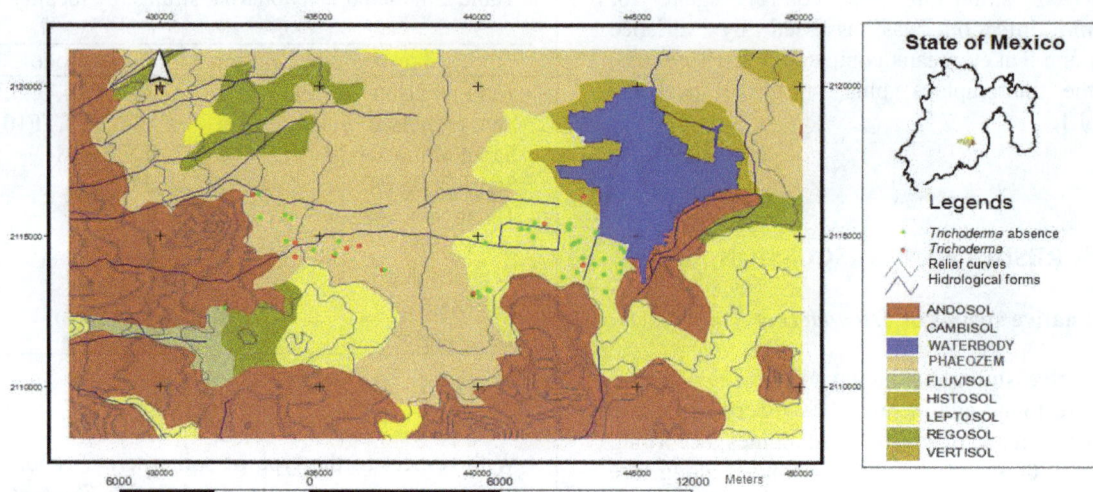

Figure 1. *Trichoderma* spp. distribution in the horticultural zone of the Valley of Toluca. Figure by: Ángel Rolando Endará Agramot

An organic matter (OM) content between 2.7 to 21.9% were found in the sampled horticultural soils (Table 3). Tukey test showed a significant difference (p<0.05) among sampling localities. These results differ to those reported by Reyes *et al*., (2002), in that work an OM content of 8.5 % in the cultivated soils of the Nevado de Toluca National Park (NTNP) was reported. The higher values that were found for OM can be attributed to the fact that certain producers add organic compounds to their plots to improve the soil, such as chicken, sheep or cow manure; records show a higher use of chicken manure, with an OM content of 20 – 40% (Estrada, 2005). Also, two sampled plots were planted for the first time; this may explains the high percentage of OM in the soil at least for those two cases. Low values of OM content found in some plots are possibly due to the fact that the constant agricultural use of the sampling sites has caused a decrease in OM content of those plots. These results agree with those of Michel-Aceves *et al*., (2001), as

in that work low values of up to 2.3% of organic matter were reported for the same zone.

Trichoderma native strains were found mainly in soils with low OM content (Table 3), this does not match up with the results of Osorio-Nila *et al*., (2005), where high organic matter content increase the establishment of *T. lignorum* in lettuce cultivation soils of the Valley of Tenango Municipality, in the State of Mexico.

Textures found in the evaluated localities were, sandy crumb, limey crumb and clayey crumb, with the greater number of isolates coming from sandy crumb. No significant relationship was found for the presence of *Trichoderma* and soil properties (pH and OM) by logistic regression analysis (Table 4). These results agree with the study by Michel-Aceves *et al*, (2001), where no significant correlation was found between the presence of *Trichoderma* and soil properties.

Table 3. Physicochemical properties of soil in the sampled localities

Town	Number of strains[v]	Type of soil	Mean pH[x]	Mean OM[y]	Texture
San Francisco Putla	3	Phaeozem	4.9a	2.7a	Sandy crumb
San Francisco Tetetla	4	Phaeozem	5.0b	5.3b	Sandy crumb
Tenango de Arista	1	Phaeozem	5.4b	8.9d	Clayey crumb
San Juan la Isla	1	Leptosol	5.7c	11.8e	Limey crumb
San Mateo Texcalyacac	0	Leptosol	6.1d	21.9g	Sandy crumb
Santa María Jajalpa	2	Histosol	6.5e	6.0c	Clayey crumb
San Pedro Techuchulco	0	Leptosol	6.6e	13.0f	Sandy crumb

The analyses correspond to 50 plots distributed among seven towns.
[v] Number of *Trichoderma* strains found by locality.
[x, y] Mean value obtained from pH, OM parameters (N=150).
The values in the same column, marked with different letters, show statistical differences (P<0.05).

Table 4. Logistic regression between the presence of *Trichoderma* and soil properties

Soil property	Regression coefficient	Standard error	Chi squared	Probability level	R^{2z}
pH	-0.1458606	0.2868994	0.26	0.611170	0.005469
OM	-0.1786044	0.1005832	3.15	0.075784	0.058225

[z] Refers to r^2 values obtained through logistic regression between the presence of *Trichoderma* and soil properties.

In vitro confrontation tests

The eleven native *Trichoderma* strains isolated in this work showed a higher growth rate than *Sclerotinia* (Table 5). These results may be related to the fact that *Trichoderma* has the capability to colonize rapidly almost any substrate (Infante *et al*, 2009). On the other hand, while *Sclerotinia* normally is a very aggressive mold, in this case it showed a growth rate that was lower than *Trichoderma* strains; this growth rate is similar to the one reported by Sanogo and Puppala, 2007, when evaluating the *in vitro* growth of *S. sclerotiorum*.

In the confrontation tests, the eleven native strains of *Trichoderma* differ statistically (p<0.005) in their potential to inhibit *Sclerotinia spp.* growth (Table 6). The TL6 strain get the lowest level of inhibition of *Sclerotinia* spp (0.08%) while TX8 (88%), TL4 (86%) and TF10 (80%) were the strains with greater inhibitory capacity of all. The differences obtained for inhibition capability among native *Trichoderma* strains may be related to the fact that the isolated strains could be different species, since they had different color pattern, as well as differences in growth patterns and inhibition mechanisms. It has also been reported that each mold may have different antagonistic potential (Bowen *et al.*, 1996; Hermosa *et al.*, 2000; Herman *et al.*, 2004; Schubert *et al.*, 2008; Komón-Zelazowska *et al*, 2007).The results obtained in the present study are similar to those reported by Michel-Aceves *et al*. (2004), where inhibition percentages were found going from 5.35 up to 42.02% in the biocontrol of *Fusarium subglutinans* by *Trichoderma;* also to those of Arzate *et al*. (2006), in the antagonistic action of *Trichoderma spp.* over *Mycospharella fijiensis,* where inhibition percentages of 14.41 to 73.48 were found in banana plants. Thus, it will be interesting to carry out further studies, including identification of native strains at molecular level, as well as the analysis of the inhibitory mechanisms involved and its bioregulatory action over *Sclerotinia spp.*

Table 5. Radial growth rate of native strains of *Trichoderma* and *Sclerotinia spp.*

Strain[w]	Radial growth (mm)$_x$	Growth rate (mm/day)[y]
TX7	90.0a	12.857
TX8	90.0a	12.857
TJ6	89.76[a]	12.824
TT6	90.0a	12.857
TL2	89.99[a]	12.852
TL4	87.57b$_z$	12.511
TL5	90.0a	12.857
TL6	90.0a	12.857
TF6	90.0a	12.857
TF8	90.0a	12.857
TF10	90.0a	12.857
Sclerotinia spp.	78.94c$_z$	12.00

[x] Radial growth (mm) of each strain, for 7 days
[y] Growth rate (mm/day) of the eleven isolated strains of *Trichoderma* and *Sclerotinia* spp, during seven days of evaluation.
[z] Values with different letters are statistically different (Tukey, p≤0.005).

Table 6. Comparison of means of *in vitro* inhibition percentages in dual cultures of native *Trichoderma* strains and *Sclerotinia spp.*

Native strain[x]	Inhibition (%)
TL6	0.08[a]
TL2	14.93b
TT6	17.81c
TL5	21.82d
TJ6	23.66e
TF6	34.35f
TX7	52.65g
TF8	59.38h
TF10	80.07i
TL4	86.18j
TX8	88.73k

Values with different letters are statistically different (Tukey, p≤0.005)

CONCLUSIONS

The number of *Trichoderma* isolates that can be found in the Valley of Toluca horticultural zone may vary. One of the important factors for the establishment of Trichoderma is the type of soil. In this study, the phaeozem favored the presence of *Trichoderma* strains. Another factor increasing the occurrence of this genus is soil acidity.

Trichoderma proved to be a successful organism that colonize the substrate rapidly and get hold of space over *Sclerotinia* spp. Confrontation tests allowed us to identify three strains with a high potential for biocontrol of *Sclerotinia*; these strains are a natural resource and an alternative to reduce the use of chemicals which have been applied in these study sites and have not shown efficient results in the control of soft rot due to this pathogen. However, it is important to carry out other studies that will allow us to show the effectiveness of these strains at the greenhouse level and on the field.

On the other hand, the molecular identification at the level of species, of the native strains that were found, will allow us to manage their potential, in an optimal and correct manner, in further studies.

ACKNOWLEDGMENTS

We wish to thank the National Council for Science and Technology (CONACyT) for the scholarship granted to Hilda Guadalupe García Núñez, in order to carry out her Masters studies during the 2008-2010 period. We also wish to thank SEP (Public Education Secretary), and PIFI Program 2007-2009, for financing of the following project: "Isolation and molecular characterization of native *Trichoderma spp* strains with potential for biocontrol of pests from horticultural soils of the Valley of Toluca" and finally thank Dr. Ángel Rolando Endará Agramot for their support in preparating the distribution map of *Trichoderma*.

REFERENCES

Adams, P.B. 1990. The potential of mycoparasites for biological control of plant diseases. Rev. Phytopathology, 28: 59-72.

Altieri, M.A. 1999. The ecological role of biodiversity in agroecosystems. Agriculture, Ecosystems and Environment, 74: 19-31.

Argumedo-Delira, R., Alarcón, A., Ferrera-Cerrato, R., Peña-Cabriales, J.J. 2009. El género fúngico *Trichoderma* y su relación con contaminantes orgánicos e inorgánicos. Rev. Internacional de Contaminación Ambiental, 25: 257-269.

Arzate-Vega, J., Michel-Aceves, A.C., Domínguez-Márquez, V.M., Santos-Emésica, O.A. 2006. Antagonismo de *Trichoderma* spp. sobre *Mycosfharella fijiensis* Morelet, agente causal de la sigakota negra del plátano (*Musa* sp.) *in vitro* e invernadero. Revista Mexicana de Fitopatología, 24: 98-104.

Bowen, J.K., Franicevic, S.C., Crowhurts, R.N., Templetom, M.D., Stewart, A. 1996. Differentiation of a specific Trichoderma biological control agent by restriction fragment length polymorphism (RFLP) analysis. New Zeland Journal of Crop and Horticultural Science, 24: 207-217.

Gashe, B.A. 1992. Cellulase production and activity by *Trichoderma sp*. A-001. Journal of Aplied Bacteriology, 73: 79-82.

Davis, R.M., Subbarao, K.V., Raid, R.N., Kurtz, E.A. 2002. Plagas y enfermedades de la lechuga. Mundi-Prensa. México.

Duffy, B.K., Ownley, B.H., Weller, D.M. 1997. Soil chemical and physical properties associated with suppression of take-all of wheat by *Trichoderma koningii*. Rev. Phytopathology, 87: 1118-1124.

Estrada, P.M. 2005. Manejo y procesamiento de la gallinaza. Revista Lasallista de Investigación, 2: 43-48.

Guigón-López, C., González-González, P.A. 2004. Selección de cepas de *Trichoderma spp*. con actividad antagónica sobre *Phytophora capsici* Leonian y promotoras de crecimiento en el cultivo de chile (*Capsicum annum* L.). Revista Mexicana de Fitopatología, 22: 117-124.

Hao, J.J., Subbarao, K.V. and Koike, S.T. 2003. Effects of brocoli rotation on lettuce drop caused by *Sclerotinia minor* and on the population density of sclerotia in soil. Plant Disease, 87:159-169.

Hao, J.J., and Subbarao, K.V. 2005. Comparative analyses of lettuce drop epidemics caused by *Sclerotinia minor* and *S. sclerotiorum*. Plant Disease, 89: 717-725.

Harman, E.G., Howell, C.R., Viterbo, A., Chet, I., Lorito M. 2004. *Trichoderma* species opportunistic, avirulent plant symbionts. Microbiology, 2: 43-56.

Hermosa, M.R., Grondona, I., Iturriaga, E.A., Díaz-Domínguez, J.M., Castro, C., Monte, E., García-Acha, I. 2000. Molecular characterization and identification of biocontrol isolates of *Trichoderma spp.* Applied and Environmental Microbiology, 66: 1890-1898.

Hubbard, J.C., Subbarao, K.V. and Koike, S.T. 1997. Development and significance of dicarboximide resistance in *Sclerotinia minor* isolates from commercial lettuce fields in California. Plant. Disease, 81: 148-153.

Infante, D., Martínez, B., González, N., Reyes, Y. 2*009.* Mecanismos de acción de *Trichoderma* frente a hongos fitopatógenos. Revista Protección Vegetal, 24: 14-21.

Kredicks, L., Antal, Z., Manczinger, L., Szekeres, A., Kevei, F., Nagy, E. 2003. Influence of enveronmental parametrers on *Trichoderma* strains with biocontrol potencial. Food Technology Biotecnology, 41: 37-42

Martínez, B., Solano T. 1994. Antagonismo de *Trichoderma* spp. frente a *Alternaria solani* (Ellis y Martin) Jones y Grout. Revista Protección Vegetal, 10: 221-225.

Martínez, B., Yusimy, Reyes., Infante. D., González, E., Baños, H., Cruz, A. 2008. Selección de aislamientos de *Trichoderma spp.* candidatos a biofungicidas para el control de *Rhizoctonia sp.* en arroz. Revista Protección Vegetal, 23: 118-125.

Matheron, M.E., Porchas, M. 2004. Activity of boscalid, fenhexamid, fluazinam, fludioxonil and vinclozolin on growth of *Sclerotinia minor* and *S. sclerotiorum* and development of lettuce drop. Plant Disease, 88: 665-668.

Michel-Aceves, A.C., Rebolledo-Domínguez, O., Lezama-Gutiérrez, R., Ochoa-Moreno, M.E., Mésima-Escamilla, J.C., Samuels, G.J. 2001. Especies de *Trichoderma* en suelos cultivados con mangoafectador por "Escoba de bruja" y su potencial inhibitorio sobre *Fusarium oxyporum* y *F. subglutinans*. Revista Mexicana de Fitopatología, 19: 154-160.

Michel-Aceves, A.C., Otero-Sánchez, M.A., Solano-Pascacio, L.Y., Ariza-Flores, R., Barrios-Ayala, A., Rebolledo-Martínez, A. 2009. Biocontrol *in vitro* con *Trichoderma* spp., *Fusarium subglutinans*, (Wollenweb y Reinking) Nelson, Toussoun y Marasas y *F. oxysporum Schlecht.*, Agentes causales de la "Escoba de bruja" del Mango (Mangifera indica L.). Revista Mexicana de Fitopatología, 27: 18-26.

Mónaco, C.I., Rollán, M.C., Nico, A. I. 1998. Efecto de micoparásitos sobre la capacidad

reproductiva de *Sclerotinia sclerotiorum.* Revista Iberoamericana de Micología, 15: 81-84.

Okoth, S.A., Roimen, H., Mutsotso, B. Muya, E., Kahindi, J., Owino, J.O., Okoth. 2007. Land use systems and distribution of *Trichoderma* species in Embu región, Kenya. Tropical and Subtropical Agroecosytems, 7: 105-122.

Osorio-Nila, M.A., Vázquez-García, L.M., Salgado-Siclán, M.L., González-Esquivel, C.E. 2005. Efecto de dos enmiendas orgánicas y *Trichoderma spp.* para controlar *Sclerotinia spp.* en lechuga (*Lactuca sativa* L.). Revista Chapingo Serie Horticultura, *11: 203-208.*

Quiroz-Sarmiento, F.V., Ferrera-Cerrato, R. 2008. Antagonismo *in vitro* de cepas de *Aspergillus* y *Trichoderma* hacia hongos filamentosos que afectan al cultivo del ajo. Revista Mexicana de Micología, 26: 27-34.

Schubert, M., Fink, S.F., Schwarze, W.M.R. Evaluation of *Trichoderma spp.* as a biocontrol agent against wood decay fungi in urban tres. Biological Control, 45: 111–123.

Subbarao, K.V. 1998. Progress toward integrated management of lettuce drop. Plant Disease, 82: 1068-1998.

Raebeendran, N., Jones, E.E., Moot, D.J., Stewart, A. 2006. Biocontrol of *Sclerotinia* lettuce drop by *Coniothyrium minita*ns and *Trichoderma hamatum*. Biological control, 39: 352-362.

Rey, M., Delgado-Jarana, J., Rincón, A.M., Limón, M. C., Benítez, T. 2000. Mejora de cepas de *Trichoderma* para su empleo como biofungicidas. Revista Iberoamericana de Micología, 17: 31-36.

Samaniego-Gaxiola, J.A. 2008. Efecto del pH en la sobrevivencia de esclerocios de *Phymatotrichopsis omnívora* (Dugg.) Hennebert expuestos a Tilt y *Trichoderma* sp. Revista Mexicana de Fitopatología, 26: 32-39.

Samuels, G.J. 2006. *Trichoderma*: Systematics, the sexual state, and ecology. Phytopathology, 96: 195-206.

Sanogo, S. and Puppala, N. 2007. Characterization of a darkly pigmented mycelial isolate of *Sclerotinia sclerotiorum* on Valencia peanut in New Mexico. Plant Disease, 91: 1077-1082.

Valenzuela, E., Leiva, S., Godoy, R. 2001. Variación estacional y potencial enzimático de microhongos asociados a la descomposición de hojarasca *Nothofagus pumilio*. Revista Chilena de Historia Natural, 74: 737-749.

Vinale, F., Sivasithamparam, K., Ghisalberti, E.L., Marra, R., Woo, S.L., Lorito, M. 2008. *Trichoderm* plant pathogen interactions. Soil Biology and Biochemistry, 40: 1-10.

Wu, B.M., and Subbarao, K.V. 2006. Analyses of lettuce drop incidence and population structure of *Sclerotinia sclerotiorum* and *S. minor*. Phytopathology, 96: 1322-1329.

7

Ruderal Plants: Temporary Hosts of Arbuscular Mycorrhizal Fungi in Traditional Agricultural Systems?

José Ramos-Zapata[1*], Denis Marrufo-Zapata[1], Patricia Guadarrama-Chávez[2], Uriel Solís-Rodríguez[1] and Luis Salinas-Peba[1]

[1]*Departamento de Ecología Tropical, Campus de Ciencias Biológicas y Agropecuarias, Universidad Autónoma de Yucatán, km 15.5 de la Carretera Mérida-Xmatkuil AP 4-116, Itzimná Mérida, Yucatán, México. E-mail: aramos@uady.mx*

[2]*Unidad Multidisciplinaria de Docencia e Investigación, Facultad de Ciencias, Universidad Nacional Autónoma de México, Sisal, Yucatán 97356, México. *Corresponding author*

SUMMARY

Ruderal plants may serve as temporary hosts of arbuscular mycorrhizal fungi (AMF), by maintaining the availability of active propagules in the soil, which in turn favors rapid colonization of roots of cultivated species during the agricultural cycle. The goals of this study were to: 1) estimate the richness of ruderal plant species in an agricultural plot and determine their mycorrhizal status, 2) quantify the number of live AMF spores in soil samples, and 3) estimate the infection potential and number of active propagules in soil samples from the agricultural site. The agricultural site used was located in Yucatan, Mexico, and consisted of a monoculture of corn subjected to low-impact agricultural practices during the last five years. A total of 20 species of ruderal plants were found at the experimental site, belonging to 11 families. All the sampled species exhibited associations with AMF, and colonization percentages ranged from 11.7±0.07 to 79.6±0.01 among species. The rhizosphere presented an average of 565±324 spores in 50 g of dry soil, of which 58.76% of the spores were alive. The inoculum potential of the soil was 50.4±0.05%, while the number of infective propagules was 193.37 (both in 50 mL of soil). Results from this study show that the presence of ruderal species in agricultural sites may promote the maintenance of AMF communities by acting as temporary hosts of these fungal species. In doing so, ruderal species can favor a higher production of infective AMF propagules and thus stronger mycorrhizal interactions with cultivated species.

Key words: Slash and burn; low-impact agriculture; corn; rhizosphere; infective propagules.

INTRODUCTION

Slash and burn agriculture is one of the most common forms of traditional agriculture used to cultivate corn in tropical regions (Palerm, 1981). In Yucatan, Mexico, this agricultural practice is very common despite the high degree of stoniness in the soil, low rainfall, and low water retention capacity of the soil (Duch, 1991; Arias, 1995). Slash and burn agriculture consists of cutting down and burning the vegetation at the onset of the rainy season, followed by a cultivation period of 2 to 3 years, after which the site remains unused (i.e. fallow) for a 15 to 20-year period during which the vegetation recovers its original structure and species composition, to then be used in another cycle (Duch, 1991). Pool and Hernández (1987), mentioned that in the state of Yucatan, Mexico, this agricultural practice has been modified by reducing the fallow period, leading to a limited regeneration of the vegetation as well as greater soil erosion. Another problem currently associated with slash and burn agriculture is the indiscriminate use of fertilizers to compensate for low nutrient availability in the soil, as well as the use of herbicides for weed control. Although both types of chemical inputs can have negative impacts on soil microbial communities, the effects on biodiversity and functionality (*v.g.* plant protection, P uptake) have been largely neglected (Arias, 1992; Ramos-Zapata *et al.*, 2012).

Ruderal plant species are an undesired component of conventional agricultural practices because they compete for resources (water, nutrients or space) with cultivated species (Hart, 1985; Caamal and Castillo, 2011). During the fallow period, ruderal species proliferate and colonize the abandoned agricultural site, given high nutrient availability. In doing so, ruderal plants play a relevant ecological role by maintaining microclimatic conditions in the soil and contributing to nutrient and mineral recycling (Altieri and Whitcomb, 1979; Gliessman, 1990). Ruderal species also contribute to the pathogen and pest control in cultivated species (Altieri and Whitcomb, 1979) and they are hosts of several groups of symbiotic microorganisms (Baumgartner *et al.*, 2005). For example, ruderal plants are hosts of many species of arbuscular mycorrhizal fungi (AMF) (Baumgartner, 2005; Hausmann and Hawkes, 2009). Previous studies have suggested that ruderal plants favor an increase in diversity and abundance of AMF in the soil of cultivated sites (Vatovec *et al.* 2005; Ramos-Zapata *et al.*, 2012).

In spite of their potentially beneficial effects through increased AMF abundance, ruderal plant species are usually removed manually or with the use of herbicides which may have negative effects on soil biota, including AMF. These negative impacts include the reduction of AMF species richness, diversity and inoculum potential (Kurle and Pfleger, 1994); other agricultural practices such as fertilization or continuo monocultures can also affect the composition and diversity of AMF (Sieverding, 1990; Jansa *et al.*, 2002; Oehl *et al.*, 2003). The use of low-impact agriculture (i.e. not mechanized), cover crops, crop rotation and fallow periods, as well as tolerance for ruderal species (up to certain levels of abundance), are practices which mitigate the negative impacts of conventional agriculture on AMF. Indeed, tolerating the abundance of ruderal species up to a given threshold may favor the maintenance of AMF communities when the crop species is not planted. In this way, the presence of ruderal species favors the early colonization of crop species by AMF. Accordingly, here we test the hypothesis that ruderal plant species may act as potential hosts of AMF and the present study has the following goals: 1) to estimate the richness of ruderal plant species in an agricultural plot and determine their mycorrhizal status, 2) to quantify live AMF spores at this agricultural site, and 3) to estimate the infection potential and number of infective propagules present in the soil of the agricultural study site.

MATERIALS AND METHODS

Site description

The study was conducted at the Campus de Ciencias Biológicas y Agropecuarias (CCBA) of the Universidad Autónoma de Yucatán (20°52' 3.86'' N; 89°37'20.05'' W). The study site is of karstic geological origin, with a high abundance of exposed rocks. The soil type is leptosol with intermediate texture, brown in color (Bautista-Zúñiga *et al.*, 2003). The climate is warm subhumid with rains during the summer and winter. The mean annual precipitation is 900 mm and the mean annual temperature is 27.5 °C (García, 1973). The dominant vegetation type corresponds to a deciduous tropical forest (Flores and Espejel, 1994) with varying levels of disturbance.

The experimental plot has an area of 1 ha, and is surrounded by secondary vegetation derived from a deciduous tropical forest. Starting in 2006, *Zea may* L. have been periodically grown as a monoculture, following traditional fallow periods but without burning or using fertilizers. The soil has a pH of 7.7, a carbon content of 5.3 mg per 100g of soil, while nitrogen and total phosphorus are 1.25 g/kg and 685.8 mg/kg, respectively.

Species richness and mycorrhizal status of ruderal plants

During the fallow period in June 2011, we characterized the community of ruderal species present in the experimental plot. Specimen sampling was conducted by randomly placing 15 quadrats of 0.5 x 0.5-m, throughout the plot. We collected all ruderal species present inside each quadrat, and identified them using specific keys and herbarium specimens. Once all plants were identified, we collected the fine roots (<2 mm in diameter) of three specimens per species.

Roots were labeled and transported to the laboratory where AMF presence and percentage of mycorrhizal colonization were determined. Roots were washed with tap water and dyed with trypane blue following Phillips and Hayman (1970), modified by Hernández-Cuevas et al. (2008). Subsequently, we prepared permanent samples with polyvinyl alcohol, lactic acid and glycerin (PVLG) and estimated the percent of total AMF colonization for each fungal structure (hyphae, vesicles, spores, arbuscules and coils) in each sample using the intersection method by McGonigle et al., (1990).

AMF propagules

Within each of the 15 sampling quadrats, we collected 1-kg soil sample from the first 15 cm of soil. The soil samples from five randomly chosen quadrats, were mixed, and finally three compound samples were drawn from this mixture. From each compound sample, a 100 g sub-sample were placed in the oven at 80°C for 24 h, in order to estimate the soil dry weight, and 50 g from these compound samples were used to isolate and quantify AMF spores, using the humid sieving method by Gerdemann and Nicolson (1963), modified by Hernández-Cuevas et al. (2008). Spores were observed under an optic microscope (40-100X), separating and counting live spores (i.e. with lipid content and no signs of damage), and dead spores (damaged wall, lacking lipid content), for each soil sample.

The infection potential of the soil samples was evaluated with a bioessay conducted in the greenhouse using 300-mL pots for each compound soil sample (three compound samples, five replicates per sample, n= 15). Pots were filled with 200 mL of soil that was previously steam sterilized, a layer of 50 mL of soil to be subsequently sampled (not mixed), and finally a layer of steam sterilized soil in order to avoid contamination. A single sorghum plant

(*Sorgum vulgare* L.) was transplanted to each pot, with plants being previously germinated with a sterile substrate. Pots were randomly distributed and rotated weekly. Plants were watered every second day during a six-week period, after which they were harvested and roots were collected to evaluate the presence of AMF structures and to determine the total percentage of AMF colonization following McGonigle et al. (1990).

The number of AMF infective propagules was estimated with a bioessay in the greenhouse, using the technique of the most probable number of infective propagules (MPN) (Porter, 1979) which consists of a series of dilutions (4^{-0} to 4^{-7}) using steam sterilized soil samples, with five replicates per dilution. Each dilution was placed in pots, as described previously using the same methodology, and after six weeks each seedling was harvested and the roots were collected to evaluate the presence of AMF structures following McGonigle et al. (1990); the MPN of infective propagules was calculated as described in Ramos-Zapata et al. (2011).

Data analyses

Ruderal species richness (S) was calculated based on the specimens collected in each of the sampled quadrats. Spore density was quantified considering both live and dead spores. The number of infective propagules in 50 mL of soil per compound sample (3 replicates) was analyzed by means of a likelihood test (Ramos-Zapata et al., 2011).

RESULTS AND DISCUSSION

A total of 20 ruderal plant species were identified during the fallow period of the study plot. These species belonged to 11 plant families, of which Euphorbiaceae and Poaceae had the highest number of species (three species each) (Table 1). The species reported for both of these families are common in agricultural systems (Villaseñor and Espinosa, 1998), in particular those in the Yucatan Peninsula (Caamal et al., 2001; Castillo-Caamal et al., 2010). However, our finding is comparatively lower in relation to the findings by Cocom et al. (2008) who reported up to 40 ruderal species during the fallow period of an agricultural site from the same area of Yucatan Peninsula. This difference in results may be related to the life history traits of the species found, in particular the competitive ability for light, water and nutrients (Holm et al., 1979; Gupta et al., 2008). In addition, the continuous application of herbicies may select for specific ruderal species which are more resistant and therefore become dominant in agricultural sites

(Caamal *et al.*, 2001). Therefore, by restricting the use of herbicides, the number of ruderal species present in agricultural sites will likely increase.

Plant species belonging to Commelinaceae, Portulacaceae and Zygophyllaceae families have been previously reported as non-mycorrhizal (Gerdemann, 1968; Trappe, 1987). However, in this study we found that several ruderal species belonging to these families had their roots colonized by AMF, so these plants may act as temporal hosts of AMF species. Our results emphasize the importance of conducting exhaustive studies which screen for the presence of mychorrhizae in roots of ruderal species belonging to families previously thought to be non-mycorrhizal, in order to reduce the use of herbicides to promote the presence of potentially mycorrhizal ruderal plants in different agriculture sites and management conditions.

Moreover, according to Baumgartner *et al.* (2005), ruderal plant species exhibit a variable mycorrhizal status depending on their abundance at an agricultural site. Indeed, although we found that all the ruderal species sampled established mycorrhizal interactions, colonization levels varied considerably (range from 11.7±0.07 in *Euphorbia ocymoidea* to 79.6±0.01 in *Andropogon virginicus*). The species *Sanvitalia procumbens*, *Sida acuta* and *Bidens pilosa* exhibited high levels of colonization (greater than 40%), which contrasts with results previously reported by Ramos-Zapata *et al.* (2012) for the same species (less than 15%) in an agricultural system with corn. Nonetheless, the comparatively lower numbers reported by this latter work may have resulted from a shorter fallow period at such agricultural site, which could have restricted the establishment of mycorrhizal interactions.

Table 1. List of ruderal species present in a traditional agricultural system with maize (Yucatan, Mexico). The agricultural system was sampled during the fallow period; values given are mean percent (± SE) of overall AMF colonization (AMF), as well as mean colonization for each fungal structure separately (H: hyphae, V: vesicles, S: spores, C: arbusculate and hyphal coils).

Family	Species	AMF	H	V	S	C
Acanthaceae	*Justicia* sp.	36.02 (±3.3)	32.52 (±2.9)	5.78 (±1.2)	1.14 (±0.7)	1.49 (±0.6)
	Ruellia nudiflora	47.56 (±4.1)	41.99 (±3.5)	20.33 (±9.2)	6.20 (±4.2)	2.77 (±0.5)
Asteraceae	*Bidens pilosa* L.	79.67 (±0.9)	71.67 (±1.8)	30.67 (±5.7)	11.33 (±4.4)	
	Sanvitalia procumbens	67.77 (±10.6)	64.79 (±11.2)	7.61 (±4.7)	3.12 (±1.02)	
Commelinaceae	*Commelina elegans*	35.33 (±9.9)	31.33 (±10.6)	12.44 (±7.8)	5.78 (±4.5)	1.33 (±1.1)
Euphorbiaceae	*Acalypha alopecuroides*	47.11 (±3)	32.19 (±1.1)	24.12 (±10.1)	9.23 (±3.3)	1.63 (±0.8)
	Euphorbia heterophylla	42.02 (±9.6)	38.35 (±9.2)	17.00 (±5.9)	6.48 (±2.5)	
	Euphorbia ocymoidea	31.49 (±6)	20.59 (±7.2)	12.68 (±0.3)	6.42 (±1.6)	
Leguminosae	*Desmanthus virgatus*	41.44 (±6.4)	39.11 (±6)	15.67 (±5.8)	3.67 (±2.7)	
Malvaceae	*Abutilon hirtum* (Lam.)	11.73 (±4.4)	11.40 (±4.3)	3.12 (±1.2)		
	Corchorus siliquosus L.	30.28 (±6)	25.68 (±5.6)	7.60 (±1.6)		
	Sida acuta Burm	49.87 (±2.2)	42.79 (±2.9)	23.87 (±8.2)	6.07 (±1.3)	
Passifloraceae	*Passiflora obovata*	64.40 (±10.3)	58.06 (±8.6)	16.21 (±12)	5.51 (±4.3)	0.67 (±0.5)
Poaceae	*Andropogon virginicus*	52.18 (±1.9)	48.46 (±1.8)	10.16 (±0.5)	3.39 (±0.3)	0.35 (±0.3)
	Cynodon dactylon (L.)	60.09 (±1.6)	46.52 (±6.4)	24.16 (±4.6)	8.08 (±2.2)	0.42 (±0.3)
	Urochloa fusca (Sw.)	69.33 (±12.4)	47.00 (±7)	46.67 (±17.7)	20.67 (±7.8)	
Portulacaceae	*Portulaca oleracea* L.	54.77 (±3.7)	50.59 (±5.4)	17.76 (±3.8)	4.43 (±1.4)	0.33 (±0.3)
	Portulaca pilosa L.	22.14 (±12.6)	15.19 (±7.2)	11.52 (±8.5)	3.57 (±3.6)	
Rubiaceae	*Moringa yucatanensis*	50.06 (±6.7)	42.56 (±10.3)	7.40 (±3)	2.94 (±1.9)	1.13 (±0.5)
Zigophyllaceae	*Kallstroemia maxima*	46.64 (±1.2)	28.01 (±6.4)	25.64 (±1.7)	8.51 (±1.8)	

Our finding that mycorrhizal colonization was widespread among all ruderal species present at the studied agricultural plot differs from previous studies (*v.g.* Janos, 1980; Pringle *et al.*, 2009). Indeed, several authors (v.g. Grime *et al.* 1987, Francis and Read 1994, Smith and Read, 2008) have suggested that high dispersal ability, high growth rates, early reproduction, and high tolerance to a wide range of abiotic conditions represent life history traits which may preclude a strong dependency on interactions with AMF. Nonetheless, it has been shown that different genotypes of the ruderal species *Ruellia nudiflora* vary in their response to mycorrhizal interaction, influencing plant survival as well as establishment (Ramos-Zapata *et al.*, 2010). Therefore, interactions among ruderal plants and AMF, and the benefits obtained by the former will depend on factors such as plant and AMF genotype as well as environmental conditions and the availability of AMF propagules as has been proposed (Annapurna *et al.* 1996, Smith and Read, 2008).

All of the ruderal species sampled presented AMF hyphae and vesicles in their roots, and some species presented spores and coils (both arbusculate and hyphal) (Table 1). The presence of coils in roots examined suggests a colonization of *Paris* type, in spite of being the *Arum* type the most common for plants with high growth rates such as ruderal species (Smith and Smith, 1997). Our results support the idea that the type of colonization depends not only on plant life-history traits, but also on identity of AMF and their availability, based on which a continuum between both types of colonization may be found as has been suggested before (v.g. Cavagnaro *et al.* 2001, Dikson, 2004). Any way, it is clear from our results that the majorities of ruderal species at least in tropical agrosystems establish a colonization of *Paris* type (but see Ramos-Zapata *et al.*, 2010), and this may be advantageous in environment which exhibit strong fluctuations in resource availability as is the case of the studied plots.

The mean number of AMF spores found was 565±324 in 50 g of dry soil, of which 58.76% were alive (332±199 spores). Spore density varies among cultivation cycles depending on the availability of potential hosts, based on which it is important to conserve a stock of ruderal species associated with a given crop, especially when the cultivated species is planted in monoculture (Feldmann and Boyle, 1999). On the other hand, the soil inoculum potential (50.4±0.05%) and the number of infective propagules (193.37 [47.33 lower limit, 637.64 upper limt]) calculated in this study in 50mL of soil, reveals that during the fallow period (when the main crop is absent) the community of AMF remains active. Levels of AMF soil inoculum may respond to the type of agricultural management scheme (Brundrett *et al.*, 1996; Ramos-Zapata *et al.*, 2012) influencing the presence of AMF both into the roots of ruderal species as well as in the rhizosphere soil. The high density of live spores of AMF and the presence of infective propagules found in our study may respond to the presence of ruderal species which act as AMF host during periods when the crop species is not planted (Baumgartner *et al.*, 2005). Therefore, we hypothesize that ruderal plants maintain the production of spores and propagules throughout long periods of time, and in doing so guarantee the maintenance of a stock of AMF to colonize crops (Jordan *et al.*, 2000; Zangaro *et al.*, 2000).

Despite the benefits of AMF in agroecosystems, for example by modifying the abundance of ruderal species (Jordan *et al.*, 2000) or suppressing aggressive weeds in agricultural systems (Vatovec *et al.*, 2005; Jordan and Huerd, 2008; Rinaudo *et al.*, 2010), much work remains to be conducted in order to understand the interactions among AMF and ruderal species in tropical agricultural systems; here with our results, we are able to suggest that the presence of ruderal species acting as temporal host of AMF may play an essential role mantaining active the AMF community until the fallow period ends and the principal crop is planted and in consequence provoking a rapid colonization of its roots.

CONCLUSIONS

Traditional agricultural systems with maize involve low-impact practices that promote the abundance and richness of ruderal plant species that are potentially colonized by AMF, at least during the fallow period of the cultivation cycle. In doing so, ruderal species that establish mycorrhizal interactions may act as temporary hosts of AMF species and contribute to the long-term establishment and stability of communities of these symbiotic microorganisms which may have positive effects on crop yield in agricultural systems.

REFERENCES

Altieri, M.A., Whitcomb, W.H. 1979. The potential use of weeds in the manipulation of beneficial insects. HortScience. 14:12-18.

Annapura, K., Tilak, K.V.B.R., Mukerji, K.G. 1996. Arbuscular mycorrhizal symbiosis recognitions and specificity. In: Mukerji, K.G. (ed.). Concepts in mycorrhizal research. Kluwer Academic Publishers. Dordrecht, The Netherlands. pp 77-90.

Arias, R.L. 1992. El proyecto dinámica de la milpa en Yaxcaba, Yucatán. In: Zizumbo, D., Rasmussen, Ch., Arias, L. M., Terán, S. (eds.). La modernización de la milpa en Yucatán: Utopía o Realidad. Centro de Investigación Científica de Yucatán. Mérida. Yucatán. México. pp. 195-201.

Arias, R.L. 1995. La producción milpera actual en Yaxcaba, Yucatán. In: Hernández, E., Bello, E., Levi, S. (comps.). La Milpa en Yucatán, un sistema de producción agrícola tradicional. Colegio de Postgraduados. México. pp. 171:200.

Baumgartner, K., Smith, R.F., Bettiga, L. 2005. Weed control and cover crop management affect mycorrhizal colonization of grapevine roots and arbuscular mycorrhizal fungal spore populations in a California vineyard. Mycorrhiza. 15:111–119.

Bautista–Zúñiga, F., Jiménez–Osornio, J., Navarro–Alberto, J. 2003. Microrelieve y color del suelo como propiedades de diagnóstico en leptosoles cársticos. Terra. 21:1–11.

Brundrett, M.C., Bougher, N., Bernie, D., Grove, T., Malajczuk, N. 1996. Working with micorrhizas in forestry and agriculture. Monografía ACIAR 32, Canberra Australia.

Caamal, J.A., Jiménez J., Torres-Barragán, A., Anaya, A. 2001. The use of allelopathic legume cover and mulch species for weed control in cropping systems. Agronomy Journal. 93(1):27-36.

Caamal, J.A., Castillo, J.B. 2011. Muestreo de arvenses In: Bautista, F., Palacio, J.L., Delfín, H. (eds.). Técnicas de muestreo para manejadores de recursos naturales. Universidad Nacional Autónoma de México. México, D. F. pp. 537-561

Castillo-Caamal, J.B., Caamal-Maldonado, J.A., Jimenéz-Osornio, J.J., Bautista-Zúñiga, F., Amaya-Castro, M.J., Rodríguez-Carrillo, R. 2010. Evaluación de tres leguminosas como coberturas asociadas con maíz en el trópico subhúmedo. Agronomía Mesoamericana. 21:39-50.

Cavagnaro, T.R., Gao, L.-L., Smith, F.A., Smith, S.E. 2001. Morphology of arbuscular mycorrhizas is influenced by fungal identity. New Phytologist. 151:469–475

Cocom, R., Avilés, L., Garrido, J., Ortiz, J. J. 2008. Plantas silvestres asociadas al cultivo de maíz en Yucatán. Universidad Autónoma de Yucatán. Fundación Produce Yucatán A. C. México.

Dickson, S. 2004. The Arum–Paris continuum of mycorrhizal symbiosis. New Phytologist. 163:187–200

Duch, G.J. 1991. Condicionamiento ambiental y modernización de la milpa en el estado de Yucatán. In: Zizumbo, D., Rasmussen, Ch., Arias, L.M., Terán, S. (eds.). La modernización de la milpa en Yucatán: Utopía o Realidad. Centro de Investigación Científica de Yucatán. Mérida. Yucatán. México. pp. 227-246.

Feldmann, F., Boyle, C. 1999. Weed-mediated stability of arbuscular mycorrhizal effectiveness in maíz monocultures. Journal of Applied Botany 73:1-5.

Flores, J.S., Espejel, I. 1994.Tipos de vegetación de la península de Yucatán. Fascículo 3. Etnoflora Yucatanense. Facultad de Medicina Veterinaria y Zootecnia. Universidad Autónoma de Yucatán. México.

García, E. 1973. Modificaciones al sistema de clasificación climática de Köppen. Universidad Nacional Autónoma de México, México.

Gerdemann, J.W. 1968. Vesicular-arbuscular mycorrhiza and plant growth. Annual Review of Phytopathlogy. 6: 397-418.

Gerdemann, J.W., Nicolson, T. 1963. Spores mycorrhizal endogone species extracted from soil by wet sleving and decanting. Transactions of the British Mycological Society. 42:235-244.

Gliessman, S. R. 1990. Agroecology. Researching the Ecological Basis for Sustainable Agriculture. Springer-Verlag. New York.

Grime, J.P, Mackey, J.M.L,, Sillier, S.H., Read, D.J. 1987. Floristic diversity in a model system using experimental microcosm. Nature. 328: 420–422.

Gupta, A., Joshi, S.P., Manahas, R.K., 2008. Multivariate analysis of diversity and composition of weeds communities of wheat fields in Doon valley India. Tropical Ecology. 49:103-112.

Hart, R.D. 1985. Conceptos básicos sobre agroecosistemas. Centro Agronómico Tropical de Investigación y Enseñanza (CATIE). Turrialba, Costa Rica.

Hausmann, N.T., Hawkes, C.V. 2009. Plant neighborhood control of arbuscular mycorrhizal community composition. New Phytologist. 183:1188–1200.

Hernández–Cuevas, L., Guadarrama–Chávez, P., Sánchez–Gallén, I., Ramos–Zapata, J. 2008. Micorriza arbuscular: colonización intrarradical y extracción de esporas. In: Álvarez–Sánchez, J., Monroy–Ata, A. (comps.). Técnicas de Estudio de las Asociaciones Micorrízicas y sus Implicaciones en la Restauración. Facultad de Ciencias. Universidad Nacional Autónoma de México. México. D.F. pp. 1–15.

Holm, L., Pancho, J., Herberger, J., Plucknett., D. 1979. A Geographical Atlas of World Weeds. John Wiley & Sons. Nueva York.

Janos, D.P. 1980. Vesicular arbuscular mycorrhizae affect lowland tropical rain forest plant growth. Ecology. 61:151-162.

Jansa, J., Mozafar, A., Anken, T., Ruh, R., Sanders, I. R., Frossard, E. 2002. Diversity and structure of AMF communities as affected by tillage in a temperate soil. Mycorrhiza. 12:225–234.

Jordan, N., Huerd, S. 2008. Effects of soil fungi on weed communities in a cornsoybean rotation. Renewable Agriculture and Food Systems. 23:108–117.

Jordan, N.R., Zhang, J., Huerd, S. 2000. Arbuscular-mycorrhizal fungi: potential roles in weed management. Weed Research. 40:397–410.

Kurle, J.E., Pfleger, F.L. 1994. The effects of cultural practices and pesticides on VAM fungi. In: Pfleger, F. L., Linderman, R. G. (eds.).

Mycorrhizae and Plant Health. APS Press, St. Paul, Minnesota. pp. 101-131.

McGonigle, T.P., Miller, M.H., Evans, D.G., Fairchild, G.L., Swan, J.A. 1990. A method which gives and objective measure of colonization of roots by vesicular-arbuscular mycorrhizal fungi. New Phytologist. 115:495-501.

Muthukumar, T., Prakash, S. 2009. Arbuscular mycorrhizal morphology in crops and associated weeds in tropical agro-ecosystems. Mycoscience. 50:233–239.

Oehl, F., Sieverding, E., Ineichen, K., Mader, P., Boller, T., Wiemken, A. 2003. Impact of land use intensity on the species diversity of arbuscular mycorrhizal fungi in agroecosystems of central Europe. Applied and Environmental Microbiology. 69:2816–2824.

Palerm, A. 1981. The agricultural basis of urban civilizations in Mesoamerica. In: Graham, J. (ed.). Ancient Mesoamerica. Selected Readings. Peek Publications, USA. Pp. 101-116.

Phillips, J., Hayman, D. 1970. Improved procedures for clearing roots and staining parasitic and vesicular–arbuscular mycorrhizal fungi for assessment of infection. Transactions of the British Mycological Society. 55:158–161.

Pool, N.L., Hernandez-Xolocotzi, E. 1987. Corn producction intensification under flash and burn agricultura in Yaxcaba, Yucatán, México. Terra. 5:149-162.

Porter, W. 1979. The Most Probable Number method for enumerating infective propagules of vesicular arbuscular mycorrhizal fungi in soil. Australian Journal of Soil Research. 17:515-519.

Pringle, A., Bever, J.D., Gardes, M., Parrent, J.L., Rilling, M.C. 2009. Mycorrhizal Symbioses and Plant Invasions. Annual Review of Ecology and Systematics. 40:699–715.

Ramos-Zapata, J., Campos, M.J., Parra-Tabla, V., Abdala-Roberts, L., Navarro, J. 2010. Genetic variation in the response of the weed *Ruellia nudiflora* (Acanthaceae) associated to arbuscural mycorrhizal fungi. Mycorrhiza. 20:275-280.

Ramos-Zapata, J.A., Guadarrama, P., Navarro-Alberto, J., Orellana, R. 2011.Arbuscular mycorrhizal propagules in soils from a tropical forest and an abandoned cornfield in Quintana Roo, Mexico: visual comparison of most-probable-number estimates. Mycorrhiza. 21:139-144.

Ramos-Zapata, J., Marrufo-Zapata, D., Guadarrama, P., Carrillo-Sánchez, L., Hernández-Cuevas, L., Caamal-Maldonado, A. 2012. Impact of weed control on arbuscular mycorrhizal fungi in a tropical agroecosystem: a long-term experiment. Mycorrhiza. DOI 10.1007/s00572-012-0443-1

Rinaudo, V., Barberi, P., Giovannetti, M., van der Heijden, M.G.A. 2010. Mycorrhizal fungi suppress aggressive agricultural weeds. Plant and Soil. 333:7-20.

Sieverding, E. 1990. Ecology of VAM fungi in tropical agrosystems. Agriculture, Ecosystems and Environment. 29:369–390.

Smith, F.A., Smith, S.E. 1997. Structural diversity in (vesicular)-arbuscular mycorrhizal symbioses. New Phytologist. 137:373-388.

Smith, S.E., Read, D.J. 2008. Mycorrhizal symbiosis. Academic Press. London.

Trappe, J.M. 1987. Phylogenetic and ecologic aspects of mycotrophy in the angiosperms from an evolutionary standpoint. In: Safir, G.R. (ed). Ecophysiology of VA Mycorrhizal Plants. CRC Press, Boca Raton. Florida. pp 5-25.

Vatovec, C., Jordan, N., Huerd, S. 2005. Responsiveness of certain agronomic weed species to arbuscular mycorrhizal fungi. Renewable Agriculture and Food Systems. 20:181–189.

Villaseñor, R.J.L., Espinosa, F.J.G. 1998. Catálogo de malezas de México. Universidad Nacional Autónoma de México. Consejo Nacional Consultivo Fitosanitario. Fondo de Cultura Económica. México, D.F.

Zangaro, W., Bononi, V.L.R., Truffen, S.B. 2000. Mycorrhizal dependency, inoculum potential and habitat preference of native woody species in South Brazil. Journal of Tropical Ecology. 16:603-622.

8

Comparative Assessment on the Nutritional and Antinutritional Attributes of the Underutilized Legumes, *Canavalia gladiata* (JACQ.) Dc, *Erythrina indica* Lam. and *Abrus precatorius* L

Pious Soris Tresina and Veerabahu Ramasamy Mohan*

Ethnopharmacology Unit, Research Department of Botany,
V.O.Chidambaram College, Tuticorin-628008, Tamil Nadu, India.
E-mail: vrmohan_2005@yahoo.com
*Corresponding author

SUMMARY

The seed samples of *Canavalia gladiata, Erythrina indica* and *Abrus precatorius* (red and black coloured seed coat and white coloured seed coat) were collected from the regions of Tamil Nadu and analyzed for their chemical composition with a view to evaluate their nutritional potential. The proximate composition revealed that, all the presently investigated seed samples were found to contain high content of crude protein and crude lipid. Mineral profiles were also analyzed in all the seed samples. All the investigated pulses appeared to be good sources of potassium, magnesium, sodium, calcium, phosphorus, iron, zinc, copper and manganese content were deficient when compared with Recommended Dietary Allowance (NRC/NAS, 1980, 1989). Vitamins (niacin and ascorbic acid) contents were found to be relatively high in the investigated pulses. The essential amino acid profiles of total seed proteins were compared favorably with FAO/WHO (1991) requirement pattern. Fatty acid profiles revealed that, all the investigated seed samples had rich unsaturated fatty acids (55.60-72.04%) and very high content of linoleic acid (24.16-34.14%). The IVPD of the investigated tribal pulses ranged from 63.31-71.36%. Antinutritional factors like total free phenolics, tannins, L-DOPA, phytic acid, hydrogen cyanide, trypsin inhibitor, oligosaccharides and phytohaemagglutinating activity were analyzed.

Key words: Legumes; nutritive value.

INTRODUCTION

Legumes are widely grown throughout the world and their dietary and economic importance is globally appreciated and recognized. Legumes not only add variety to diet but also serve as an economical source of supplementary proteins for a large human population. In India, they provide the only high protein component of the average diet and over 10 million tones are consumed annually (Sood *et al.*,

2002). They have high protein content (20-26%) and can be considered as a natural supplement to cereals. Next to fish (dry) which provides 335g protein per kg, grain legumes provide 220-250g protein per kg. Hence legumes are considered as "poor man's meat" (Kakati *et al.*, 2010). The availability and consumption of protein food in India will remain inadequate due to population explosion and urbanization leading to Protein Energy Malnutrition (PEM). The PEM problem can be alleviated by finding alternative cost effective sources of proteins (Waterlow, 1994).

With an increasing interest in new food resources, the seeds of wild plants including the tribal pulses receive more attention, because they are highly resistant to disease and pests and exhibit good nutritional qualities (Janardhanan *et al.*, 2003a). The underutilized legumes/wild tribal pulses have tremendous potential for commercial exploitation but remain ignored. They offer good scope to meet the ever-increasing demands for vegetable protein. Although they have high protein content and good nutritional value, their utilization is limited by the presence of some antinutritional/antiphysiological/toxic substances (Pugalenthi *et al.*, 2004).

Hence, the present study deals with the nutritional and antinutritional aspects of three underexploited legumes/little known/underutilized tribal pulses viz, *Canavalia gladiata*, *Erythrina indica* and *Abrus precatorius*.

Canavalia gladiate

The sword bean, *Canavalia gladiata* (Jacq.) DC (local name: Koliawarai) is a representative of the family Fabaceae and is distributed in North and Penninsular India. Fruits of sword bean are consumed by the Indian ethnic/tribal peoples of Arunachal Pradesh, Nagaland, Manipur, Mizoram, Tripura and Meghalaya (Borthakur, 1996). The Kaddar, Mannan and Muthuvan tribal sects of Kerala state, South India, consume unripe fruits of sword bean (Radhakrishnan *et al.*, 1996). Cooked young pods and seeds are known to be consumed by the Palliyar tribals living in Grizzled Giant Squirrel Wildlife Sanctuary, Srivilliputhur, South-Eastern slopes of Western Ghats, Tamil Nadu, India (Arinathan *et al.*, 2007). This species is also used as a cover crop and roasted seeds are ground to prepare a coffee-like drink in Guatemala (Bressani *et al.*, 1987).

Erythrina indica

Erythrina indica Lam. is commonly known as Kalyana murungai, and grows throughout India, mostly in Assam, Bengal, Konkan, Tamil Nadu and Kerala. It is a deciduous tree with orange red flowers and grows to a height of 15m with rough bark. The young fruits and seeds are taken as vegetable by the tribes of Susale island, Pune district (Vartak and Suryanarayana, 1995). Despite, being edible, seeds of *E. indica* are also medicinally valuable. Seed paste is used to massage in case of paralysis and to kill worms (Pal and Jain, 1988).

Abrus precatorius

Abrus precatorius L., an underutilized food legume, is commonly known as Gundumani. It is indigenously found throughout India, even at altitudes upto 1200m on the outer Himalayas. It is now naturalized in all tropical countries (Dwivedi, 2004) and Andaman Islands, having good nutritional properties receive more attention as an alternative protein source. It is a twining herb, with pinkish white or pink coloured flowers. The seeds are brightly coloured, elliptic to sub globose, smooth, glossy, and shinning. The seeds of *A. precatorius* are very similar in weight. The seeds are much valued in making jewelry for their bright colouration. The cooked seeds have been consumed by Onges of Andaman and Katharis of Pune District, Maharashtra, India during extreme famine (Janardhanan *et al.*, 2003b; Pugalenthi *et al.*, 2007). The seeds of *A. precatorius* have medicinal values. The seeds are used to treat diabetes and chronic nephritiss. Dry seeds of *A. precatorius* are powdered and taken one teaspoonful once a day for two days to cure worm infection (Rain-tree, 2004).

MATERIALS AND METHODS

Collection of seed samples

The seeds of *C. gladiata* (Jacq.) DC were collected from Arumugamangalam, Tuticorin district, Tamil Nadu, during August 2010, *E. indica* Lam. seeds were collected from Spic Nagar, Tuticorin district, Tamil Nadu during July, 2010. The seeds of *A. precatorius* L. red and black coloured seed coat were collected from Sathankulam, Tuticorin district and white coloured seed coat of *A. precatorius* L. were collected from Vadavalli, Coimbatore district, Tamil Nadu during August 2010. The collected pods were thoroughly dried in the sun; the pods were thrashed to remove seeds. The seeds, after thorough cleaning and removal of broken seeds, foreign materials and immature seeds were stored in airtight plastic jars at room temperature (25°C).

Proximate composition

The moisture content was determined by drying 50 transversely cut seed in an oven at 80°C for 24 hr and is expressed on a percentage basis. The air-dried samples were powdered separately in a Wiley mill (Scientific Equipment, Delhi, India) to 60-mesh size and stored in screw capped bottles at room temperature for further analysis.

The nitrogen content was estimated by the micro-Kjeldahl method (Humphries, 1956) and the crude protein content was calculated (N x 6.25). Crude lipid content was determined using Soxhlet apparatus (AOAC 2005).The ash content was determined by heating 2g of the dried sample in a silica dish at 600°C for 6hr (AOAC 2005). Total dietary fibre (TDF) was estimated by the non-enzymatic-gravimetric method (Li and Cardozo, 1994). To determine the TDF, duplicate 500 mg ground samples were taken in separate 250 ml beakers. To each beaker 25 ml water was added and gently stirred until the samples were thoroughly wetted, (i.e. no clumps were noticed). The beakers were covered with Al foil and allowed to stand 90 min without stirring in an incubator at 37°C. After that, 100 ml 95% ethanol was added to each beaker and allowed to stand for 1 hr at room temperature (25±2°C). The residue was collected under vacuum in a pre-weighed crucible containing filter aid. The residue was washed successively with 20 ml of 78% ethanol, 10 ml of 95% ethanol and 10 ml acetone. The crucible containing the residue was dried ≥ 2 hr at 105°C and then cooled for ≥ 2 hr in a desiccator and weighed. One crucible containing residue was used for ash determination at 525°C for 5 hr. The ash-containing crucible was cooled for ≥ 2hr in a desiccator and weighed. The residue from the remaining duplicate crucible was used for crude protein determination by the micro-Kjeldahl method as already mentioned. The TDF was calculated as follows.

$$TDF\% = 100 \text{ x } \frac{Wr - [(P+A)/100] \ Wr}{Ws}$$

Where: Wr is the mg residue, P is the % protein in the residue; A is the % ash in the residue, and Ws is the mg sample.

The nitrogen free extract (NFE) was obtained by difference (Muller and Tobin, 1980). The energy value of the seed (kJ) was estimated by multiplying the percentages of crude protein, crude lipid and NFE by the factors 16.7, 37.7 and 16.7 respectively (Siddhuraju *et al.*, 1996).

Minerals and vitamins analysis

Five hundred mg of the ground legume seed was digested with a mixture of 10ml concentrated nitric acid, 4ml of 60% perchloric acid and 1ml of concentrated sulphuric acid. After cooling, the digest was diluted with 50ml of deionised distilled water,

filtered with Whatman No. 42 filter paper and the filtrates were made up to 100ml in a glass volumetric flask with deionised distilled water. All the minerals except phosphorus were analysed from a triple acid-digested sample by an atomic absorption spectrophotometer – ECIL (Electronic Corporation of India Ltd., India) (Issac and Johnson, 1975). The phosphorus content was determined colorimetrically (Dickman and Bray, 1940)
.

Ascorbic acid and niacin contents were extracted and estimated as per the method given by Sadasivam and Manickam, 1996. For the extraction of ascorbic acid, 3g air-dried powdered sample was ground with 25ml of 4% oxalic acid and filtered. Bromine water was added drop by drop to 10ml of the filtrate until it turned orange-yellow to remove the enolic hydrogen atoms. The excess of bromine was expelled by blowing in air. This filtrate was made up to 25ml with 4% oxalic acid and used for ascorbic acid estimation. Two millilitres of the extract was made up to 3ml with distilled H_2O in a test tube. One millilitre of 2% 2, 4-dinitrophenyl hydrazine reagent and a few drops of thiourea were added. The contents of the test tube were mixed thoroughly. After 3hr incubation at 37°C, 7ml of 80% H_2SO_4 was added to dissolve the osazone crystals and the absorbance was measured at 540nm against a reagent blank.

For the extraction of niacin, 5g air-dried powdered sample was steamed with 30ml concentrated H_2SO_4 for 30min. After cooling, this suspension was made up to 50ml with distilled H_2O and filtered. Five ml of 60% basic lead acetate was added to 25ml of the filtrate. The pH was adjusted to 9.5 and centrifuged to collect the supernatant. Two ml of concentrated H_2SO_4 was added to the supernatant. The mixture was allowed to stand for 1hr and centrifuged. The 5ml of 40% $ZnSO_4$ was added to the supernatant. The pH was adjusted to 8.4 and centrifuged again. Then the pH of the collected supernatant was adjusted to 7 and used as the niacin extract. For estimation, 1ml extract was made up to 6ml with distilled water in a test tube; 3ml cyanogen bromide was added and shaken well, followed by addition of 1ml of 4% aniline. The yellow colour that developed after 5min was measured at 420nm against a reagent blank. The ascorbic acid and niacin contents present in the sample were calculated by referring to a standard graph and expressed as mg per 100 g of powdered samples.

Extraction and estimation of total proteins and protein fraction

The total (true) protein was extracted by the method of Basha *et al* (1976) with slight modification (ethanol treatment was omitted to save prolamin fraction). The extracted proteins were purified by precipitation with cold 20% trichloroacetic acid (TCA) and estimated by the method of Lowry et al (1951). The albumin and globulin fractions of the seed protein were extracted and separated according to the method of Murray (1979). The prolamin fraction was extracted from the residual pellet by treating the pellet with 80% ethanol (1:10w/v) overnight. After centrifugation (20,000g for 20 min at room temperature) the supernatant containing prolamins was air-dried and dissolved in 0.1N NaOH. The resulting pellet was extracted with 0.4N NaOH (1:10w/v) overnight and centrifuged as above. The supernatant was designated as glutelins. All four fractions so obtained were precipitated and washed with cold 10% TCA. All samples were redissolved in 0.2M NaOH and protein content was determined by the method of Lowry et al (1951).

Amino acid analysis

The total seed protein was extracted by a modified method of Basha et al., (1976). The extracted proteins were purified by precipitation with cold 20% trichloroacetic acid (TCA). A protein sample of 30mg was hydrolysed by 6N HCL (5ml) in an evacuated sealed tube, which was kept in an air oven maintained at 110 °C for 24 hr. The sealed tube was broken and the acid removed completely by repeated flash evaporation after the addition of de-ionized water. Dilution was effected by means of citrate buffer pH 2.2 to such an extent that the solution contained 0.5 mg protein ml^{-1}. The solution was passed through a millipore filter (0.45μM) and derivitized with O-phthaldialdehyde by using an automated pre-column (OPA). Aminoacids were analysed by a reverse – phase HPLC (Method L 7400, HITACHI, Japan) fitted with a denali C_{18} 5 micron column (4.6X 150mm). The flow rate was 1 ml min^{-1} with fluorescence detector. The cystine content of protein sample was obtained separately by the Liddelle and Saville, (1959) method. For the determination of tryptophan content of proteins, aliquots containing known amounts of proteins were dispersed into glass ampoules together with 1 ml 5M NaOH. The ampoules were flame sealed and incubated at 110°C for 18 hr. The tryptophan contents of the alkaline hydrolysates were determined colorimetrically using the method of Spies and Chambers (1949) as modified by Rama Rao et al (1974).

The contents of the different amino acids were expressed as g100g-1 proteins and were compared with FAO/WHO (1991) reference pattern. The essential amino acid score (EAAS) was calculated as follows:

$$EAAS = \frac{\text{Essential a.a. g /100g total protein}}{\text{Essential a.a. g/100g FAO/WHO (1991) reference pattern}} \times 100$$

Lipid extraction and fatty acid analysis

The total lipid was extracted from the seeds according to the method of Folch et al., (1957) using chloroform and methanol mixture in ratio of 2: 1 (v/v). Methyl esters were prepared from the total lipids by the method of Metcalfe et al., (1966). Fatty acid analysis was performed by gas chromatography (ASHMACO, Japan; Model No: ABD20A) using an instrument equipped with a flame ionization detector and a glass column (2mX3mm) packed with 1% diethylene glycol succinate on chromosorb W. The temperature conditions for GC were injector 200°C and detector 210°C. The temperature of the oven was programmed from 180°C and the carrier gas was nitrogen at a flow rate of 30ml/min. Peaks were identified by comparison with authentic standards, quantified by peak area integration and expressed as weight percentage of total methyl esters; the relative weight percentage of each fatty acid was determined from integrated peak areas.

Analysis of antinutritional compounds

The antinutritional compounds, total free phenolics (Bray and Thorne, 1954), tannins (Burns, 1971), the non-protein amino acid, L-DOPA (3, 4-dihydroxyphenylalanine) (Brain, 1976), phytic acid (Wheeler and Ferrel, 1971) and hydrogen cyanide (Jackson, 1967) were quantified. Trypsin inhibitor activity was determined by the enzyme assay of Kakade et al., (1974) using benzoil-DL-arginin-p-nitroanilide (BAPNA) as a substrate. One trypsin inhibitor unit (TIU) has been expressed as an increase of 0.01 absorbance units per 10ml of reaction mixture at 410nm. Trypsin inhibitor activity has been defined in terms of trypsin units inhibited per mg protein.

Extraction, TLC separation and estimation of Oligosaccharides

Extraction of oligosaccharides was done following the method of Somiari and Balogh, (1993). Five grams each of all the samples of seed flours were extracted with 50 ml of 70% (v/v) aqueous ethanol and kept on an orbital shaker at 130 rpm for 13 hr and then filtered through Whatman No. 1 filter paper. Residue was further washed with 25 ml of 70% (v/v) ethanol. The filtrates obtained were pooled and vacuum-dried at 45°C. The concentrated sugar syrup was dissolved in five ml of double-distilled water.

Separation of oligosaccharides was done by TLC. Thirty g of cellulose-G powder was dissolved in 45 ml of double distilled water and shaken well to get homogeneous slurry. TLC plates were coated with the slurry and air-dried. Spotting of the sugar samples was done by using micropipettes. Five μl aliquots of each sample were spotted thrice separately. The plates were developed by using a solvent system of n-propanol, ethyl acetate and distilled water (6:1:3), and dried (Tanaka *et al.,* 1975). The plates were sprayed with α –naphthol reagent (1%, w/v). Plates were dried in a hot-air oven. The separated spots were compared with standard sugar spots. A standard sugar mixture containing raffinose, stachyose and verbascose was procured from sigma chemical co., St. Louis, USA. Separated sugars that appeared were verbascose, stachyose and raffinose. The sugar spots were scrapped, eluted in 2 ml of distilled water kept overnight and filtered through Whatman No. 1 filter paper. The filtrates were subjected to quantitative estimation.

The eluted individual oligosaccharides were estimated by the method of Tanaka *et al.,* (1975). One ml of the eluted and filtered sugar solution was treated with one ml of 0.2 M thiobarbituric acid and one ml of concentrated HCL. The tubes were boiled in a water bath for exactly 6 min. After cooling, the oligosaccharide contents were quantified in an Elico UV-Spectrophotometer model SL 150 at 432 nm. Average values of triplicate estimations were calculated and the content of oligosaccharides was expressed on dry weight basis

Quantitative determination of phyto-haemagglutinating (Lectin) activity

Lectin activity was determined by the method of Almedia *et al.,* (1991). One g of air-dried seed flour was stirred with 10ml of 0.15N sodium chloride solution for 2hours and the pH was adjusted to 4.0. The contents were centrifuged at 10,000 X g for 20min. and the supernatants were collected separately. The protein content was estimated by the Lowry *et al.,* (1951) method. Human blood (blood groups A, B and O) was procured from the blood bank of Jothi Clinical Laboratory, Tuticorin. Blood erythrocyte suspensions were prepared by washing the blood samples separately with phosphate-buffered saline and centrifuged for 3min at low speed. Supernatants were removed with Pasteur pipettes. The washing procedure was repeated three times. The washed cells were diluted by one drop of cells with 24 drops of phosphate – buffered saline.

The determination of lectin was done by the method of Tan *et al.,* (1983). Clear supernatant (50μl) was poured into the depression (pit) on a micro-titration plate and serially diluted 1:2 with normal saline. The human blood erythrocyte (A, B and O blood groups) suspensions (25μl) were added to each of the twenty depressions. The plates were incubated for 3 hours at room temperature. After the incubation period, the titre values were recorded. One haemagglutinating unit (HU) is defined as the least amount of haemagglutinin that will produce positive evidence of agglutination of 25μl of a blood group erythrocyte after 3hr incubation at room temperature. The phytohaemagglutinating activity was expressed as haemagglutinating units (HU) / mg protein.

Determination of *in vitro* protein digestibility (IVPD)

This was determined using the multi-enzyme technique (Hsu *et al.,*1977). The enzymes used for IVPD were purchased from Sigma Chemical Co., St. Louis, MO, USA. Calculated amounts of the control (casein) and sample were weighed out, hydrated in 10ml of distilled water and refrigerated at 5°C for 1h. The samples containing protein and enzymes were all adjusted to pH 8.0 at 37°C. The IVPD was determined by the sequential digestion of the samples containing protein with a multi-enzyme mixture (trypsin, ∝-chymotrypsin and peptidase) at 37°C followed by protease at 55°C. The pH drop of the samples from pH 8.0 was recorded after 20min of incubation. The IVPD was calculated according to the regression equation $Y= 234.84 - 22.56\ X$, where Y is the % digestibility and X the pH drop.

RESULTS AND DISCUSSION

Crude protein

Proteins are required for maintenance (replacement of wear and tear of tissues) in adults for the growth of infants and children, for foetal development in pregnancy and secretion of milk during lactation. The requirement of proteins for the later groups is higher than the adults (Rao *et al.,* 1989). Legume seeds are valuable source of protein, oil, carbohydrates, minerals and vitamins. They are playing an important role in human nutrition mainly in developing countries (Yanez *et al.,* 1995). In the present study, *C. gladiata* has high content of crude protein than the other pulses investigated (Table 1). The crude protein content of *C. gladiata*, *E. indica* and *A. precatorius* (both samples) was higher than the other tribal pulses like *Dolichos trilobus, Entada rheedi, Rhynchosia cana, R. suaveolens, R. filipes, Vigna radiata* var. *sublobata, V. unguiculata* subsp. *cylindrica, Atylosia scarabaeoides, Neonotonia wightii* var. *coimbatorensis* and *V. unguiculata* subsp. *unguiculata* (Arinathan *et al.,* 2003; 2009). To meet the protein demands in developing countries where animal protein is grossly inadequate, considerable attention is being paid to less consumed protein sources, especially in legumes (Balogun and Fetuga, 1986) which are considered as protein tablets (Salunkhe, 1982). The crude protein levels of the studied samples suggest its usefulness as alternative source of protein.

Crude lipid

Fat is an important component of diet and serves a number of functions in the human body. Fat is a concentrated source of energy and supplies per unit weight more than twice the energy furnished by either proteins or carbohydrates. *Abrus precatorius* (both samples) contained a high level of crude lipid content (Table 1). This value is found to be more or less equal to that of earlier reports in the same species (Mohan and Janardhanan, 1995a). Crude lipid content of *Canavalia gladiata* is found to be higher than that of *Canavalia ensiformis* (Doss *et al.*, 2011). Crude lipid content of *Erythrina indica* is found to be higher than that of same species (Pugalenthi *et al.*, 2004) and the tribal pulses *Vigna aconitifolia* and *V. unguiculata* subsp. *unguiculata* (Tresina and Mohan, 2011a).

Total Dietary Fibre and ash content

The presence of fibre in the diet is necessary for digestion and for elimination of wastes. The contraction of muscular walls of the digestive tract is stimulated by fibre, thus counteracting constipation (Rao *et al.*, 1989). The World Health Organization (WHO) has recommended an intake of 22-23 kg of fibre for every 1000 Kcal. of diet (Kanwar *et al.*, 1997). *Abrus precatorius* (red and black coloured seed coat) contained the highest percentage of TDF (Table 1) among the legumes of this study. However, the TDF level of presently investigated tribal pulses seems to be higher compared to certain legumes like cowpea and kidney bean (Singh *et al.*, 2000); different varieties of *Vigna mungo* (Tresina *et al.*, 2010); Co_9, Co_{11} and Co_{12} varieties of *Lablab purpureus* (Kala *et al.*, 2010a).

The ash content of the presently investigated tribal pulses (4.11-5.21%) (Table 1) would be important to the extent that it contains the nutritionally important mineral elements, which are presented in Table 2. In the present study, all the investigated pulses exhibited the highest level of ash when compared with an earlier report in different varieties of *Vigna mungo* (Tresina *et al.*, 2010).

Nitrogen Free Extractives and Calorific value

Among the presently investigated tribal pulses, *A. precatorius* (white coloured seed coat) exhibited higher level of Nitrogen Free Extractives (NFE) than *C. gladiata*, *E. indica* and *A. precatorius* (red and black coloured seed coat) (Table 1). These values are found to be higher than that of some of the earlier investigated tribal pulses like, three accessions of *Mucuna pruriens* var. *pruriens* (Kala and Mohan, 2010a; Fathima *et al.*, 2010); *M. pruriens* var. *utilis* and *M. deeringiana* (Kala and Mohan, 2010b); five accessions of *M. atropupurea* (Kala *et al.*, 2010b); five accessios of *M. pruriens* var. *pruriens* (Kalidass and Mohan, 2011) and *V. aconitifolia* (Tresina and Mohan, 2011a). The high NFE content of these legumes act as a good source of calories which would be antimarasmus, especially infant nutrition (Vadivel and Janardhanan, 2000). The range in calorific values exceeds the energy values of different varieties of *Lablab purpureus* (Kala *et al.*, 2010a) which are in the range of 1524-1589 kJ $100g^{-1}$DM respectively.

Mineral composition

The determination of minerals and trace elements in food stuffs is an important part of nutritional and toxicological analyses. Copper, chromium, iron and zinc are essential micronutrients for human health. In addition, these elements play an important role in human metabolism, and interest in these elements is increasing together with reports of relationships between trace elements status and oxidative diseases. Legumes supply adequate protein while being a good source of vitamins and minerals (Fennema, 2000). In the present study, all the investigated tribal pulses exhibited lower level of sodium when compared to Recommended Dietary Allowances (RDA) of NRC/NAS (1980).

In the present study, all the investigated tribal pulses registered a higher level of potassium (Table 2) when compared to earlier reports in chick pea (Alajaji and Ed-Adawy, 2006); *L. purpureus* varieties (Kala *et al.*, 2010a) and *V. mungo* varieties (Tresina *et al.*, 2010) and Recommended Dietary Allowance value (RDA) of infants and children (< 1550mg) (NRC/NAS, 1980). The high content of potassium can be utilized beneficially in the diets of people who take diuretics to control hypertension and suffer from excessive excretion of potassium through the body fluid (Siddhuraju *et al.*, 2001).

Table: 1 Proximate composition of *Canavalia gladiata, Erythrina indica* and *Abrus precatorius* (g 100g⁻¹)[a]

Components	Canavalia gladiata	Erythrina indica	Abrus precatorius (red and black coloured seed coat)	Abrus precatorius (white coloured seed coat)
Moisture	5.65 ± 0.14	7.04 ± 0.12	7.38 ± 0.21	6.76 ± 0.34
Crude protein (Kjeldahl N x 6.25)	26.35 ± 0.56	22.56 ± 0.48	20.40 ± 0.78	19.34 ± 1.10
Crude lipid	6.31 ± 0.11	6.58 ± 0.18	8.34 ± 0.31	8.56 ± 0.26
Total Dietary Fibre (TDF)	7.24 ± 0.09	7.40 ± 0.27	7.86 ± 0.14	6.24 ± 0.11
Ash	5.21 ± 0.07	4.14 ± 0.11	4.30 ± 0.08	4.11 ± 0.07
Nitrogen Free Extractive (NFE)	54.89	59.32	59.10	61.31
Calorific value kJ 100g⁻¹DM	1594.595	1615.46	1642.67	1669.57

[a] All values are means of triplicate determination expressed on dry weight basis; ± denotes standard error.

All the presently investigated tribal pulses contained more calcium content when compared to *Cicer arietinum* (Alajaji and Ed-Adawy, 2006); *Phaseolus vulgaris, Cajanus cajan* (Sangronis and Machado, 2007); *Rhynchosia cana* and *R. suaveolens* (Arinathan *et al.*, 2009). But, all the presently investigated legumes were deficient in calcium content compared to RDA's of infants (NRC/NAS, 1980).

Among the presently examined tribal pulses, *Erythrina indica* registered the highest level of magnesium content (Table 2) when compared to some legumes like, *P. vulgaris* (black and white beans), *C. cajan* (Sangronis and Machado, 2007); *Cicer arietinum* (Alajaji and Ed-Adawy, 2006); *Dolichos trilobus, R. cana, R. suaveolens, Vigna radiata* var. *sublobata, V. unguiculata* subsp. *cylindrica* (Arinathan *et al.*, 2009); *V. mungo* varieties (Tresina *et al.*, 2010) and *Vigna* species (Tresina and Mohan, 2011a). *Erythrina indica* is found to contain more than adequate level of magnesium compared to RDA's of NRC/NAS (1980).

Among the presently studied tribal pulses, *C. gladiata, E. indica* and *A. precatorius* (white coloured seed coat) registered the highest level of phosphorus content than that of earlier reports in *C. arietinum* (Alajaji and Ed-Adawy, 2006); *Atylosia scarabaeoides, Neonotonia wightii* var. *coimbatorensis, R. cana, R. suaveolens, V. radiata* var. *sublobata, V. unguiculata* subsp. *cylindrica, D. trilobus* (Arinathan *et al.*, 2003; 2009) and *Vigna* species (Tresina and Mohan, 2011a). But the phosphorus content of presently studied species is deficient according to RDA's (NRC/NAS, 1980).

Among the presently investigated tribal pulses, *E. indica* registered high level of iron (Table 2) and this value seems to be lower than that of an earlier report in the same species (Pugalenthi *et al.*, 2004). Among the presently investigated tribal pulses, *E. indica* exhibited the highest level of zinc and manganese; and the copper level is low in all the presently studied pulses. But all the presently investigated pulses were deficient in Fe, Cu, Zn and Mn content when compared to children RDA's of (NRC/NAS, 1989).

The ratios of sodium to potassium (Na/K) and calcium to phosphorus (Ca/P) are also shown in Table 2. Na/K ratio in the body is of great concern for prevention of high blood pressure Na/K ratio less than one is recommended. Hence, in the present study, all the seed samples would probably reduce high blood pressure disease because they had Na/K less than one. Modern diets which are rich in animal proteins and phosphorus may promote the loss of calcium in the urine (Shills and Young, 1988). This had led to the concept of the Ca/P ratio. If the Ca/P ratio is low (low calcium, high phosphorus intake), more than the normal amount of calcium may be lost in the urine, decreasing the calcium level in bones. Food is considered "good" if the ratio is above one and "poor" if the ratio is less than 0.5 (Nieman *et al.*, 1992). The Ca/P ratio in the present study ranged between 0.74 to 1.19 indicating they would serve as good sources of minerals for bone formation.

Table 2. Mineral composition of *Canavalia gladiata, Erythrina indica* and *Abrus precatorius* (mg 100g^{-1})[a]

Components	Canavalia gladiata	Erythrina indica	Abrus precatorius (red and black coloured seed coat)	Abrus precatorius (white coloured seed coat)
Sodium	78.04 ± 0.17	44.08±0.78	33.16 ± 0.54	25.08 ± 0.16
Potassium	1938.51±2.34	1838.12±1.24	1970.30 ± 1.56	1864.14 ± 2.52
Calcium	341.14 ± 0.76	294.04±0.61	246.10 ± 0.38	213.08 ± 0.24
Magnesium	106.06 ± 0.36	324.68 ± 0.54	94.26 ± 0.14	114.16 ± 0.11
Phosphorus	313.68 ± 0.53	396.02 ± 0.36	206.13 ± 0.32	256.53 ± 0.46
Iron	4.74 ± 0.11	5.37 ± 0.11	3.48 ± 0.15	4.11 ± 0.13
Zinc	1.56 ± 0.03	0.67 ± 0.03	0.74 ± 0.03	0.82 ± 0.02
Copper	0.76 ± 0.01	6.34 ± 0.17	2.44 ± 0.07	2.58 ± 0.01
Manganese	0.59 ± 0.02	0.56 ± 0.01	0.31 ± 0.01	0.338 ± 0.01
Na/K	0.04	0.02	0.02	0.01
Ca/P	1.09	0.74	1.19	0.83

[a]All values are means of triplicate determination expressed on dry weight basis; ± denotes standard error.

Table 3 Vitamins (niacin and ascorbic acid) content of *Canavalia gladiata, Erythrina indica* and *Abrus precatorius* (mg 100g^{1})[a]

Components	Canavalia gladiata	Erythrina indica	Abrus precatorius (red and black coloured seed coat)	Abrus precatorius (white coloured seed coat)
Niacin	33.20 ± 0.38	11.36 ± 0.08	9.42 ± 0.11	10.12 ± 0.07
Ascorbic acid	27.37 ± 0.11	14.36 ± 0.13	17.38 ± 0.21	14.07 ± 0.32

[a]All values are means of triplicate determination expressed on dry weight basis. ± denotes standard error.

Vitamins

Legumes constitute an important part of the human diet in many parts of the world and are sources of vitamins (Shills *et al.*, 1999). Niacin or nicotinic acid is a true anti-pellagra vitamin. It is an essential vitamin needed for our health. The presently investigated tribal pulse exhibited highest level of niacin content (Table 3) which was found to be higher than that of an earlier report in *Cajanus cajan, D. lablab, D. biflorus, M. pruriens,, P. mungo, Vigna catjang* and *Vigna* species (Rajyalakshmi and Geervani, 1994); *V. unguiculata* subsp. *unguiculata* (Arinathan *et al.*, 2003); *C. arietinum* (Alajaji and Ed-Adawy, 2006) and *V. mungo* varieties (Tresina *et al.*, 2010).

Ascorbic acid is an essential nutrient for humans, but they lack the capacity to synthesize it. It is involved in collagen synthesis, bone and teeth calcification. The presently investigated tribal pulses registered higher level of ascorbic acid content (Table 3) than *C. arietinum, P. aureus, D. biflorus* (Khatoon and Prakash, 2006); *P. vulgaris* (white and black beans); *C. cajan* (Sangronis and Machado, 2007), *V. radiata, V. mungo* (Kakati *et al.*, 2010) and *V. mungo* varieties (Tresina *et al.*, 2010).

Total protein and protein fractionation

Proteins are indispensable for normal growth and metabolism of human. The storage or reserve proteins constitute the major portion of the proteins in grains. Among the studied tribal pulses, *C. gladiata* exhibited highest level of total proteins than the other pulses (Table 4). *Canavalia gladiata* is found to contain more total protein than that of *L. purpureus* (Co$_2$, Co$_9$, Co$_{11}$, Co$_{12}$ varieties) (Kala *et al.*, 2010a) and *V. mungo* (TMV-1 and Vamban-1 varieties) (Tresina *et al.*, 2010).

In general, the globulin constitutes the major seed storage protein in legumes. In all the presently investigated tribal pulses, globulins constitute the major bulk of the proteins (Table 4). This is in consonance with some earlier reports in *Lablab purpureus* (Kala *et al.*, 2010a) and *V. mungo* varieties (Tresina *et al.*, 2010).

Fatty acid composition

Linoleic and linolenic acids are the most important essential fatty acids required for growth, physiological functions and maintenance. The fatty acid composition of the total seed lipids of presently investigated tribal pulses were given in Table 6. The current data revealed that, all the seed lipids were rich in unsaturated fatty acids (55.60-72.04%) and had very high contents of linoleic acid (24.16-34.14%). These values are nutritionally desirable.

The palmitic acid content of *C. gladiata* was higher than the other legumes such as *L. purpureus* (Kala *et al.*, 2010a), *M. pruriens* var. *pruriens* (Fathima *et al.*,

2010; Kala and Mohan, 2010a; Kalidass and Mohan, 2011) and *Vigna* species (Tresina and Mohan, 2011a). Oleic acid was found to be higher than the pulse crop commonly consumed in India such as *C. cajan* (Salunkhe, 1982); tribal pulses *M. atropurpurea* (Kala *et al.*, 2010b) and *Vigna* species (Tresina and Mohan, 2011a). Similarly, the level of stearic acid detected in the samples investigated in the present study seems to be higher than the samples of *L. purpureus* (Kala *et al.*, 2010a); *V. mungo* and *V. aconitifolia* (Tresina *et al.*, 2010; Tresina and Mohan, 2011a).

The detected level of antinutritional fatty acid, behenic acid in *E. indica* (1.04%) is in agreement with earlier reports in Shenbagathoopu accession of *M. pruriens* var. *pruriens* (Kala and Mohan, 2010a) and Seithur accession of *M. pruriens* var. *pruriens* (Kalidass and Mohan, 2011). The presence of behenic acid has been implicated with antherogenic properties (Kritchevsky *et al.*, 1973).

In vitro protein digestibility (IVPD)

Among the presently investigated tribal pulses, *Canavalia gladiata* registered the highest level of IVPD (71.36%) than the other pulses (Table 7), and their protein digestibility was found to be higher than that of *V. mungo* varieties (Tresina *et al.*, 2010) and comparable with some edible legumes (Table 7).

Table 4 Data on total (true) protein and protein fractions of seed flour of *Canavalia gladiata*, *Erythrina indica* and *Abrus precatorius* [a]

Components	Canavalia gladiata		Erythrina indica		Abrus precatorius (red and black coloured seed coat)		Abrus precatorius (white coloured seed coat)	
	g/100g seed flour	g/100g seed protein	g/100g seed flour	g/100g seed protein	g/100g seed flour	g/100g seed protein	g/100g seed flour	g/100g seed protein
Total protein	22.34±1.26	100	17.48±0.74	100	16.30±0.94	100	15.26±0.41	100
Albumins	5.98 ± 0.31	26.77	5.21±0.34	29.81	5.16±0.21	29.81	4.86±0.08	31.85
Globulins	13.61±0.65	60.92	10.28±0.56	58.81	8.74±0.76	58.81	8.77±0.16	57.47
Prolamins	0.78±0.06	3.49	0.84±0.03	4.81	0.94±0.01	4.81	0.62±0.03	4.06
Glutelins	1.97±0.04	8.82	1.15±0.06	6.57	1.46±0.03	6.57	1.01±0.05	6.62

[a]All values are means of triplicate determination expressed on dry weight basis; ± denotes standard error.

Table 5. Amino acid profiles of acid- hydrolysed, purified seed proteins of *Canavalia gladiata, Erythrina indica* and *Abrus precatorius* (g 100g^{-1})

Amino acid	*Canavalia gladiata*	EAAS	*Erythrina indica*	EAAS	*Abrus precatorius* (red and black coloured seed coat)	EAAS	*Abrus precatorius* (white coloured seed coat)	EAAS	FAO WHO 1991
Glutamic acid	14.38		14.28		14.82		13.84		
Aspartic acid	13.36		11.02		11.14		12.56		
Serine	3.78		4.34		2.41		2.74		
Threonine	3.53	103.82	3.30	97.06	2.86	84.12	2.31	67.94	3.4
Proline	4.08		2.94		3.84		3.72		
Alanine	4.16		4.14		5.46		5.84		
Glycine	4.26		4.38		5.08		5.20		
Valine	4.06	116	4.58	130.86	6.18	176.57	5.98	170.86	3.5
Cystine	1.21		0.38		0.42		0.56		2.5
Methionine	0.48	67.60	1.04	56.80	1.68	84	1.34	76	
Isoleucine	7.12	254.29	2.78	99.29	4.34	155	4.68	167.14	2.8
Leucine	6.28	95.15	6.96	105.45	7.14	108.18	6.96	105.45	6.6
Tryosine	2.78		3.41		2.81		3.04		6.3
Phenylalanine	3.38	97.78	5.04	134.13	4.30	112.86	4.61	121.43	
Lysine	5.64	97.24	6.28	108.28	5.86	101.03	5.92	102.07	5.8
Histidine	3.14	165.26	2.14	112.63	1.34	70.53	1.76	92.63	1.9
Tryptophan	0.78	70.91	0.64	58.18	0.76	69.09	0.84	76.36	1.1
Arginine	4.36		6.18		5.43		5.88		

EAAS-Essential amino acid score

Table 6. Fatty acid profile of lipids of *Canavalia gladiata, Erythrina indica* and *Abrus precatorius* [a]

Components	*Canavalia gladiata*	*Erythrina indica*	*Abrus precatorius* (red and black coloured seed coat)	*Abrus precatorius* (white coloured seed coat)
Myristic acid (C14:0)		2.14	-	-
Palmitic acid (C16:0)	32.14	12.68	20.48	21.43
Stearic acid (C18:0)	12.26	11.30	8.41	34.41
Oleic acid (C18:1)	21.48	27.46	32.08	7.08
Linoleic acid (C18:2)	24.16	34.14	29.40	26.61
Linolenic acid (C18:3)	9.36	9.40	8.14	8.88
Behenic acid (C22:0)	-	1.04	-	-
Others		1.84	1.49	1.59

[a]Average values of two determinations.

Antinutritional factors

The problem of protein digestibility has been attributed to the interplay of several factors, including protease inhibitors, phytates, oxalates, goitrogens and other antinutritional factors. Naturally, in societies where legumes are consumed, rather than much more expensive animal foods, there is great concern over the level of antinutrients present in the diet. For this reason, a preliminary evaluation of some of the antinutritional factors in raw seeds of *Canavalia gladiata, Erythrina indica* and *Abrus precatorius* (both samples) was examined (Table 8).

Total free phenolics and tannins

Phenolic compounds inhibit the activity of digestive as well as hydrolytic enzymes such as amylase, trypsin, chymotrypsin and lipase (Salunkhe *et al*, 1992) and decrease the digestibility of proteins, carbohydrates and availability of vitamins and minerals (Rao and Deosthale, 1982). However, recent researchers report that the phenolic compounds are the main human dietary antioxidant and have a decreased incidence of chronic diseases. A number of polyphenolic compounds are present, which contribute towards the defense mechanism of plants. Although these are considered earlier as antinutritional compounds, under the present nomenclature, phenols fall under the category of nutraceuticals, offering many nutritional advantages to man (Shanthakumari *et al*., 2008). The content of total free phenolics of currently investigated tribal pulses appeared to be higher than the earlier reports in *L. purpureus* (Kala *et al*., 2010a), *V. mungo* (Tresina *et al*., 2010) and lower than those of other tribal pulses such as *Dolichos lablab* (Ramakrishnan *et al*., 2006), *D. trilobus, Entada rheedi, M. atropurpurea, Rhynchosia cana, R. suaveolens, Tamarindus indica, Teramnus labialis, V. radiata* var. *sublobata, V. unguiculata* subsp. *cylindrica* (Arinathan *et al*., 2009) and *V. aconitifolia* (Tresina and Mohan, 2011a). The tannin content of the investigated pulses were relatively lower than red gram, bengal gram, lentil (Salunkhe *et al*., 2006), *C. arietinum* (Alajaji and Ed-Adawy, 2006), *P. vulgaris* (white and black beans), *C. cajan* (Sangronis and Machado, 2007), *Pisum sativum* (Nikolopoulou *et al*., 2007); *V. radiata, V. mungo* (Kakati *et al*., 2010) and *V. aconitifolia* (Tresina and Mohan, 2011a).

L-DOPA

L-DOPA (3,4- dihydroxyphenylalanine) is a non-protein amino acid which causes skin eruptions and increases body temperature in the consuming people when present in high concentrations (Jebedhas, 1980). L-DOPA, a compound chiefly used in the treatment of Parkinson's disease, has been reported to cause hallucinations, in addition to causing gastrointestinal disturbance such as nausea, vomiting and anorexia (Reynolds, 1989). All the presently investigated tribal pulses contained low level of L-DOPA (Table 8) when compared with *M. atropurpurea* (Fathima and Mohan, 2009; Kala *et al*., 2010a); *M. pruriens* var. *pruriens* (Kala and Mohan, 2010a; Kalidass and Mohan, 2011). Among the investigated tribal pulses, *C. gladiata* exhibited the highest percentage of L-DOPA. The level of L-DOPA is significantly reduced by repeated soaking and boiling of seeds (Jebedhas, 1980). It is also observed that, drying effects substantial lose in content of L-DOPA (Longo *et al*., 1974; Larher *et al*., 1984). Repeated boiling of seeds in water and decanting the water for seven times resulted in significant reduction in the level of L-DOPA (Janardhanan, 1982). Dry heat treatment also has been found to be more effective in reducing the L-DOPA content (Siddhuraju *et al*., 1996).

Table 7. IVPD of seeds of *Canavalia gladiata, Erythrina indica* and *Abrus precatorius* (two samples) compared with some pulses

Pulses	(IVPD %)	Sources
Canavalia gladiata	71.36	
Erythrina indica	69.73	
Abrus precatorius (red and black coloured seed coat)	64.34	
Abrus precatorius (white coloured seed coat)	63.31	
Dolichos diflorus	71.00	Rajyalakshmi and Geervani (1990)
Vigna radiata	80.05-85.33	Reddy and Gowramma (1987)
Vicia faba	57.20-72.07	Moneam (1990)
Vigna umbellata	73.48-74.30	Laurena *et al*., (1991)
Lablab purpureus	64.36-70.30	Kala *et al*., (2010a)
Vigna species	69.32-74.21	Tresina and Mohan, (2011a)

Table 8 Data on antinutritional factors of *Canavalia gladiata, Erythrina indica* and *Abrus precatorius*

Components	Canavalia gladiata	Erythrina indica	Abrus precatorius (red and black coloured seed coat)	Abrus precatorius (white coloured seed coat)
Total free phenolics[a] g 100 g^{-1}	1.21±0.03	0.94±0.07	0.76±0.11	0.84±0.11
Tannins[a] g 100g^{-1}	0.41±0.01	0.36±0.03	0.54±0.02	0.61±0.07
L-DOPA[a] g 100g^{-1}	2.50±0.21	2.48 ±0.16	1.12±0.07	0.98±0.03
Phytic acid[b] mg 100g^{-1}	354.30±1.76	386.24±1.04	348.14±2.21	366.32±1.34
Hydrogen cyanide[a] mg 100g^{-1}	0.31±0.01	0.09±0.01	0.12±0.01	0.09±0.01
Trypsin inhibitor (TIU mg^{-1} protein)	25.54±0.48	36.24±0.21	34.16±0.14	35.30±0.02
Oligosaccharide[a] g100g^{-1} Raffinose	0.64±0.06	1.24±0.04	1.38±0.05	1.21±0.04
Stachyose	1.64±0.11	0.98±0.06	1.12±0.03	1.08±0.03
Verbascose	2.11±0.10	5.86±0.74	3.34±0.56	3.96±0.24
Phytohaemagglutinating activity[b] (Hu mg^{-1} protein) A group	138	138	58	64
B group	94	68	178	168
O group	36	16	28	32

[a] All values are of means of triplicate determination expressed on dry weight basis; [b] All values of two independent experiments; ± Standard error

Phytic acid

Phytic acid, a major phosphorus storage form in plants and its salts are known as phytates, which regulates the various cellular functions such as DNA repair, chromatin remodelling, endocytosis, nuclear messenger and potential hormone signalling which is important for plants and seed development (Zhou and Erdman, 1995), as well as animal and human nutrition (Vucenik and Shamsuddin, 2006). It is often regarded as an antinutrient because of strong mineral, protein and starch binding properties, thereby decreasing their bioavailability (Weaver and Kannan, 2002). Phytic acid content of investigated seed samples was found to be low when compared to that of some commonly consumed legumes,TMV-1 variety of *V. mungo* (Tresina *et al.*, 2010); *V. radiata, V. mungo* (Kakati *et al.*, 2010) tribal *pulses M. pruriens* var. *pruriens* (Fathima *et al,* 2010; Kala and Mohan, 2010a); *M. pruriens* var. *utilis* and *M. deeringiana* (Kala and Mohan, 2010b) and *V. aconitifolia* (Tresina and Mohan, 2011a). It is worthwhile to note that, the phytate content in *Mucuna* beans could be substantially eliminated by processing methods such as soaking and cooking (Vijayakumari *et al.*, 1996).

Hydrogen cyanide (HCN)

Hydrogen cyanide is known to cause acute or chronic toxicity. A lot of HCN (known to inhibit the respiratory chain at the cytochrome oxidase level) is lost during soaking and cooking (Kay *et al.*, 1977). The content of HCN level in the presently investigated tribal pulses was far below the lethal level i.e., 36mg/100g (Oke, 1969) and comparable with those of *V. mungo* (Tresina *et al.*, 2010), *L. purpureus* (Kala *et al.*, 2010a) and certain tribal pulses investigated in our laboratory (Arinathan, 2003; 2009; Tresina and Mohan, 2011b).

Trypsin inhibitor activity

The presence of protease inhibitors such as trypsin and chymotrypsin inhibitors in the diet leads to the formation of irreversible trypsin enzyme-trypsin inhibitor complexes, causing a decrease in trypsin in the intestine and decrease in the digestibility of dietary protein, thus leading to slower animal growth.

As a result, the secretary activity of the pancreas increases, which could cause pancreatic hypertrophy and hyperplasia (Liener, 1994). The trypsin inhibitor activities of all the studied samples were higher than that of *P. vulgaris* Roba variety (4.59mg/g) (Shimelis and Rakshit, 2007); *C. arietinum* (11.90mg/g) (Alajaji and Ed-Adawy, 2006); *P. vulgaris* (white and black beans), *C. cajan* (4.13-4.75mg/g) (Sangronis and Machado, 2007) and they seem to be lower than that of Co$_5$ and TMV-1 varieties of *V. mungo* (38.74-41.36mg/g) (Tresina *et al.*, 2010) *M. pruriens* var. *pruriens* (40.40-48.24%) (Kala and Mohan, 2010a; Kalidass and Mohan, 2011). Trypsin inhibitor activity has greater impact on the IVPD of the legumes where the trypsin inhibitor activity was known to be heat labile.

Oligosaccharides

Ingestion of large quantities of beans is known to cause flatulence in humans and animals. The raffinose family sugars (raffinose, stachyose and verbascose) are important contributors of flatus. These are not digested by man due to lack of α-galactosidase enzyme (Gitzelmann and Aurricctuo, 1965). The microflora in the lower intestine metabolizes these oligosaccharides and produces flatus gases.

Among the presently investigated tribal pulses, the seed sample of *E. indica* is found to contain the highest level of total oligosaccharides (Table 8). In the present study, all the investigated samples contained verbascose as the major oligosaccharide. This is in agreement with earlier reports in *C. cajan* (Mulimani and Devendra, 2010); *M. pruriens* var. *utilis* (Janardhanan *et al.*, 2003a; Kala and Mohan, 2010b), *M. pruriens* var. *pruriens* (Kala and Mohan, 2010a; Kalidass and Mohan, 2011).

Haemagglutinins (lectins)

Lectins are toxic glycoproteins that have the ability to bind with carbohydrate moieties on the surface of the human red blood cells (RBC) and cause them to agglutinate. Lectins can combine with intestinal mucosal cells and cause interference with the absorption of available nutrients (Liener, 1994).

Regarding phytohaemagglutinating activity, among the presently investigated tribal pulses, *C. gladiata* and *E. indica* registered higher haemagglutinating activity with respect to the 'A' blood group of human erythrocytes; whereas, the seed samples of *A. precatorius* registered higher haemagglutinating activity with respect to the 'B' blood group of human erythrocytes. All the presently investigated seed samples showed low levels of phytohaemagglutinating activity with respect to the 'O' blood group. Haemagglutinating activity of *C. gladiata* and *E. indica* is in agreement with earlier reports in *Mucuna* (Kala and Mohan, 2010a; Fathima *et al.*, 2010; Kalidass and Mohan, 2011). Similarly haemagglutinating activity of *A. precatorius* (both samples) is in good agreement with earlier reports in the tribal pulse *Vigna* species (Tresina and Mohan, 2011a).

Lectins are highly sensitive to heat treatment (Singh *et al.*, 1988). Haemagglutinating activity decreases during germination in *Glycine max, P. vulgaris, Vicia faba* and *V. radiata* (Valdebouze *et al.*, 1980). A significant reduction in lectin activity has been noticed when the seeds of certain pulses were subjected to dry heat treatment and autoclaving (Siddhuraju *et al.*, 1996).

CONCLUSION

On the basis of the comparative assessment, *C. gladiata, E. indica* and *A. precatorius* (red and black coloured seed coat and white coloured seed coat) which are referred as underutilized legumes are a valuable source of nutrition due to high proteins and carbohydrates with an adequate quantity of minerals, vitamins, essential amino acids and unsaturated fatty acids. The currently investigated tribal pulses possess a variety of antinutritional factors that cause adverse effects on consumers. The presence of antinutritional factors identified in the current report should not pose a problem for humans, if the beans are not properly processed. The overall interpretation of this present investigation may offer a scientific basis for increased and versatile utilization of these protein-rich underutilized legumes as a food and protein supplement.

REFERENCES

Alajaji, S.A., El-Adawy, T.A. 2006. Nutritional composition of chick pea (*Cicer arietinum* as affected by microwave cooking and other traditional cooking methods. Journal of Food Composition and Analysis. doi:10.1016/j.jfca.2006.03.015.

Almedia, N.G., Calderon de la Barca, A.M., Valencia, M.E. 1991. Effect of different heat treatments on the anti-nutritional activity of *Phaseolus vulgaris* (variety ojode Carbra) lution. Journal of Agricultural Food Chemistry, 39: 1627–1630.

AOAC. 2005. Official Methods of Analysis (18th edn.). Association of Official Analytical Chemists. Washington. DC.

Arinathan, V., Mohan, V.R., De Britto A.J., Murugan, C. 2007. Wild edibles used by Palliyars of the Western Ghats, Tamil Nadu, India. Journal of Traditional Knowledge, 6: 163-168.

Arinathan, V., Mohan, V. R., Maruthupandian, A., Athiperumalsami, T. 2009. Chemical evaluation of raw seeds of certain tribal pulses in Tamil Nadu, India. Tropical and Subtropical Agroecosystems, 10: 287 – 294.

Arinathan, V., Mohan, V.R., John de Britto, A. 2003. Chemical composition of certain tribal pulses in South India. International Journal of Food Science and Nutrition, 54: 209 – 217.

Balogun, A.M., Fetuga, B.L. 1986. Chemical composition of some underexploited leguminous crop seeds in Nigeria. Journal of Agricultural Food Chemistry, 34: 189-192.

Basha, S.M.M., Cherry, J.P., Young, C.T. 1976. Changes in free amino acids, carbohydrates and proteins of maturity seeds from various peas (Arachis hypogaea) cultivars. Cereal Chemistry, 53: 583 – 597.

Borthakur, S.K. 1996. Wild edible plants in markets of Assam, India- an ethnobotanical investigation. In: Ethnobiology In Human Welfare S. K. Jain (Ed.) pp31-36, New Delhi, India: Deep Publications.

Brain, K.R. 1976. Accumulation of L-DOPA in cultures from Mucuna pruriens. Plant Science Letters, 7: 157-161.

Bray, H.G., Thorne,W.V. 1954. Analysis of phenolic compounds methods. Biochemical Analyst, 1: 27-52.

Bressani, R., Brenes, R. S., Gracia, A., Elias, L. G. 1987. Chemical composition, amino acid content and protein quality of Canavalia spp. seeds. Journal of Science Food and Agriculture 40: 17–23.

Burns, R.B., 1971. Methods of estimation of tannin in the grain, sorghum. Agronomy Journal, 63: 511 -512.

Dickman, S.R., Bray, R.H. 1940. Colorimetric determination of phosphate. Industrial Engineering Chemistry Analytical Education, 12: 665-668.

Doss, A., Pugalenthi,M., Vadivel,V., Subhashini, G., Subash, R.A. 2011. Effects of processing technique on the nutritional composition and antinutrient content of under-utilized food legume Canavalia ensiformis L. DC. International Food Research Journal, 18: 965-970.

Dwivedi, R.S. 2004. Unnurtured and untapped super sweet nonsacchariferous plant species in India. http://www.ias.ac.in/currsci/ jun10/ articles19.htm (viewed on 18.09.2011).

FAO/WHO. 1991. Protein quality evaluation, (p 66). Rome, Italy: Food and Agricultural Organization of the United Nations.

Fathima, K.R., Mohan,V.R. 2009. Nutritional and antinutritional assessment of Mucuna atropurpurea DC: an under utilized tribal pulse. African Journal of Basic and Applied Science. 1: 129-136.

Fathima, K. R., Tresina Soris, P., Mohan, V. R. 2010. Nutritional and antinutritional assessment of Mucuna pruriens (L.) DC var. pruriens an underutilized tribal pulse. Advances in Bioresearch, 1: 79-89.

Fennema, O.R. 2000. Food Chemistry, New York. Marcel Dekker.

Folch, J., Lees, M., Solane-Stanly, G.M. 1957. A simple method for the isolation and purification of total lipids from animal tissues. Journal of Biological Chemistry, 226:497 – 506.

Gitzelmann, R., Aurricchio, S. 1965. The handling of soy alpha-galactosides by a normal and galactosemic child. Pediatrics, 36: 231-235.

Hsu, H.W., Vavak, D.L., Satterlee, L.D., Miller, G.A. 1977. A multi-enzyme technique for estimating protein digestibility. Journal of Food Science, 42: 1269 – 1271.

Humphries, E.C. 1956. Mineral composition and ash analysis In: Peach K. and M.V. Tracey (eds.) Modern Methods of Plant Analysis Vol.1, Springer-Verlag, Berlin, pp: 468-502.

Issac, R.A., Johnson, W.C. 1975. Collaborative study of wet and dry techniques for the elemental analysis of plant tissue by Atomic Absorption Spectrophotometer. Journal of Association of Official Analytical Chemist, 58: 376-38.

Jackson, M-L. 1967. Cyanide in Plant tissue. In: Soil Chemical Analysis. Asia Publishing House New Delhi India. pp. 337.

Janardhanan, K. 1982. *Studies on seed development and germination in Mucuna utilis* Wall. ex. Wt. (Papilionaceae). Ph.D. Thesis, Madras Univ., Madras, India.

Janardhanan, K., Gurumoorthi, P., Pugalenthi, M. 2003a. Nutritional potential of five accessions of a South Indian tribal pulse, *Mucuna pruriens* var. *utilis* I. The effect of processing methods on the content of L-DOPA, phytic acid and oligosaccharides. Tropical and Subtropical Agroecosystems, 1:141 – 152.

Janardhanan, K., Vadivel, V., Pugalenthi, M. 2003b. Biodiversity of Indian underexploited / tribal pulses. In: *Improvement strategies for leguminosae Biotechnology*. (Eds.). Jaiwal, P.K. and Singh, R.P. Great Britain: Kluwer Academic Publishers. pp 353-405.

Jebadhas, A.W. 1980. *Ethnobotanical studies on some hill tribes of South India.* Ph.D. Thesis, Madras Univ., Madras, India.

Kakade, M.L., Rackis, J.J., McGhce, J.E., Puski, G. 1974. Determination of trypsin inhibitor activity of soy products: a collaborative analysis of an improved procedure. Cereal Chemistry, 51: 376 -38.

Kakati, P., Deka S.C., Kotoki, D, Saikia, S. 2010. Effect of traditional methods of processing on the nutrient contents and some antinutritional factors in newly developed cultivars of green gram [*Vigna radiata* (L.) Wilezek] and black gram [*Vigna mungo* (L.) Hopper] of Assam, India. International Food Research Journal, 17: 377-384.

Kala, B.K., Mohan, V.R. 2010a. Nutritional and antinutritional potential of three accessions of itching bean (*Mucuna pruriens* (L.) DC var. *pruriens*): an under-utilized tribal pulse. International Journal of Food Science and Nutrition, 61: 497-511.

Kala, B.K., Mohan, V.R. 2010b. Chemical composition and nutritional evaluation of lesser known pulses of the genus, *Mucuna.* Advances in Bioresearch, 1: 105-116.

Kala, K.B., Tresina Soris, P., Mohan, V.R., Vadivel,V. 2010a. Nutrient and chemical evaluation of raw seeds of five varieties of *Lablab purpureus* (L.) Sweet. Advances in Bioresearch, 1: 44-53.

Kala, K.B., Kalidass, C., Mohan, V.R. 2010b. Nutritional and antinutritional potential of five accessions of a South Indian tribal pulse *Mucuna atropurpurea* DC. Tropical and subtropical Agroecosystems, 12: 339-352.

Kalidass, C., Mohan, V.R. 2011. Nutritional and antinutritional composition of itching bean (*Mucuna pruriens* (L.) DC var. *pruriens*): An underutilized tribal pulses in Western Ghats, Tamil Nadu. Tropical and Subtropical Agroecosystems, 14: 279-279.

Kanwar, K.C., Kanwar, V., Shah, S. 1997. Friendly fibres. Science Reports, 34: 9-14.

Kay, T., Ogunsona, V.A., Eka, O.U. 1977. The prevention of beany taste development and the elimination of better taste in preparing soya bean food in the rural community in Nigeria. Samuru Agricultural News Letter, 19: 11.

Khatoon, N., Prakash, J. 2006. Nutritional quality of microwave-cooked and pressure-cooked legumes. International Journal of Food Science and Nutrition, 55: 441-448.

Kritechevsky, D., Tepper, S.A., Vesselinovitch, D., Wissler, R.W.1973. Cholesterol vehicle in experimental antherosclerosis, 13. Randomised peanut oils. Atherosclerosis, 17: 225 – 237.

Larher, I., Gerad, J., Gerant sauvage, D., Hamelin, J., Briens, M. 1984. An assessment of potential *Vicia faba* minor in the storage of the C-form of 3,4 Dihydroyphenyl alanine. Journal of Plant Physiology, 116: 171-180.

Laurena, A.C., Rodrigues, F.M., Sabino, N.G., Zamora, A.F., Mendoza, E.M.T. 1991. Amino acid composition, relative nutritive value and *in vitro* protein digestibility of several Philippine indigenous legumes. Plant Foods for Human Nutrition, 41: 59-68.

Li, B.W., Cardozo, M.S. 1994. Determination of total dietary fiber in foods and products with little or no starch, non-enzymatic gravimetric method: collaborative study. Journal of Association of Official Analytical Chemists International, 77: 687 -689.

Liddelle, H.F., Saville, B. 1959. Colorimetric determination of cysteine. Analyst, 84: 133 - 137.

Liener, I.E. 1994. Implications of anti-nutritional components in soybean foods. CRC Critical Reviews of Food Science and Nutrition, 34: 31-67.

Longo, R., Castelloni, A., Sberze, P., Tibolla, M. 1974. Distribution of L-Dopa and related amino acid in *Vicia*. Phytochemistry, 13: 161-171.

Lowry, O.H., Rorebrough, N.J., Farr, A.L., Randall, R.J. 1951. Protein measurement with folin phenol reagent. Journal of Biological Chemistry, 193: 265 – 275.

Metcalfe, L.D., Schemitz, A.A., Pelka, J.R. 1966. Rapid preparation of fatty acid esters from lipids for gas chromatographic analysis. Analytical Chemistry, 38: 514 – 515.

Moneam, N.M.A. 1990. Effect of pre-soaking on faba bean enzyme inhibitors and polyphenols after cooking. Journal of Agricultural Food Chemistry, 38: 1476-1482.

Mohan, V.R., Janardhanan, K. 1995a. Chemical determination of nutritional and antinutritional properties in tribal pulses. Journal of Food Science and Technology. 32: 465-469.

Muller, H.G., Tobin, G. 1980. Nutrition and food processing, London : Croom Helm Ltd.

Mullimani, V.H., Devendra, S. 2000. Effect of soaking and germination on oligosaccharide content of red gram. ICPN. 7: 69-72.

Murray, D.R. and Roxburg, C. M. 1984. Amino acid composition of the seed albumins from chickpea. Journal of Science Food and Agriculture, 35: 893-896.

Nieman, D.C., Batterworth, Nieman, C.N. 1992. Nutrition pp 237-312. Wmc. Brown Publishers, Dubugue, USA.

Nikolopoulou, D., Grigorakis, K., Stasini, M., alexis, M.N., Iliadis, K. 2007. Differences in the chemical composition of field pea (*Pisum sativum*) cultivars: Effect of cultivation area and year. Food Chemistry, 103: 847-852.

NRC/NAS, 1980. National Research Council Committee on Dietary Allowances. Recommended Dietary Allowances 9th edn. National Academy of Science Press. Washington, DC. USA.

NRC/NAS, 1989. National Research Council Committee on Dietary Allowances. Recommended Dietary Allowances 10th edn. National Academy of Science Press. Washington, DC. USA.

Oke, O.L. 1969. The role of hydrocyanic acid in nutrition. World Review of Nutrition and Dietetis, 11: 118-147.

Pal, D.C. and Jain, S.K. (eds) 1988. Ethnomedicine. Naya Prakesh Publications, 206 Bidhan, Sarani, Calcath. Pp1-317.

Pugalenthi, M., Vadivel, V., Janaki, P. 2007. Comparative evaluation of protein quality of raw and differentially processed seeds of an under-utilized food legume, *Abrus precatorius* L. Livestock Research for Rural Development. 19. Article# 168, retrieved on November 02, 2007.

Pugalenthi, M., Vadivel, V., Gurumoorthi, P., Janardhanan, K. 2004. Comparative nutritional evaluation of little known legumes, *Tamarindus indica*, *Erythrina indica* and *Sesbania bispinosa*. Tropical and Subtropical Agroecosystems, 4: 107-123.

Radhakrishnan, K, Pandurangan A.C., Pushpangadan D. 1996. Ethnobotany of the wild edible plants of Kerala,India. In*: Ethnobotany in Human Welfare* (Ed.) Jain SK. Deep Publications, New Delhi, India, pp 48 – 51.

Rain-tree, 2004. http://www.rain-tree.com/abrus.htm (viewed on 18.09.2011).

Rajyalakshmi, P., Geervani, P. 1994. Nutritive value of the foods cultivated and consumed by the tribals South India. Plant Foods for Human Nutrition, 46: 53 -61.

Ramakrishnan, V., Jhansi Rani, P., Ramakrishna Rao, P. 2006. Antinutritional factors during germination in Indian bean (*Dolichos lablab* L.) seeds. World Journal of Diary and Food Sciences, 1: 6-11.

Rama Rao, M.V., Tara, M.R., Krishnan, C.K. 1974. Colorimetric estimation of tryptophan content of pulses. Journal of Food Science and Technology, 11: 13– 216.

Rao, B.S.N., Deosthale, Y.G., Pant, K.C. 1989. Nutritive value of Indian goods. Pp 47-91. Hyderabad, India. National Institute of Nutrition. Indian Council of Medical Research.

Rao, P.U., Deosthale, Y.G. 1982. Tannin content of pulses varietals differences and effects of germination and cooking. Journal of Science Food and Agriculture, 33: 1013 – 1016.

Reddy, P.R.C., Gowramma, R.S. 1987. Cooking characters and *in vitro* protein digestibility of green gram varieties. Mysore Journal of agricultural Science, 21: 50-53.

Reynolds, J.E.F. 1989. Martindale: The Extra Pharmacopoeia. The Pharmaceutical Press: London pp 1015-1020.

Sadasivam, S., Manickam, A. 1996. Biochemical methods, New age International (P) limited publishers, New Delhi, India.

Salunke, B.K., Patil, K.P.,Wani, M.R., Maheshwari, V.L. 2006. Antinutritional constituents of different grain legumes grown in North Maharastra. Journal of Food Science and technology, 43: 519–521.

Salunkhe, D.K. 1982. Legumes in human nutrition: current status and future research needs. Current Science, 57: 387-394.

Salunkhe, D.K., Jadhaw, S.J., Kadam, S.S., Chavan, J.K. 1982. Chemical, biochemical and biological significance of polyphenol in cereals and legumes. Critical review of Food Science and Nutrition, 17: 277-305.

Sangronis, E., Machado, C.J. 2007. Influence of germination on the nutritional quality of *Phaseolus vulgaris* and *Cajanus cajan*. *LWT*, 40: 116-120.

Shanthakumari, S., Mohan, V.R. Britto J. De. 2008. Nutritional evaluation and elimination of toxic principles in wild yam (*Dioscorea* spp.). Tropical and Subtropical Agroecosystems, 8: 313-319.

Shills, M.E., Olson, J.A., Shrike, M., Ross. A.C. 1999. Modern nutrition in health and disease. Battimore. M.D: Williams and Wilkins.

Shills, M.E.G., Young,V.R. 1988. Modern nutrition in health and disease. In Nutrition, D.C. Nieman, D.E. Buthepodorth and C.N. Nieman (eds). Pp: 276-282 WmC. Brown publishers Dubugue, USA.

Shimelis, E.A. and Rakshit, S.K. 2007. Effect of processing on antinutrients and *in vitro* protein digestibility of kidney bean (*Phaseolus vulgaris* L.) varieties grown in east Africa. Food chemistry, 103:161-172.

Siddhuraju, P., Becker, K., Makkar, H.S. 2001. Chemical composition, protein fractionation, essential amino acid potential and antimetabolic constituents of an unconventional legume, Gila bean (*Entada phaseoloides* Merrill.) seed kernel. Journal of Science Food and Agriculture, 82: 192 -202.

Siddhuraju, P., Vijaykumari, K., Janardhanan, K. 1996. Chemical composition and protein quality of the little known legume, velvet bean [*Mucuna pruriens* (L.) DC.]. Journal of Agriculture and Food Chemistry, 44: 2636 – 2641.

Singh, J., Sharma, A., Ranjan, V. , Kumar, S. 1988. Plant drugs for liver disorders. Journal of Medicinal and Aromatic Plant Science, 20: 673-678.

Singh, J.N., Kumar, R., Kumar, P., Singh, P.K. 2000. Status of dietary fibres in new millennium- A review. Indian Journal of Nutrition and Dietetics, 37: 261-273.

Sood, M., Malhotra, S.R., Sood, B.C. 2002. Effect of processing and cooking on proximate composition of chick pea varieties. Journal of Food science and Technology, 39: 69-71.

Somiari, R.T., Balogh, E. 1993. Effect of soaking, cooking and alpha- galactoside treatment on the oligosaccharide content of cowpea flours. Journal of Science Food and Agricultural, 61: 339 – 343.

Spies, J.R., Chamber, D.C. 1949. Chemical determination of tryptophan in proteins. Analytical Chemistry, 21: 1249 – 1266.

Tan, N.H., Rahim, Z.H.A., Khor, H.T., Wong, K.C. 1983. Winged bean (*Psophocarpus etragonolobus*). Tannin level, phytate content and Haemagglutinating activity. Journal of Agricultural Food Chemistry, 31: 916 – 917

Tanaka, M., Thanankul, D., Lee, T.C., Chichester, L.O. 1975. A simplified method for the quantitative determination of sucrose, raffinose and stachyose in legume seeds. Journal of Food Science, 40: 1087 – 1088.

Tresina, P.S., Mohan V.R. 2011a. Chemical analysis and nutritional assessment of two less known pulses of genus *Vigna*. Tropical and Subtropical Agroecosystems, 14: 473-484.

Tresina, P.S., Mohan V.R. 2011b. Effect of gamma irradiation on physicochemical properties, proximate composition, vitamins and antinutritional factors of the tribal pulse, *Vigna unguiculata* subsp. *unguiculata*. International Journal of Food Science and Technology, 46: 1739-1746.

Tresina, P.S., Kala, K.B., Mohan, V.R., Vadivel, V. 2010. The biochemical composition and nutritional potential of three varieties of *Vigna mungo* (L.) Hepper. Advances in Bioresearch, 1: 6-16.

Vadivel, V., Janardhanan,K. 2000. Chemical composition of the underutilized legume *Cassia hirsuta* L. Plant Foods for Human Nutrition, 55: 369-381.

Valdebouze, P., Bergeron, E., Gaborit, T., Delort – Laval, J. 1980. Content and distribution of trypsin inhibitors and haemagglutinins in some legume seeds. Canadian Journal of Plant Science, 60: 695-701.

Vartak, V.D., Suryanarayanan, M.C. 1995. Enumeration of wild edible plants from Susale Island, Malsi reservoir, Pune district.

Journal of Economic Taxonomic Botany, 19: 565-569.

Vijayakumari, K., Siddhuraju, P., Janardhanan, K. 1996. Effect of different post-harvest treatments on anti-nutritional factors in seeds of the tribal pulse, *Mucuna pruriens* (L.) DC. International Journal of Food Science and Nutrition, 47: 263-272.

Vucenik, I., Shamsuddin, A.M. 2006. Protection against cancer by dietary IP6 and inositol. Nutrition and Cancer. 55: 109.

Waterlow, J.C. 1994. Childhood malnutrition in developing nations: Cooking back and looking forward. Annual Review of Nutrition, 14: 1-19.

Weaver, C.M., Kannan, S. 2002. Phytate and Mineral Bioavailability. In: Food Phytates, (Eds.). Reddy, N.R. and S.K. Sathe CRC Press, Boca Raton Florida, ISBN: 156676-867-5.

Wheeler, E.L., Ferrel, R.E. 1971. A method for phytic acid determination in wheat and wheat fractions. Cereal Chemistry, 48: 312 – 320.

Yanez, E., Zacarias, I., Aguayo, M. Vasquez, M ., Guzman, E. 1995. Nutritive value evaluated on rats of new cultivars of common beans (*Phaseolus vulgaris*) released on Chile. Plant Foods for Human Nutrition, 47: 301-307.

Zhou, J.R., Erdman, J.W. 1995. Phytic acid in health and disease. Critical Reviews of Food Science and Nutrition, 35: 495-508.

9

Seedling Growth of Rainforest Species Inoculated with Arbuscular Mycorrhizal Fungi: An Analysis of the Size Fragment Effect

N. Zamarripa, A.M. Patterson, I. Sánchez-Gallen* and J. Álvarez-Sánchez

Departamento de Ecología y Recursos Naturales, Facultad de Ciencias,
Universidad Nacional Autónoma de México
Circuito Exterior Ciudad Universitaria 04510 México, D.F. México.
e-mail: irene_sgallen@ciencias.unam.mx
*Corresponding author

SUMMARY

Deforestation is a process that brings as a consequence strong environmental problems in tropical rain forests. Restoration of damaged areas can accelerate succession process and improve seedling performance. One way to reach this objective is to inoculate them with native arbuscular mycorrhizal fungi. This study analyzed the effect of mycorrhizae inoculation on seedling survivorship and growth of two tree species, *Pleuranthodendron lindenii* (light demanding) and *Pimenta dioica* (shade tolerant) in shaded greenhouse and field conditions in the region of "Los Tuxtlas", Veracruz. We applied three inoculation treatments, without mycorrhizal inoculum (control), mycorrhizal inoculum from small fragments, and inoculum from large fragments. We analyzed survivorship and relative growth rates for height and diameter. For both species, significant differences ($p<0.05$) in growth rates in height and diameter were found for inoculum origin and time, as well as their interaction. The highest mean values corresponded to plants with inoculum from small fragments. Differences in survival among arbuscular mycorrhizal fungi treatments were significant only under shaded greenhouse conditions. The results are discussed in terms of life history traits and environmental conditions.

Keywords: Arbuscular mycorrhizal fungi; fragmentation; growth; tropical rain forest; *Pleuranthodendron lindenii; Pimenta dioica.*

INTRODUCTION

Fragmentation of tropical forests due to deforestation has carried out severe biodiversity loss and ecological interaction breakage, as well as depletion in economic and social activities related to these forests (Laurance, 1999; Kareiva and Marvier, 2007).

Deforestation disrupts the continuous forest in several patches, or fragments (Guevara et al., 2004b). Fragmentation of landscape is a complex process that produces abiotic and biotic changes that depend on fragment size, shape and arrangement in landscape (Bennett and Saunders, 2010). These changes lead to reduction and isolation of original populations, resulting in extinction population rate increase and colonization population rate decrease (MacArthur and Wilson, 1967), as well as ecological interaction shifts (Lienert, 2004; Mangan et al., 2004).

Fragmentation effects have been well studied on animal and plant communities (Debinski and Holt, 2000; Matthies et al., 2004; Helm et al., 2006). In contrast, in Neotropical forests there are few works that analyze how fragmentation modifies arbuscular mycorrhizal fungi (AMF) community structure and functionality (Mangan et al., 2004).

AMF form an important ecological relationship with a great number of plants (Siqueira et al., 1998; Zangaro et al., 2003). This mutualistic association has been proven to increase plant biomass and mineral concentration in plant tissue (Smith and Read, 2008; Varma, 2008), to stimulate flower, fruit, and pollen production in some plant species (Daft and Okusanya, 1973), and to increase cytokinins on leaves and roots (Allen et al., 1980), in exchange fungi receive the required carbohydrates for their existence.

Incipient data point at AMF colonizing capability differs depending on their origin site because AMF distribution in time and space is not haphazard, and it can change according to physical, biological and ecological characteristics of each place (Johnson and Wedin 1997; Allen et al., 1998; Picone, 2000; Mangan et al., 2004; Violi et al., 2008). As a consequence, we hypothesized that AMF communities from small fragments should be different in structure and functionality, as well as their effects on plants, compared to those from large fragments.

Given seedling vulnerability, inoculation with AMF can be essential to improve survival and growth, as it has been demonstrated in several studies (Kiers et al., 2000; van der Heijden, 2004; Zangaro et al., 2005). However, plant species response to AMF species will depend upon plant and AMF identity (van der

Heijden, 2002), and plant life history (Siqueira et al., 1998), as a consequence, each species of AMF can exhibit different affinities and impacts on seedling fitness (van der Heijden et al., 1998a; Kiers et al., 2000; Bever et al., 2010).

Plant life histories in tropical rain forest have been classified according to their role in natural regeneration and their light requirements. There are basically two large groups, light demanding and shade tolerant species. The former comprises plants that need high light intensities during its whole life cycle and have relative growth rates and resource requirements higher than those of shade tolerant species. On the contrary, shade tolerant species can germinate and live under shade for several years, their relative growth rates and resource requirements are lower than those of light claimants (Martínez-Ramos, 1994).

Following the previous ideas, our main goal was to assess the effects of different origin AMF inocula on seedling growth of two native tropical rain forest species with contrasting life histories under shaded greenhouse and field conditions. Because the responses of plants to AMF presence is determined by their life history and identity of AMF species, we hypothesized that growth and survival of shade tolerant species (TS) will be higher with inoculant from large forest fragments, while species light demanding (LD) will respond more positively to inoculant from small fragments.

This study contributed to the development of more accurate models of restoration in deforested areas, in order to develop strategies to accelerate the regeneration process of Los Tuxtlas tropical rain forest. Our hypothesis is that, in the long term, changes in AMF community from small fragments will occur, and this will have an effect on their functionality, favoring those plant species with the highest functional complementarity or affinity, i.e. light demanding species in small fragments, and shade tolerant species in large fragments.

MATERIAL AND METHODS

The Los Tuxtlas region is located in the coastal plain of the Gulf of Mexico, at the south part of the State of Veracruz (Guevara et al., 2004a). At low altitudes, tropical rain forest dominates (Miranda and Hernandez-X., 1963), most of it is concentrated into remnant fragments of different sizes which are mainly surrounded by a matrix of secondary vegetation and/or cattle ranching lands (Ibarra-Manriquez et al., 1997). Af(m)w"(i')g is the main climate type with a mean annual precipitation of 4,084 mm, and 25 °C as mean annual temperature (data taken from the Los

Tuxtlas Tropical Biology station, belonging to the period from 1996 to 2008). In particular, our shaded greenhouse was located at Los Tuxtlas Tropical Biology station (LTTBS), a research center of tropical biology under the Universidad Nacional Autónoma de México protection and we carried out our field transplants in two small fragments, 5 and 7 ha in size (Figure 1). All sites are part of the Los Tuxtlas National Biosphere Reserve.

Selected species

Pleuranthodendron lindenii (Turcz.) Sleumer belongs to Tiliaceae family. It is an evergreen native tree species. It is common in secondary vegetation at Los Tuxtlas forest, mainly associated with gaps; we classified it as a light demanding (LD) species (Pennington and Sarukhán, 2005).

Pimenta dioica (L.) Merr. belongs to Myrtaceae family. It is an evergreen native tree that grows under shade; we classified it as a shade tolerant (ST) species (Pennington and Sarukhán, 2005).

Inoculum collection

In March 2005, we collected soil from the interior and most conserved zone of two large fragments (211 and 640 ha in size) and two small ones (5 and 7 ha)

(Figure 1). Large fragments are little disturbed sites. Particularly, 640 ha fragment is the LTTBS and has never been managed or cleared. In contrast, small fragments have a higher LD species presence (Sánchez-Gallen, 2011). All four fragments were surrounded by cattle ranching lands.

In each of these fragments, we chose twenty five sampling points in a one-way field trip; each point was separated every two meters. Soil samples were taken from the first 20 cm in depth where we can find colonized roots by AMF, as well as mycelium and spores. After, we mixed all soil samples from the same size fragment group. Rocks and pieces of coarse roots were removed.

Subsequently, we placed collected soil in pots containing trap plants of various forest species, and after six month we characterized inoculum from each of the fragment groups (Luna, 2009). Identification was conducted by MSc. Laura Hernández-Cuevas, she based her results on current descriptions of different manuals (Schenk and Pérez 1988) and the International culture collection of arbuscular mycorrhizal fungi website (http://invam.caf.wvu.edu/Myc_Info/Taxonomy.speci es.html), following Schüßer and Walker (2010) classification.

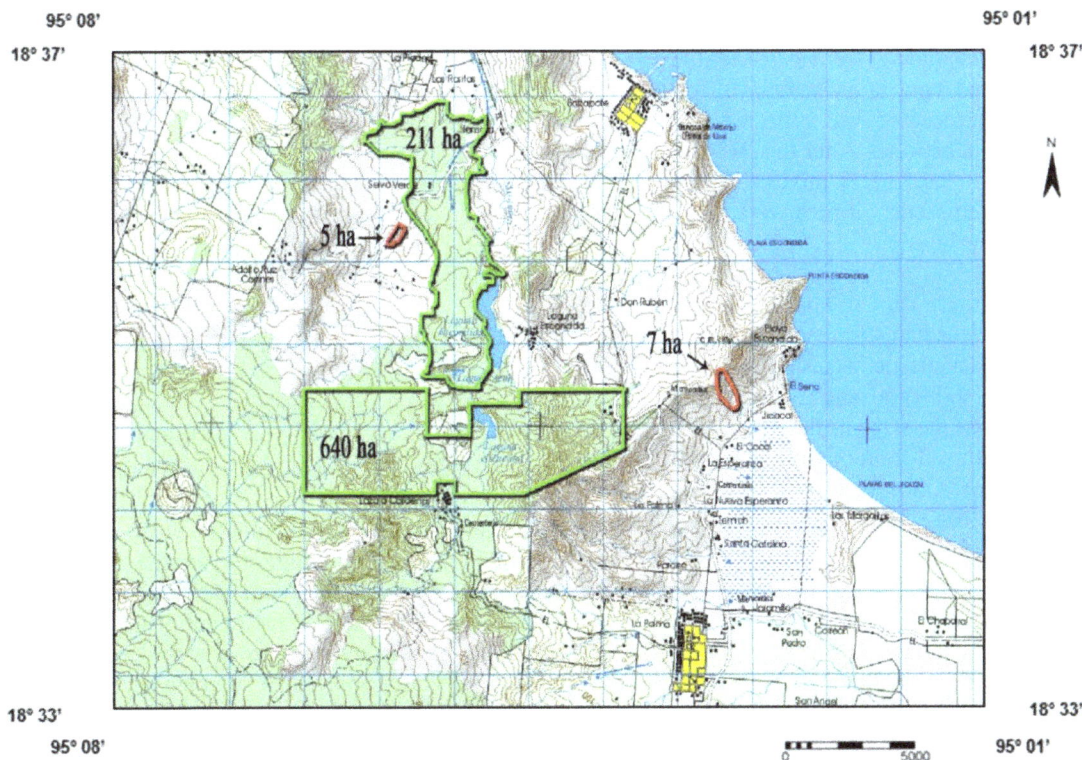

Figure 1.
Fragment location at Los Tuxtlas region, Veracruz, Mexico (modified from the INEGI (2000) topographic chart, key e15 a63, New Victory scale 1: 50 000). 640 ha Fragment = LTTBS. We show fragments 5 ha and 7 ha, where plants were transplanted.

Small fragment inoculum had an average of 8.7 spores g^{-1}, we identified a total of 11 AMF species. *Glomus tenebrosum* was the predominant one and *Acaulospora morrowiae*, *G. flavisporum*, *Sclerocystis rubiformis,* and *A. mellea* were only found in this type of inoculum. Whereas, in large fragments the average number of spores per gram was 9.7. We recorded a total of 15 species of AMF, with an abundance of *A. scrobiculata* and eight exclusive species (*A. foveata*, *G. aggregatum*, *Claroideoglomus claroideum*, *Sclerocystis sinuosa*, *Scutellospora gilmorei*, *A. laevis*, *Ambispora leptoticha*, and *Gigaspora decipiens*). Shared species by both inocula were *A. scrobiculata*, *A. spinosa*, *G. tenebrosum*, *Funneliformis geosporum*, *G. microaggregatum*, *F. verruculosum*, and *Redeckera fulvum*.

Experimental

The experiment considered the inoculum origin factor with three levels: 1) without AMF (M-), 2) with AMF from large fragments (MLF), and 3) with AMF from small fragments (MSF).

Prior to starting the experiment, in December 2004 we obtained seeds of the two above mentioned species by collecting fruits from 10 different trees located inside the forest of Los Tuxtlas. We washed all seeds with tap water and submerged them in a 3% chlorine solution during 10 min to disinfect them.

Simultaneously, in laboratory, we steam-sterilized soil from the forest and sand in an autoclave, for 1 hour at 90 °C. This procedure was repeated twice with a 24-hour period in between. After this, we filled pots with a sterile soil-sand (1:1) mixture and we sowed in the disinfected seeds, all pots were kept in the LTTBS shaded greenhouse.

After three months of growth, in March 2005, we randomly selected 420 seedlings for each species and transplanted to 2 kg black plastic bags. The bags were filled with sterilized soil and sand mixture (3:1). Sterilization process was the same as described above. We placed 50 g of fresh soil from the trap pots as the inoculum around each seedling belonging to large or small fragments per bag. The bags were tagged according to species name and inoculum type (treatment).

Three months later, in July 2005 (rainy season) we transplanted a total of 90 seedlings to the field, in the border zone of two small fragments (5 and 7 ha in size), 45 for each species, 15 plants per treatment. The mean *Pleuranthodendron* seedling height (± SD) was 17.7 cm (±3.4), 21.4 cm (±5.3), and 21.7 cm (±4.6), for M-, MLF and MSF, respectively. Mean height (± SD) for *Pimenta* per treatment was 9.9 cm

(±1.9) (M-), 9.8 cm (±1.6) (MLF), and 9.4 cm (±1.8) (MSF).

Simultaneously, we left 150 seedlings for each species, 50 for each treatment in the shaded greenhouse of the LTTBS, they were randomly placed into three blocks for each species, following the applied inoculation treatments (M-, MLF, or MSF) to avoid any contamination among treatments. Also, we rotated them every month to assure that seedlings were growing under similar light conditions, environmental temperature and humidity were not controlled, and they were consistent with those outside the shaded greenhouse.

All surviving plants were harvested nine months later (March 2006). And to confirm AMF colonization, we took roots from two plants of every treatment and species, we stained them following Phillips and Hayman technique (1970) modified by Hernández-Cuevas et al. (2008) and estimated total colonization percentage.

Data analysis

Every month, we measured total height and diameter at the base of 15 randomly chosen seedlings per species and treatment, in shaded greenhouse and field. Similarly, survival was recorded counting the number of total seedlings that remained alive every month, per species per treatment. We repeated these measures until January 2006.

Growth analysis

We calculated relative growth rates based in height (cm cm^{-1} day^{-1}) and diameter (mm mm^{-1} day^{-1}) following a functional growth analysis (Hunt, 1982):

$$RGR = \frac{H \ or \ D_{t2} - H \ or \ D_{t1}}{t_2 - t_1}$$

Where H= Heigh, D= diameter, t_1 = number of days elapsed after previous data collection, and t_2 = number of days elapsed after next data collection.

We transformed RGR diameter data by natural logarithm and RGR height by arcsine in order to reach normal distribution and variance homogeneity (Zar, 2009). We analyzed treatment mean differences by using repeated measure analysis of variance (ANOVA). And, only when we obtained significant differences, in order to discriminate among different treatment levels we applied a Tukey analysis (Montgomery, 1991). An analysis of covariance was not necessary to perform because RGRs allow to weight initial differences among individuals. We used

STATISTICA 8 (Statsoft Inc., 2000) software for all analyses of variance.

Survivorship

We compared survivorship among inoculum treatments, separating field and shaded greenhouse sites with Peto and Peto test (Pyke and Thompson, 1986).

RESULTS

Growth analysis

Average relative growth rates (RGR) for *Pleuranthodendron lindenii* fluctuated between 0.01 and -0.001 cm cm^{-1} day^{-1}, and 0.07 and -0.004 mm mm^{-1} day^{-1}, for height and diameter, respectively. In *Pimenta dioica*, values were between 0.008 and -0.005 cm cm^{-1} day^{-1}, and 0.06 and -0.002 mm mm^{-1} day^{-1} for height and diameter, respectively.

Inoculum origin and time factors, for both species and response variables had significant differences among levels, while their interaction was not significant; on the contrary, zone factor was only significant for *Pleuranthodendron* (Table 1).

Both species, *Pleuranthodendron*, and *Pimenta* had the highest significant mean values in height and diameter when are inoculated with fungi from large fragments (MLF) (Table 1).

However, even when inoculum origin is an important factor for both species growth, when we analyzed interactions among factors, shaded greenhouse interacting with any date of collection and type of inoculum had the highest plant response (Figure 2 and 3) for both species.

Mean total colonization percentage was higher in MLF treatment for both species (*Pleuranthodendron* 52% (±5) and *Pimenta* 46% (±7)). While M- had *Pleuranthodendron* 15% (±5) and *Pimenta* 23% (±7).

Survivorship

Survival individual number comparisons among AMF inoculum treatments had significant differences (p<0.05) only under shaded greenhouse conditions. For both species, we observed the lowest survivorship value with inoculum from MSF, while control and large fragment inoculum treatments had the same trend (Figure 4).

DISCUSSION

Our hypothesis was that growth, measured in terms of height and diameter, and survival of *P. lindenii* (light demanding species) would reach their highest values with inoculum from small fragments (MSF). On the contrary, we expected that *P. dioica* would have higher relative growth rates and survival with inoculum from large fragments (MLF) than with MSF. Our results partially supported this hypothesis, since both *Pleuranthodendron* and *Pimenta* growth was benefited, at some time, from both inocula.

We did not find a clear relationship among inoculum origin and species life history. This indicates that AMF presence is important for both species growth but not all the time, this perhaps is related with changes of plant physiological needs or energetic costs involved with the relationship, because even when several studies characterized arbuscular mycorrhiza as a mutualistic interaction, sometimes, depending on environmental conditions it could turn parasitic, where fungi parasites plant due to an excess of plant carbon investment (Johnson *et al.*, 1997), and this can be happening in the field, where environmental conditions are so stressing that plants can barely produce carbon and allocate it to its main functions.

The fact that presence of AMF, regarding origin, favored higher growth rates has been shown in other studies with tropical forest native species (Allen *et al.*, 2003; Fischer *et al.*, 1994; Kiers *et al.*, 2000; Zangaro *et al.*, 2000). Fischer *et al.* (1994) found that inoculum from abandoned grasslands resulted in higher growth, compared with inoculum from lowland secondary forest, and Zangaro *et al.* (2000) found that early successional species had higher growth rates when they were inoculated with inoculum from an area dominated by pioneer species. Although Kiers *et al.* (2000) found that fungi are important for species growth, they reported that a tree pioneer species and a mature forest tree species grew much more with an inoculum from a late successional plant species (which in our case would correspond to MLF inoculum) and with inoculum from an early successional plant species (equivalent to our MSF inoculum), respectively. In the same terms, Allen *et al.* (2003) showed that early late successional seedlings had a higher benefit when they were inoculated with soil from an early successional site of deciduous tropical forest instead of using inoculum from a mature forest soil of the same forest.

Table 1. Summary of analyses of variance. We show F-value and significance value (p). We also highlight the highest and lowest mean values of the response variable for each factor.

Species	Response variable	Factor	F	p	Highest mean	Lowest mean
Pleuranthodendron lindenii	RGR in height	Inoculum origin (IO)	275.292	<0.001	MLF	MSF
		Zone	28.768	<0.001	Shaded greenhouse (SG)	Field
		Time	5.398	<0.001	Jul-Aug	Dec-Jan
		IO×Zone	11.370	<0.001	MLF×SG	MSF×Field
		IO×Time	0.788	>0.05	MLF×Jul-Aug	MLF×Dec-Jan
		Zone×Time	9.416	<0.001	SG×Jul-Aug	Field×Dec-Jan
		IO×Zone×Time	2.535	<0.001	MLF×SG×Jul-Aug	MLF×Field×Sep-Oct
	RGR in diameter	Inoculum origin (IO)	284.22	<0.001	MLF	MSF
		Zone	14.08	<0.001	Shaded greenhouse (SG)	Field
		Time	24.11	<0.001	Aug-Sep	Sep-Oct
		IO×Zone	8.96	<0.001	MLF×SG	M-×SG
		IO×Time	1.46	>0.05	MSF×Aug-Sep	MLF×Sep-Oct
		Zone×Time	4.60	<0.001	SG×Aug-Sep	Field×Sep-Oct
		IO×Zone×Time	1.55	>0.05	MLF×SG×Aug-Sep	MLF×Field×Sep-Oct
Pimenta dioica	RGR in height	Inoculum origin (IO)	432.159	<0.001	MLF	M-
		Zone	1.825	>0.05	Shaded greenhouse (SG)	Field
		Time	3.819	<0.001	Aug-Sep	Dec-Jan
		IO×Zone	4.188	<0.01	MLF×SG	M-×Field
		IO×Time	1.304	>0.05	M-×Oct-Nov	M-×Nov-Dec
		Zone×Time	4.675	<0.001	SG×Jul-Aug	Field×Sep-Oct
		IO×Zone×Time	0.597	>0.05	M-×SG×Jul-Aug	M-×Field×Oct-Nov
	RGR in diameter	Inoculum origin (IO)	423.53	<0.001	MLF	M-
		Zone	2.90	>0.05	Shaded greenhouse (SG)	Field
		Time	7.89	<0.001	Aug-Sep	Sep-Oct
		IO×Zone	4.44	<0.01	MLF×SG	M-×Field
		IO×Time	0.87	>0.05	M-×Oct-Nov	M-×Nov-Dec
		Zone×Time	1.80	>0.05	SG×Aug-Sep	Field×Dec-Jan
		IO×Zone×Time	1.75	<0.05	MLF×SG×Aug-Sep	MLF×Field×Dec-Jan

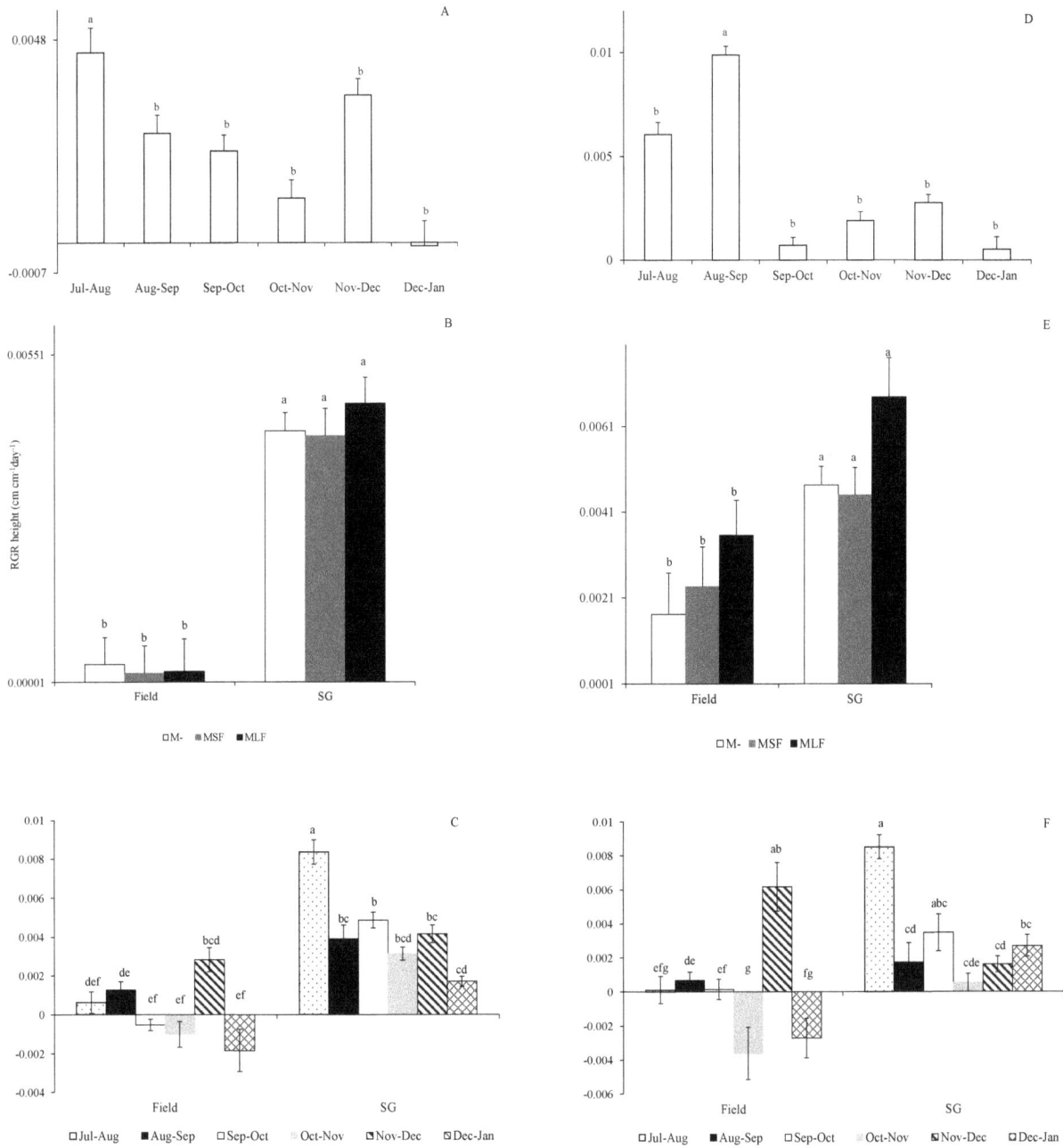

Figure 2. Mean height relative growth rates according to time (A, and D), and interactions among factors, inoculum origin (IO)×Zone (B, and E), and Zone×Time (C, and F). A, B, and C correspond to *Pleuranthodendron lindenii* while D, E and F correspond to *Pimenta dioica*. Different letters indicate significant differences according to Tukey test (p < 0.05). Zone: field and shaded greenhouse (SG).

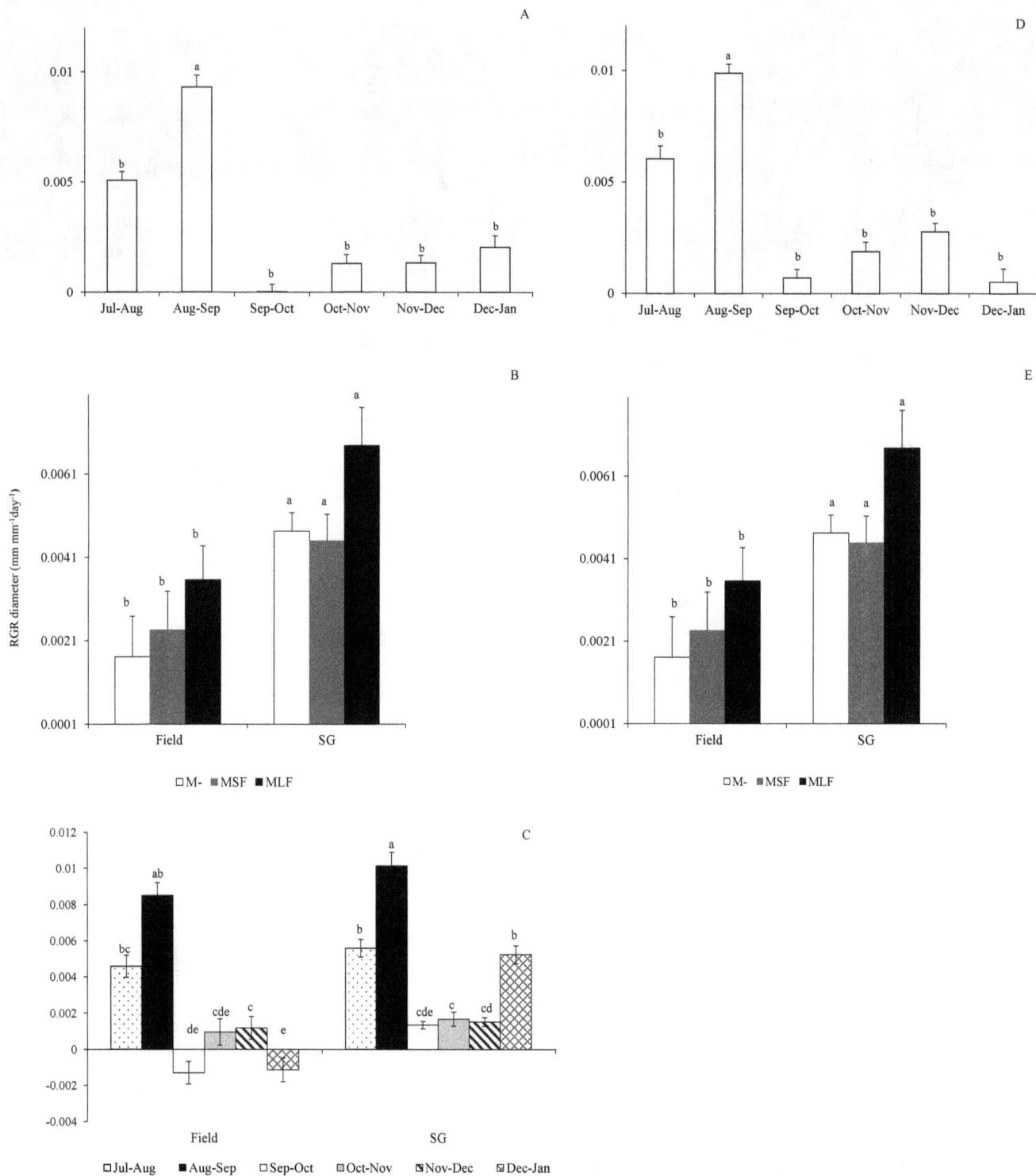

Figure 3. Mean diameter relative growth rates according to time (A, and D), and interactions among factors, inoculum origin (IO)×Zone (B, and E), and Zone×Time (C). A, B, and C correspond to *Pleuranthodendron lindenii* while D, and E correspond to *Pimenta dioica*. Different letters indicate significant differences according to Tukey test (p < 0.05). Zone: field and shaded greenhouse (SG).

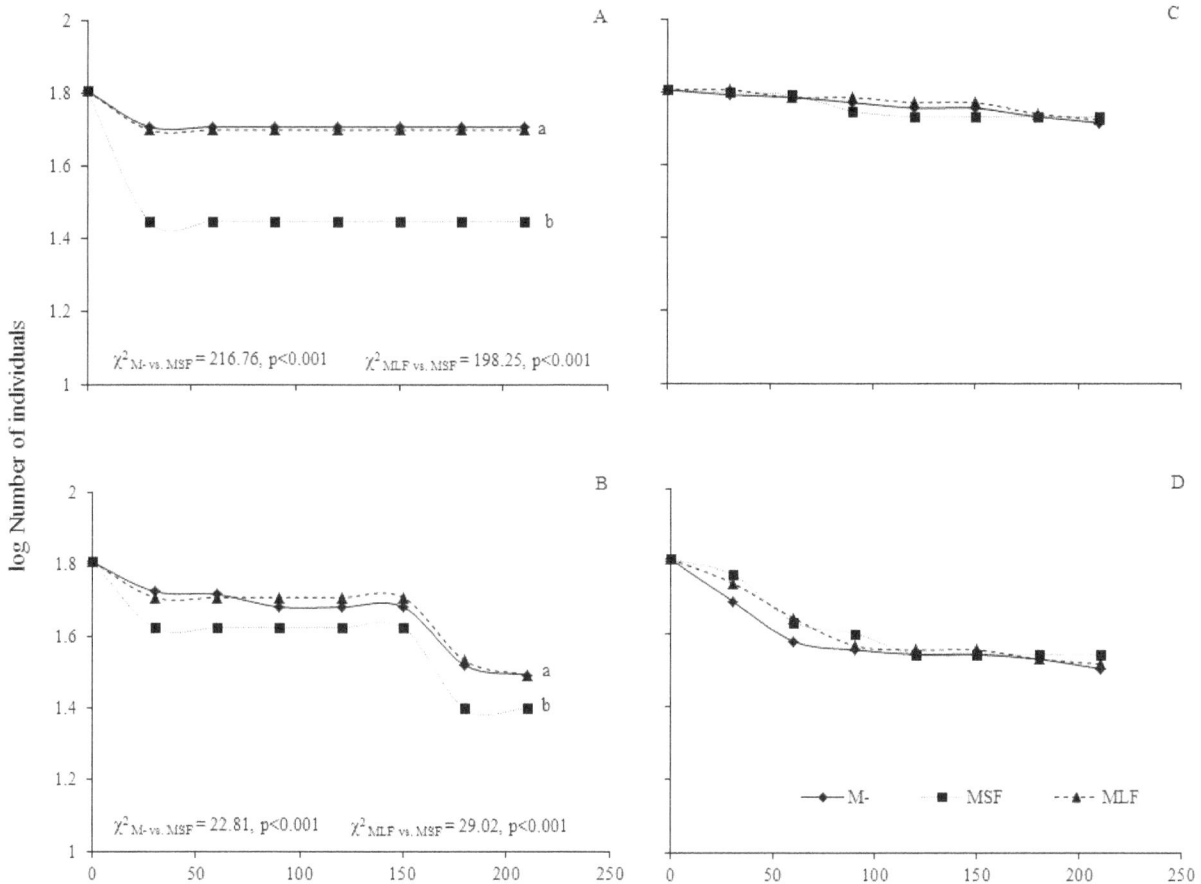

Figure 4. Survivorship curves of both species growing under shaded greenhouse (A and B) and field (C and D) conditions, (A and C, *Pleuranthodendron lindenii*; B and D, *Pimenta dioica*).

If we thoroughly analyze species response to AMF, we found that in general *Pleuranthodendron* in field had the highest mean values growing with MLF treatment while *Pimenta* pattern is not clear, but in general, MLF together with no inoculum had the highest growth and survival responses. The above suggests life history traits (Martínez-Ramos, 1994) are critical for determining plant response to inoculation, predominantly in the case of the former species because as a light demanding, its successful when being transplanted to field beneath tree shade is low, and *Pimenta* as shade tolerant can stand low light field conditions and allocate more resources to keep growing in natural conditions into the forest; this performance has been observed in other shade tolerant species (Álvarez-Sánchez *et al*., 2009).

The AMF did not have a significant effect after transplantation to the field, which indicates that for any of the two life history traits there were not effect of the inoculum for increase survival according to the size of the fragment,. This is contrary to that found by Guadarrama *et al*. (2004); they reported survival differences for the genus *Heliocarpus*, light demanding specie. It is clear that for species used in this study there is not a survival-growth trade off in terms of the benefit by the AMF.

On the other hand, AMF effects on plants can have different quality; some AMF species enhance phosphorus absorption, while other species protect plants from pathogens and parasites (Klironomos *et al*., 2000). This occurs since each AMF species can have a different effect depending on the plant species identity, this is explained by AMF multifunctionality (van der Heijden *et al*., 1998a, 1998b) supported by the fact that there is a great genetic diversity in AMF (Clapp *et al*., 2002). Our inocula shared close to 70% of species, but both inocula (from small and large fragments) had exclusive species (Luna, 2009), that could explain differential plant responses. Whether this can be happening at Los Tuxtlas tropical rain forest is a question that has to be solved in the short time because it is patent that native forest plants need AMF to grow and survive, now the question is who needs whom, answering it will give major success to restoration actions.

CONCLUSIONS

Despite there was no relationship between plant life history and response to AMF inoculum origin, we found that AMF inoculum factor by itself had a significant and positive effect on relative growth rates of both species, and this result is more evident when inoculum belonged to large fragments.

However, we also found that these positive effects depend on the site conditions where seedlings are growing; AMF inoculum origin and site interaction showed that arbuscular mycorrhiza effects are clearly positive only when seedlings are growing under shaded greenhouse conditions, but when these are transplanted to field, AMF positive effects on seedling height and diameter can not overcome harsh field environmental conditions. And this might be due to the high energetic costs this mutualistic interaction has for seedlings, regardless inoculum origin.

Nevertheless, we widely recommend the use of native AMF to inoculate seedlings before field transplants, but taking care of them, at least the first two years.

Acknowledgements.

To the project SEMARNAT-CONACyT-2002-c01-668 by funding and scholarships granted to ANZ and AP. Guadalupe Barajas help us in the statistical analysis. Also, we thank Marco Antonio Romero Romero for his support in computer issues. We thank all help received from LTTBS staff, especially Gregorio, Gilberto Quinto, Rosamond Coates and Jorge Perea. To José Palacios and Domingo Toto for allowing us to work in their lands.

REFERENCES

Allen, M.F., Moore Jr., T.S. and Christensen, M., 1980. Phytohormone changes in *Bouteloua gracilis* infected by vesicular-arbuscular mycorrhizae. I. Cytokinin increases in the host plant. Canadian Journal of Botany 58:371-374.

Allen, E.B., Rincón, E., Allen, M.F., Pérez-Jiménez, A. and Huante, P., 1998. Disturbance and seasonal dynamics of mycorrhizae in a tropical deciduous forest in Mexico. Biotropica 30:261-274.

Allen, M.F., Swenson, W., Querejeta, J.I., Egerton-Warburton, L.M. and Treseder, K.K. 2003. Ecology of mycorrhizae: a conceptual framework for complex interactions among plants and fungi. Annual Review of Phytopathology 41:271-303.

Álvarez-Sánchez, J., Sánchez-Gallen, I. and Guadarrama, P. 2009. Ecophysiological Traits of Tropical Rain Forest Seedlings Under Arbuscular Mycorrhization: Implications In Ecological Restoration. In: Symbiotic fungi: Principles and practice. Varma, A. and. Kharkwal, A.C. (eds.). Springer-Verlag, Berlin. pp. 293-305.

Bennett, A.F. and Saunders, D.A., 2010. Habitat fragmentation and landscape change. In: Sodhi, N.S. and Ehrlich, P.R. (eds.). Conservation biology for all. Oxford University Press, UK. pp. 88-106.

Bever, J.D., Dickie, I.A., Facelli, E., Facelli, J.M., Klironomos, J., Moora, M., Rillig, M.C., Stocks, W.D., Tibbett, M. and Zobel, M., 2010. Rooting theories of plant community ecology in microbial interactions. Trends in Ecology and Evolution 25:468-478.

Clapp, J.P., Helgason, T., Daniell, T.J. and Young, J.P.W., 2002. Genetic studies of the structure and diversity of arbuscular mycorrhizal fungal communities. In: Mycorrhizal ecology. van der Heijden, M. G. A. and Sanders, I. R. (eds.). Springer-Verlag, Nueva York. pp. 243-266.

Daft, M.J. and Okusanya, B.O., 1973. Effect of *Endogone* mycorrhiza on plant growth V. Influence of infection on the multiplication of viruses in tomato, petunia and strawberry. New Phytologist 72:975-983.

Debinski, D.M. and Holt, R.D., 2000. A survey and overview of habitat fragmentation experiments. Conservation Biology 14:342-355.

Fischer, C.R., Janos, D.P., Perry, D.A. and Linderman, R.G., 1994. Mycorrhizal inoculum potentials in tropical secondary succession. Biotropica 26:369-377.

Guadarrama, P., Álvarez-Sánchez, F.J. and Estrada-Torres, A., 2004. Phosphorus dependence in seedlings of a tropical pioneer tree: the role of arbuscular mycorrhizae. Journal of Plant Nutrition 27:1-6.

Guevara, S., Laborde, J. and Sánchez-Ríos, G., 2004a. Introducción. In: Guevara, S., Laborde, J. and Sánchez-Ríos, G. (eds.). Los Tuxtlas. El paisaje de la sierra. Instituto de Ecología, A.C., Unión Europea. Xalapa Ver.. pp. 18-26.

Guevara, S., Laborde, J. and Sánchez-Ríos, G., 2004b. La fragmentación. In: Guevara, S., Laborde, J. and Sánchez-Ríos, G. (eds.). Los Tuxtlas. El paisaje de la sierra. Instituto de Ecología, A.C. , Unión Europea. Xalapa Ver. pp. 111-134.

van der Heijden, M.G.A., Boller, T., Wiemken, A. and Sanders, I.R., 1998a. Different arbuscular mycorrhizal fungal species are potential determinants of plant community structure. Ecology 79:2082-2091.

van der Heijden, M.G.A., Klironomos, J.N., Ursic, M., Moutoglis, P., Streitwolf-Engel, R., Boller, T., Wiemken, A. and Sanders, I.R., 1998b. Mycorrhizal fungal diversity determines plant biodiversity, ecosystem variability, and productivity. Nature 396:69-72.

van der Heijden, M.G.A., 2002. Arbuscular mycorrhizal fungi as determinant of plant diversity: In search for underlying mechanisms and general principles. In: van der Heijden M.G.A. and Sanders, I.R. (eds.). Mycorrhizal ecology. Springer-Verlag, New York. pp. 243-266.

van der Heijden, M.G.A., 2004. Arbuscular mycorrhizal fungi as support systems for seedling establishment in grassland. Ecology Letters 7:293-303.

Helm, A., Hanski, I. and Pärtel, M., 2006. Slow response of plant species richness to habitat loss and fragmentation. Ecology Letters 9:72-77.

Hernández-Cuevas, L. V., Guadarrama-Chávez, P., Sánchez-Gallen, I. and Ramos-Zapata, J., 2008. Micorriza arbuscular. Colonización intrarradical y extracción de esporas del suelo. In: Álvarez-Sánchez, J. and Monroy, A. (eds.). Técnicas de estudio de las asociaciones micorrízicas y sus implicaciones en la restauración. Universidad Nacional Autónoma de México, Facultad de Ciencias. México, D. F. pp. 1-15.

Hunt, R., 1982. Plant growth curves: The functional approach to plant growth analysis. Edward Arnold, London.

Ibarra-Manríquez, G., Martínez-Ramos, M., Dirzo, R. and Núñez-Farfán, J., 1997. La vegetación. In: González-Soriano, E., Dirzo, R. and Vogt, R. (eds.). Historia Natural de Los Tuxtlas. Universidad Nacional Autónoma de México–Comisión Nacional para el Conocimiento y Uso de la Biodiversidad, México, D.F. pp. 61-85.

Johnson, N.C. and Wedin, D.A., 1997. Soil carbon, nutrients, and mycorrhizae during conversion of dry tropical forest to grassland. Ecological Applications 7:171-182.

Johnson, N.C., Graham, J.H. and Smith, F.A., 1997. Functioning of mycorrhizal associations along the mutualism-parasitism continuum. New Phytologist 135:575-585.

Kareiva, P. and Marvier, M., 2007. Conservation for the people. Scientific American 297:50-57.

Kiers, E.T., Lovelock, C.E., Krueger, E.L. and Herre, E.A., 2000. Differential effects of tropical arbuscular mycorrhizal fungal inocula on root colonization and tree seedling growth: implications for tropical forest diversity. Ecology Letters 3:106-113.

Klironomos, J.N., McCune, J., Hart, M. and Neville, J. 2000. The influence of arbuscular mycorrhizae on the relationship between plant diversity and productivity. Ecology Letters 3:137-141.

Laurance, W.F., 1999. Reflections on the tropical deforestation crisis. Biological conservation 91:109-117.

Lienert, J., 2004. Habitat fragmentation effects on fitness of plant populations - a review. Journal for Nature Conservation 12:53-72.

Luna, W.B., 2009. Diversidad y potencial de inóculo de hongos micorrizógenos arbusculares en fragmentos de diferentes tamaños de selva alta perennifolia. Thesis dissertation. Facultad de Ciencias, Universidad Nacional Autónoma de México, México, D.F.

MacArthur, R.H. and Wilson, E.O., 1967. The theory of island biogeography. Princeton University Press, Princeton, New Jersey.

Mangan, S.A., Eom, A.H., Adler, G.H., Yavitt, J.B. and Herre, E.A., 2004. Diversity of arbuscular mycorrhizal fungi across a fragmented forest in Panama: insular spore communities differ from mainland communities. Oecologia 141:687-700.

Martínez-Ramos, M., 1994. Regeneración natural y diversidad de especies arbóreas en selvas húmedas. Boletín de la Sociedad Botánica de México 54:179-224.

Matthies, D., Bräuer, I., Maibom, W. and Tscharntke, T., 2004. Population size and the risk of local extinction: empirical evidence from rare plants. Oikos 105:481-488.

Miranda, F. and Hernández-X, E., 1963. Los tipos de vegetación de México y su clasificación. Boletín de la Sociedad Botánica de México 28:29-179.

Montgomery, D.C., 1991. Diseño y análisis de experimentos. Grupo Editorial Iberóamerica, México, D. F., México.

Pennington, T.D. and Sarukhán, J., 2005. Árboles tropicales de México. Manual para la identificación de las principales especies. 3rd. Ed. Universidad Nacional Autónoma de México, Fondo de Cultura Económica, México.

Phillips, J.M. and Hayman, D.S., 1970. Improved procedure for clearing roots and staining parasitic and vesicular-arbuscular mycorrhizal fungi for rapid assessment of infection. Transactions of the British Mycological Society 5:158-161.

Picone, C., 2000. Diversity and abundance of arbuscular-mycorrhizal fungus spores in tropical forest and pasture. Biotropica 32:734-750.

Pyke, D.A. and Thompson, J.N., 1986. Statistical analysis of survival and removal rate experiments. Ecology 67:240-245.

Sánchez-Gallen, I., 2011. Análisis de la comunidad de plántulas, en relación con la de hongos micorrizógenos arbusculares, en fragmentos de vegetación remanente de una selva húmeda. PhD. Dissertation. Posgrado en Ciencias Biológicas, Universidad Nacional Autónoma de México, Facultad de Ciencias, México, D.F.

Schenk, N. and Perez, Y., 1988. Manual for the identification of VA mycorrhizal fungi. 2nd. Ed. INVAM. University of Florida, Gainesville, Florida.

Schüßer, A. and Walker, C., 2010. The Glomeromycota. A species list with new families and new genera. Gloucester, UK. In: http://www.genetik.biologie.uni-muenchen.de/research/schuessler/publications/papers_schuessler/schuessler_walk_2010.pdf Consulted May 3, 2012.

Siqueira, J.O., Carneiro, M.A.C., Curi, N., da Silva Roçado, S.C. and Davide, A.C., 1998. Mycorrhizal colonization and mycotrophic growth of native woody species as related to successional groups in Southeastern Brazil. Forest Ecology and Management 107:241-252.

Smith, S.E. and Read, D.J., 2008. Mycorrhizal symbiosis. 3rd. Ed. Academic Press, UK.

Statsoft, Inc., 2000. STATISTICA for Windows (Computer program manual). Statsoft Inc., Tulsa, Oklahoma, EU.

Varma, A., 2008 (ed.). Mycorrhiza. State of the art, genetics and molecular biology, eco-function, biotechnology, eco-physiology, structure and systematics. 3rd. Ed. Springer-Verlag, Berlín.

Violi, H.A., Barrientos-Priego, A.F., Wright, S.F., Escamilla-Prado, E., Morton, J.B., Menge, J. A. and Lovatt, C.J., 2008. Disturbance changes arbuscular mycorrhizal fungal phenology and soil glomalin concentrations but not fungal spore composition in montane rainforests in Veracruz and Chiapas, Mexico. Forest Ecology and Management 254:276-290.

Zangaro, W., Bononi, V.L.R. and Trufen, S.B., 2000. Mycorrhizal dependency, inoculum potential and habitat preference of native woody species in South Brazil. Journal of Tropical Ecology 16:603-622.

Zangaro, W., Nisizaki, S.M.A., Domingos, J.C.B. and Nakano, E.M., 2003. Mycorrhizal response and successional status in 80 woody species from south Brazil. Journal of Tropical Ecology 19:315-324.

Zangaro, W., Nishidate, F.R., Camargo, F.R.S., Romagnoli, G.G. and Vandresen, J., 2005. Relationships among arbuscular mycorrhizas, root morphology and seedling growth of tropical native woody species in southern Brazil. Journal of Tropical Ecology 21:529-540.

Zar, J.H., 2009. Biostatistical Analysis. 5th ed., Prentice Hall Inc. Englewoods Cliffs, New Jersey.

10

Participatory Evaluation of Sustainable Land use and Technology Adoption in Two Agroecosystems

Jaime Ruiz-Vega*, Rafael Pérez-Pacheco, Teodulfo Aquino-Bolaños and María Eugenia Silva-Rivera

Centro de Investigación Interdisciplinaria para el Desarrollo Integral Regional (CIIDIR-IPN-OAXACA). Calle Hornos 1003, Santa Cruz Xoxocotlán, Oax. CP 71 230, México
e-mail: jvega@ipn.mx
*Corresponding Author

SUMMARY

In order to identify the main agroecosystems, their limiting factors and adequate technological options, participatory approaches, such as community ranking, were used in a micro-hydrological basin in Central Oaxaca, Mexico. This area is characterized by small farm size (1-2 ha), low input agriculture and low standards of living. The results of a pretested survey were presented at community meetings and were subjected to discussion to rank the problems found in order of importance. Overall, the main production constraints were: low soil fertility, insect pests and plant diseases, lack of rain and soil erosion. After field evaluations of several sustainable technologies, the following was found: a) organic mulching can reduce soil erosion, weeds and conserve soil moisture, b) intercropped green manures with maize could be a mean to improve soil fertility while still allowing producing this staple crop, c) composting of crop residues with weeds and farmyard manure was also promoted amongst the peasants, but only a few of them adopted this practice due mostly to high labor requirements and d) even though it is an expensive technology, the use of floating row covers to produce tomatoes and hot peppers was quickly adopted by the peasants. It was concluded that the best way to convince the peasants to adopt a technological innovation is to show them that it works under their own circumstances.

Key words: participation; coal; oxen teams; maize; beans.

INTRODUCTION

An agroecosystem can be defined as a spatially and functionally coherent entity of agricultural activity, and includes the living and nonliving components, as well as their interactions (Gliessman, 2000). However, an agroecosystem not only affects the site of agricultural activity, but also the region where is located. Water is one of the main production factors in most agroecosystems and moves from the micro catchment level to the hydrological basin level; therefore, the micro-hydrological basin is the minimum study area to be used. This leads to the concept of hierarchical systems, where each level is represented by a set of functional subsystems and where the products of one can be the inputs of other subsystems (Hart, 1984).

Many traditional agroecosystems are considered sustainable, but there are not many scientific evaluations to support this affirmation (Brunett *et al.*, 2005); some are considered sustainable because they have passed the test of time. A sustainable use implies that the resources will be managed in such a way that they can still provide goods to future generations and a sustained yield is achieved (Gliessman, 2000). Furthermore, the system's production has to be distributed as equitably as possible in order to guarantee a sustainable development.

As previously mentioned, the identification of the main crop-production constraints in the agroecosystem is one of the critical steps for agroecosystem's improvement. An approach for production system`s diagnosis is the so called "farmer first" approach, a form of Participatory Rural Appraisal (Chambers, 1994). While many considered farmer first thinking as a step in the right direction, some argued that the approach failed to consider the socio-cultural and political economic dimensions of knowledge creation, innovation, transmission and use within rural societies and scientific organizations. Guijt and Cornwall (1995), found that the methodology was not applied correctly in many cases, the main issue was that the people did not actually participate in priority setting or in the subsequent actions to be taken.

Therefore, when applied in a simplistic, populist manner, participatory strategies encounter the same sorts of problems as other interventionist approaches. No matter how firm the commitment, the concept of powerful outsiders helping powerless insiders is always present (Scoones *et al.*, 2007). Since embracing participatory methods from the late 1980s, scientists and a variety of public and private agencies have encountered both successes and failures. However, recent evidence shows that participatory methods can generate quantitative data which are useful to determine local priorities and potential for innovation (Mayoux and Chambers, 2005).

Ashley *et al.* (1989) and Pretty (1997), have proposed that the peasants themselves evaluated the technology and decide which was the most suitable. Thus, a menu of sustainable technologies could be subject to validation in the farmer´s fields.

Many of the technological innovations generated to increase the well being of peasant peasants and for conservation of their production resources have been frequently not adopted in Mexico. Some examples in Oaxaca are fertilization of rain-fed maize (Ruiz, 1987), the introduction of improved varieties in low rainfall areas (Ruiz, 1990) and the construction of bench terraces with heavy machinery. These technologies were not adopted because: a) they did not address the most limiting factor and/or b) the peasants did not receive proper training on the new technology or b) the new technologies did not fit well within the farmer's agroecosystem.

However, there are a few successful cases. One of these is credited to Bunch and Lopez (1995), who found that contour grass barriers, use of organic fertilization and crop rotation were successfully introduced in the San Martin Jilotepeque area in the Guatemalan highlands. Contour ditches and side-dressing of nitrogen fertilizer in maize were used as starting technologies to motivate people. This is an example of how technology that addresses the main limiting factors and fits into the agroecosystem is readily adopted. The agriculture practiced in San Martin Jilotepeque is considered traditional and for self-consumption.

The State of Oaxaca, where this study was carried out, is characterized by peasant agriculture, low use of modern technologies and low general development. In this state, there are 7210 human settings, but about 80 % of them have less than 500 inhabitants (INEGI, 2009). These peasants carry out a variety of activities, including the cultivation of staple crops (*Zea mayz, Phaseolus vulgaris* and *Cucurbita pepo*), forest exploitation, small cattle rising and non-agricultural activities. The Central Region is crossed from N to S by The Atoyac River, which flows mostly during summer and fall. The SW basin of the river is represented mostly by the San Bernardo River and The Valientes River, forming the SW micro-hydrological basin of the Atoyac River (SWAR). This area covers some 70, 000 ha, and includes about 7,000 ha of temperate forests.

According to Valdés-Rodríguez *et al.* (2011), to facilitate the process of technology adoption, the UNDP has developed a methodology that allows for the design and evaluation of sustainable livelihoods through five steps, which include a participatory appraisal to determine the adaptive strategies of the people and a study of the potential of technology and science to complement indigenous knowledge. This study included these parts of the UNDP approach.

Using field surveys and Participatory Rural Appraisal (PRA) methodologies, this study was carried out in the SWAR with the following objectives: the characterization of the main agroecosystems (AE) present, the identification of their production constraints, and the evaluation and promotion of sustainable production practices (Pretty, 1995) for AE's improvement.

MATERIALS AND METHODS

Considering several criteria; including population size land tenure conflicts and road accessibility, two communities out of twelve present in the SW microhydrological basin of The Atoyac River were selected. These were San Lucas Tlanichico and Magdalena Mixtepec.

To give an idea of the environmental conditions in the study area, some agroecological parameters of these communities are presented in Table 1.

A pretested survey was applied to 10 % of the homesteads (N = 290), and the results were used to identify the main agricultural activities, local technology, and production problems. In most cases, the informers were the peasants, which were located at work in nearby fields. According to their registered frequency, the production problems were ranked. Afterwards, these problems, and others not perceived during the survey were ranked again during at least two community meetings per site. After the results of the survey were presented, the peasants were invited to propose other problems and to participate in the ranking process; every farmer was allowed to vote only once. These meetings were promoted one day in advance and lasted about 90 minutes per session. Even though everybody in the community was invited, women participation was low.

In order to tackle some of the main production constraints, field experiments on recognized sustainable technologies were carried out in both localities. The feasibility of intercropping green manures (soybeans, chickpeas and wheat) with maize was evaluated, as well as the use of maize stubble for mulching rain-fed peanuts. Composting of crop residues with farmyard manure was also promoted, and the use of floating row covers (Agribon e™, PGI-Bonlam México, SA de CV) was evaluated in tomatoes and hot peppers. These experiments were established in randomized block designs with 6 replicates and the data were subjected to analysis of variance and Tukey's test for comparison of means.

RESULTS AND DISCUSSION

Agroecosystems present

In Magdalena Mixtepec, all the peasants surveyed planted staple crops (maize, beans and squash) and also exploited the forest to get firewood and to make charcoal. About 86 % were producing passion fruit *Pasiflora edulis* and 50 % of then had some peach trees *Prunus persica* at home. About 15 % of the peasants produce horticultural crops and 40 % of them declared to have goats and sheep. This agroecosystem (AE) was called Staple crops-Firewood and charcoal-Passion fruit (S-F-P) and was present in at least eight communities of the SWAR.

In the S-F-P agroecosystem only 27 % of the peasants applyied farmyard manure to their land. The main reason was the scarcity of the product, as there are not many oxen teams due to the steep slopes of the land. The scarce manure produced is kept in the open. Most peasants have practiced slash and burn agriculture in communal lands. About 60 % of the peasants recognized to have some degree of soil erosion for using this practice. A few peasants have built stone faced terraces, which can be considered as a local technology. This practice was more common in land under irrigation.

From the standpoint of income, the decreasing order of importance of the different sub agroecosystems is: charcoal and firewood, staple crops and passion fruit production. This means that there is a year round extraction of wood, which has resulted in more time to reach the areas where it is possible to find *Quercus* sp. to make charcoal. Also, some peasants practice clandestine logging, as it is not permitted to cut whole pine trees, except for construction purposes. However, once in a while, the community may decide to sell a few hundred trees to local logging companies.

In San Lucas Tlanichico, all the surveyed peasants planted peanuts and staple crops, and 80 % have an oxen team for their draught-power requirements. About 50 % of them was selling minor cattle (poultry, pigs and goats) occasionally. This FS was named the Peanuts and staple crops - Oxen teams - Minor cattle agroeco system (P-O-M). This AE is common in at least six communities of light textured soils of the SWAR.

Table 1. Agroecological parameters of the communities selected.

Community name	Altitude (m)	Mean yearly Temp. (° C)	Yearly rainfall (mm)	Slope (%)	Soil Units*
Magdalena	2100	19.5	670	15-100	Cambisol
San Lucas	1585	21.0	930	2-15	Regosol

* Soil classification, FAO (1998).

In the P-O-M agroecosystem most peasants used sustainable practices such as application of farmyard manure and crop residues, and crop rotation. Most of the manure comes from oxen teams and crop residues of peanuts or beans are either brought back to the field or fed to cattle. Some peasants, however, burn these residues in the field. The most common crop rotation is maize-peanuts-maize, as they know that continuous maize shows decreased yields after two years. Terracing on gently slopes has been promoted over the years by means of mechanic plowing or by the use of wooden plows drawn by oxen teams.

The decreasing order of importance of the different subsystems is: Oxen teams, peanuts and staple crops, and minor cattle. Peanuts represent the cash crop, as most is sold to local dealers. The oxen teams are sold every 2-3 years to the butcher as they become badly tempered.

In both localities it was common that during the dry months (December to March), the peasants look for work in Oaxaca city or they even emigrate to the USA.

Validation of Problems and ranking

Discussions and voting about the main problems during community meetings showed that in the S-F-P agroecosystem the most important problems were: insect pests, low soil fertility, drought and diseases affecting horticultural crops. According to the survey, the second most important problem was soil erosion. Even though they admitted that the land yields scarcely after three years of maize cultivation, they did not recognized soil erosion as a problem. During the field visits, however, it was evident that the soil was being washed away. After talking with individual peasants latter, it was clear that some were concerned about soil losses and showed interest on the evaluation of a planting method where the slashed vegetation were left unburned.

In the P-O-M agroecosystem there was a closer agreement between the results of the survey and the ranking carried out at the community meeting. The top three problems were drought, insect pests and cattle diseases. Again, the research team was concerned about soil erosion, but the peasants ranked it fifth, while poor soils and crop disease where both ranked in

the fourth place. The reason for a closer match in rankings may be that the community surveyed for this system is more open and developed than the one surveyed in the other AE.

Evaluation and promotion of sustainable technologies

Several demonstrations about composting of crop residues and manure were carried out with peasants in both AE. In some cases, weeds and forest litter were also used. After the demonstrations, however, only one farmer in each community had the initiative to make his own compost. The compost produced was used mostly to fertilize their backyard garden, where they produce a variety of crops. Some of the reasons for the low impact of this technology include the scarcity of crop residues, high labor requirements for cutting crop residues for better decomposition, and relative availability of inorganic fertilizers.

Mulching peanuts with grinded maize stubble gave higher soil water contents, especially during a year with early season drought. This resulted in a higher number of pods per plant, seed weight, total plant weight and yield per plant (Table 2). In terms of kg/ha of shelled seed, the plot with mulch out-yielded the un-mulched one by 428 kg/ha.

Regarding the production of green manures in intercropped with maize, it was seen that dry matter yields were 30 % smaller than the observed when these were grown as monocrops. The legumes tested yielded more when intercropped at the time of maize planting, but wheat produced higher yields when intercropped during the first weeding in a shallow soil. This effect was attributed to cooler temperatures.

All these crops produced twice as much when planted during the first weeding than when planted during the second weeding, Showing a general trend of decreasing dry matter yields as planting was delayed (Table 3). Thus, early planting is advisable for maximum growth. The amount of nitrogen fixed ranged from 10.4-19.9 kg/ha and from 2.8-4.7 kg/ha of phosphorus.

Table 2. Peanut yield and yield components under mulching and without mulching in San Lucas Tlanichico, Oax.

Mulching ?	Pods/plant	Total dry weight (g)	Weight/seed (g)	Seed weight per plant (g)
No	18.7 b	13.1 b	0.78 b	19.8 b
Yes	23.5 a	20.0 a	0.96 a	25.9 a

* Different letters indicate significant differences according to Tukey's test ($\alpha = 0.05$).

Table 3. Dry matter yields (kg/ha) of intercropped green manure crops during three opportunities in a deep soil. San Lucas Tlanichico, Oax.

Green manure crop	Seeded at planting	Seeded 1st weeding	Seeded 2nd weeding
Wheat	1240 b	666 b	324 b
Chickpeas	1762 a	718 b	343 b
Soybeans	1294 b	1277 a	582 a

* Different letters indicate significant differences according to Tukey's test (α = 0.05).

Preliminary estimates of maize yields under intercropping showed that, at most, a decrease of 10 % in maize yields could be expected as a result of competition with green manure crops.

Bunch and Lopez (1993) consider that intercropping a multipurpose crop, instead of a green manure crop, is more likely to be accepted, as peasants rarely will turn them under. The planting of such crops should not represent any opportunity cost. All plantings were done immediately after normal operations such as seeding and mechanical weeding.

One of the concerns of peasants about the so called green manure crops was the availability of seed. Fortunately, there is a range of open pollinated varieties of leguminous and minor cereals.

Producing cash crops such as tomatoes and hot peppers is a good way to use scarce water and land, providing that the selling price is worth harvesting. Using floating row covers in these crops proved to be a safe way to counteract serious viral diseases transmitted by white flies, reducing insecticide use by half. This cloth, however, is expensive for the peasants. Even so, as the peasants saw that this was an effective technology, which reflected immediately in higher crop yields (Table 4), they started to implement it by themselves. At least 20 peasants in several communities have bought Agribon to produce tomatoes and hot peppers. Several short courses were implemented to train the peasants in using this technology.

Table 4. Tomato crop yields under Agribon protection and with and without insecticide in Magdalena Mixtepec, Oax.

Treatment	Infected plants (%)	Fruit size (g)	Fruit yield (ton ha[-1])
Agribon	12.3 c*	53.6 a	43.5 a
With insecticide	38.7 b	28.4 b	14.9 b
Blank	54.5 a	25.0 b	6.8 c

* Different letters indicate significant differences according to Tukey's test (α = 0.05).

According to Bunch and Lopez (1995) what matters is the sustainability of increasing yields or the sustainability of the development process. They believe that the peasants will keep doing certain practices as long as they get increased yields, decreased costs or decreased risks. Therefore, technologies should be chosen for their ability to produce such effects in a relatively short time.

Why farmers adopt production technologies seems elementary and apparent; new production technologies are adopted when the techniques are perceived as being in farmers' best interests (Nowak, 1992). However, this is not a guarantee that the technology will be permanently adopted; other factors such as conflict, socio-cultural organization and empowerment can lead to a new survivorship strategy (Aguilar-Cordero, 2008).

CONCLUSIONS

In both agroecosystems it was possible to find indigenous technologies with a high degree of sustainability, such as stone faced terraces and crop rotation. Composting was not well accepted by the peasants mostly because of high labor demands. Since they prefer to apply manure directly to the fields, it is necessary to increase its quantity and quality. The use of multipurpose crops as green manures may not be possible due to cultural and practical reasons, such as not harvesting an immature crop and cutting it without damaging the maize crop. Thus, these crops must be grown to maturity to use the seed as human food and then use their residues as fodder. The peasants adopted technologies of immediate impact, such as floating row covers, in spite of their cost. Such technologies can be used as effective lures to hook the peasants to use other sustainable technologies.

REFERENCES

Aguilar-Cordero, W.J. 2008. Toma de decisiones en la elección y adopción de opciones productivas en unidades domésticas de dos grupos de productores campesinos del Municipio de Hocabá, Yucatán, México". Edición electrónica gratuita. Texto completo disponible en www.eumed.net/tesis/2008/wjac/, obtenido en septiembre 13, 2011.

Ashley, J., Quiroz, C., Rivers, Y. 1989. Farmer participation in technology development: Work with crop varieties. *In:* Chambers, R., Pacey A., Thrupp, L.A. (eds.) Farmer First: Farmer Innovation and Agricultural Research. IT Publications, London. pp: 115-122.

Brunett, P.L., González, E.C., García, H.L.A. 2005. Evaluación de la sustentabilidad de dos agroecosistemas campesinos de producción de maíz y leche, utilizando indicadores. *Livestock Research for Rural Development. Volume 17, Article #78,* sin paginación. Available from http://www.cipav.org.co/lrrd/lrrd17/7/pere17078 .htm, retrieved March 23, 2005.

Bunch, R. 1993. The Use of Green Manures by Villager Peasants, What we have learned to Date. Technical Report No. 3, Second Edition. CIIDICIO, Tegucigalpa, Honduras.

Bunch, R., G. López. 1995. Soil Recuperation in Central America: Sustaining Innovation After Intervention.Gatekeeper Series No. 55, IIED, London.

Chambers, R. 1994. Participatory rural appraisal (PRA): Challenges, potentials and paradigm. World Development 22(10): 1437-1454.

Mayoux, L., Chambers, R. 2005. Reversing the paradigm: quantification, participatory methods and pro-poor impact assessment. Journal of International Development 17(2): 271-298.

FAO. 1998. World Reference Base for Soil Resources. Food and Agriculture Organization, Rome, Italy. Available from: http://www.fao.org/docrep/W8594E/W8594E00 .htm, retrieved September 20, 2011.

Gliessman, S.R. 2000. Agroecology: ecological processes in sustainable Agriculture. Lewis Publishers. 357 pp.

Guijt, I., Cornwall, A. 1995. Critical reflections on the practice of PRA. Participatory Learning and Action Notes 24, IIED, London. pp: 2-7.

Hart, R.D. 1984. The effect of inter level hierarchical system communication on agricultural system input output relationships. Options Mediterraneennes Ciheam IAMZ-84/1: 111-123. International Association for Ecology Series Study. Available from: http://ressources.ciheam.org/om/pdf/s07/CI0108 40.pdf, retrieved September 10, 2011.

INEGI. 2009. Anuario estadístico de Oaxaca. Available from: http://www.inegi.org.mx/est/contenidos/espanol/ sistemas/sisnav/default.aspx?proy=aee&edi=20 09&ent=20, retrieved June 15, 2011.

Nowak, P. 1992. Why farmers adopt production technology: Overcoming impediments to adoption of crop residue management techniques will be crucial to implementation of conservation compliance plans. Journal of Soil and Water Conservation 47(1):14-16.

Pretty, J.N. 1995. Regenerating Agriculture. Earthscan Editors, London, first edition. 320 pp.

Pretty, J.N. 1997. The sustainable intensification of agriculture. Natural Resources Forum, 21: 247–256.

Ruiz, V.J., Laird, R.J. 1979. Definición de fórmulas de fertilización con base en probabilidades de sequía utilizando el método CP para diagnóstico de agrosistemas. *In:* Memorias del XII Congreso Nacional de la Sociedad Mexicana de la Ciencia del Suelo, Morelia, Michoacán, México. p. 78.

Ruiz, V.J. 1987. Rainfall probabilities and corn yields in the Central Valleys of Oaxaca, Mexico. PSMP Report Series No. 25, World Meteorological organization, Geneva. pp. 25-27.

Ruiz, V.J. 1990. Evaluación de variedades de maíz, fertilización y densidades de siembra bajo riego de auxilio. *In:* Memoria de Investigación, XX Aniversario del Campo Experimental Valles Centrales de Oaxaca, Sto. Domingo Barrio Bajo, Etla, Oax. México. Publicación especial No. 3. INIFAP-SARH. p. 57.

Scoones, I., Thompson, J., Chambers, R. 2007. Farmer first – retrospects and prospect. Reflections on the Changing Dynamics of Farmer Innovation in Agricultural Research and Development in Preparation for the Farmer First Revisited Workshop Institute of Development Studies, University of Sussex 12-14 December 2007. Available from: http://www.future-agricultures.org/farmerfirst/files/Farmer_First_R evisited_Post_Workshop_Summary_Final.pdf, retrieved June 23, 2005.

Valdés-Rodríguez, O. A., Pérez-Vázquez, A. 2011. Sustainable livelihoods: an analysis of the methodology. Tropical and Subtropical Agroecosystems, 14: 91 – 99.

11

Margins of Sheep Meat Marketing in Capulhuac, State of Mexico

J. Mondragón-Ancelmo[1], I. A. Domínguez-Vara[1*], S. Rebollar-Rebollar[2],
J. L Bórquez-Gastélum[1], J. Hernández-Martínez[2]

[1]*Departamento de Nutrición Animal. Facultad de Medicina Veterinaria y Zootecnia.
Universidad Autónoma del Estado de México (UAEMéx.).Campus Universitario "El
Cerrillo" Toluca, México. C.P. 50090.*
[2]*Centro Universitario UAEM Temascaltepec. Universidad Autónoma del Estado de
México (UAEMéx.).*
Email: igv92@hotmail.com
Corresponding Author

SUMMARY

The objective of this investigation was to identify the channels and analyze the incorporation of the marketing margins of different participants in the productive chain of sheep meat in the municipality of Capulhuac, State of Mexico, during the period 2009- 2010. The marketing channel most widely used by the participants of this market was identified, and absolute and relative profit margins were calculated through their equivalent values of the costs and profits of each participating actor. Of the 31 sheep farms or producers interviewed, 58.3 % carried out the sale of live sheep at farm. Considering the absolute margin of marketing in the final price to the consumer per kg of raw meat, the producer obtained 2 USD/kg (40 %) and the intermediaries 3 USD/kg (60 %). Profit in selling cooked final product (barbacoa typical dish), was obtained 4.5 USD/kg (25.7 %) for the producers and 13 USD/kg (74.3 %) for intermediaries. Evidence was found that the most common marketing channel was producer, intermediary, barbacoa seller and final consumer, in which the barbacoa seller obtained the highest benefit-cost ratio.

Key words: sheep; farms; marketing; margins; meat; barbacoa; producer.

INTRODUCTION

In Mexico, sheep farming faces various problems that limit the development of sheep meat production, including technological lag, undervalued activity, practiced on small farms and traditional consumption (barbacoa) (Trejo, 2008). Investigation in Mexico is still isolated or is removed from the real production needs (Samaniego, 2000; Tomillo, 2001; FAO, 2010). This has been characterized by generating technologies for those producers that have the economic resources necessary for their application, without attending the producers with low economic incomes (Tomillo, 2001; Góngora *et al.*, 2010). Furthermore, the different actors of the sheep production chain (sheep farms or producers, marketers, processors, barbacoa sellers and consumers), as well as the investigators, technicians and governmental sectors, have not recognized the need for integration to achieve strategies that contribute to the improvement of sheep meat production (Samaniego, 2000; Montossi, 2002; FAO, 2010). Other countries have taken advantage of this opportunity to introduce sheep meat to the national market, due to the demand that exists of this meat; for example, meat imported from New Zealand, Australia, Canada, Chile and the United States is used to elaborate barbacoa.

In Mexico sheep carcass production increased 54 % (2.8 % annually), from 24, 695 to 53, 737 t (1990-2009); in 2009 the participation of this kind of meat was 0.9 % of the total meat production of farm animals (153.8 million t), which has not undergone significant changes since 1980 to 2009 with respect to other meats (Figure 1). The principal production entities were: the State of Mexico, Hidalgo, Veracruz, Puebla, Jalisco and Zacatecas with 14.7, 12.8, 9.3, 6.6 and 5.4 % (SIAP, 2010). These entities contributed with slightly more than half (55.4 %) of the national production, the rest (44.6 %) was covered by other entities.

The per capita consumption of sheep meat in Mexico from 2008-2009 increased from a range of 0.5-0.8 to 1 kg (SIAP, 2010). In 2008 the national apparent consumption (NAC) registered 90,000 t of meat, and the participation of imports in total consumption was approximately 45 %. The price of national sheep meat carcass was attractive (USD 4.2/kg), which was above the price of imported meat (USD 2.7/kg) (Trejo, 2008). Suárez and Sagarnaga (2000) mentioned that per capita availability of sheep meat was very low compared with meat from other farm species, indicating that consumption of this kind of meat by Mexicans is low due to high cost of the product (barbacoa). Three important aspects are considered to affect this low consumption such as a reduction in the growth rate of human population (0.9 %) in the period 2005-2009 (INEGI, 2009), reduction in importation of sheep meat (6.8 %) from 2004-2007, as well as an increase of sheep meat production (9.5 %). Although, production has not been high enough to satisfy the national demand, which represents an opportunity of production and marketing for sheep producers.

In Mexico in 2009, the State of Mexico occupied first place as producer of sheep meat carcass. In this entity, the sheep were distributed among the eight Rural Development District (RDD): Atlacomulco, Toluca, Texcoco, Zumpango, Valle del Bravo, Jilotepec, Coatepec Harinas and Tejupilco, which participated with 28.6, 25.5, 12.9, 12.0, 7.3, 7.3, 5.6 and 0.9 % of sheep meat carcass. Furthermore, in 2009 five of the 24 municipalities of the RDD of Toluca supplied 62 % of sheep meat, outstanding Temoaya, Zinacantepec, Almoloya de Juárez, Chapultepec and Otzolotepec (SIAP, 2010).

Due to its great importance in supplying and marketing sheep meat, the municipality of Capulhuac of the State of Mexico participated with only 2 % of the sheep meat production in the RDD of Toluca in 2009. However, this municipality is of great importance in the marketing of barbacoa meat, to the extent that it occupies first place in barbacoa production and as sheep stocking center in Mexico (Aguilar, 2007). Furthermore, it is outstanding in the importation of frozen meat from Australia, New Zealand and U.S.A. This is because the high demand of barbacoa in the central states of Mexico and Federal District. Presently, the routes or commercialization channels and the appropriation of the margins of sheep meat marketing in this municipality and in the region are not documented.

The channels of distribution or marketing are the routes followed by a product to reach the final consumer (agents or actors involved in obtaining the product and taking it to the consumer or meat transformers) (Caldentey, 1979; Bustamante, 2001). The agents may be whole sellers or retailers and can have influence on the management of the product. The marketing margin is the remuneration established and received by each one of the marketing agents (intermediary, transformer, stocker and distributer). This is represented by the retributions derived from the investments they make, the costs they incur in, plus a profit for each one of them to carry out the marketing; it is determined by the differences between the prices to the consumer, retailer, whole seller and producer. By measuring the differences among prices, the percentages of these differences are estimated (NAFIN, 1998).

The marketing margin can be divided into its components; a) price of the primary product (price to the producer), and b) price of marketing from the farm to the final consumer. The part that covers commercialization is also known as marketing margin, which is the difference between what the consumer pays and what is received by the farm producer (Wohlken, 1991). The marketing margin includes all of the expenses made to add value to the product, such as storage, conditioning, transportation and offering it to the consumer; it also includes the profits of the agents of transformation, storage, distribution and marketing (Schwentesius and Gómez, 2004). The objective of this investigation was to identify the channels and to analyze the appropriation of the marketing margins by the different actors in the sheep meat production chain in the municipality of Capulhuac, State of Mexico, during the period 2009-2010, so that it may be used as a base in identification of opportunities of chain integration, evaluation of sheep production system, and opportunity to develop sheep husbandry in the region.

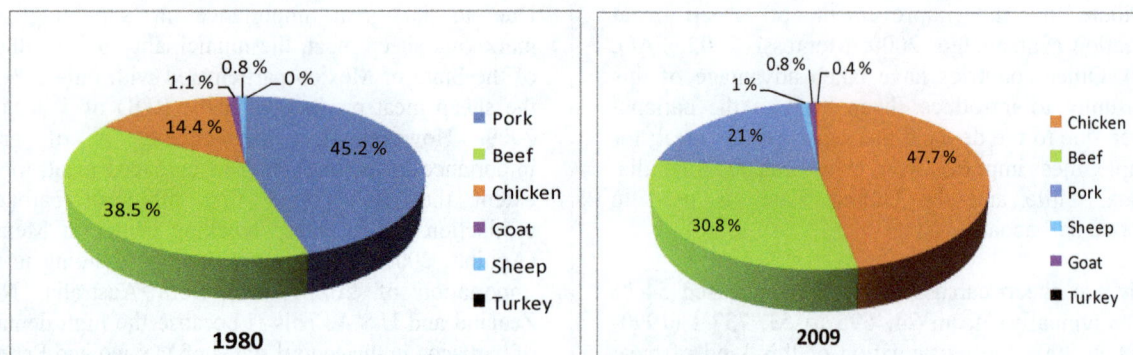

Figure 1. Meat production of the principal farm species in Mexico in 1980 and 2009 years.

MATERIALS AND METHODS

The field work was carried out from 2009-2010 in the municipality of Capulhuac, State of Mexico. The information was obtained by observational investigation method, proposed by Lovelock *et al.* (2004) and application of questionnaires through personal interviews (Cochran, 1984). Were carried out 31 interviews to sheep farms, 35 to barbacoa sellers and two to introducers; the interviews to introducers were reduced, due to the mistrust of these actors.

The coefficient yield of meat carcass was obtained from a sample of 64 male sheep, with a live weight of 46.4±3.8 kg. The coefficient yield of cooked carcass meat (barbacoa) and cooked viscera byproducts (pancita), was directly obtained from the slaughter of the animals *in situ* (at home of barbacoa producer). Sheep losses during transport from production unit to *in situ* slaughter was obtained from 64 animals; body weight losses to the sale points in the region, was found to be 3.8 kg (8 %) during transport.

The prices of carcass meat were obtained from 35 barbacoa producers who slaughtered animals *in situ*. The prices of direct sale to the consumer of barbacoa and pancita per kilogram and in the form of tacos were obtained from direct interview to these same barbacoa sellers. The parity of Mexican pesos to American dollars prices was consulted in http://www.sat.gob.mx.To calculate the number of tacos per kilogram of barbacoa and pancita, an Ohaus precision scale was used with maximum capacity of 610 g. Twenty six barbacoa sellers were interviewed, located in the municipality of Capulhuac, State of Mexico where two tacos were bought and weighed in each stand. The amount of consommé per plate was also measured with a graduated test tube with a capacity of 500 ml.

The prices used to calculate the marketing margins were as follows: prices of live sheep in the production unit, prices at slaughter and at municipal plaza for live animals sale, carcass price at slaughter, price of viscera byproducts (pancita) and prices to the final

consumer of cooked meat as barbacoa (Tables 1 and 2).

The comparison of prices in each marketing level was obtained by calculating the equivalent value to the producer at arriving to slaughter (*in situ*) and of meat carcass, as well as the barbacoa and pancita to the consumer. The marketing margins were calculated from the difference between the sale price of one unit of product by each marketing agent and the payment made in the purchase of the equivalent amount to the unit sold (Caldentey, 2007; García *et al.*, 1990). To calculate the gross absolute margins (M) and total relative margins (m), the formula used was $M = Pc-VEP$, and $m = (M/Pc)*100$ and was adjusted to each stage of the marketing process with different prices (Table 3). The data were processed by means of a numerical matrix created as data base in the program Microsoft Excel, according to the methodology proposed by Caldentey (1979), and Rebollar *et al.* (2007).

RESULTS AND DISCUSSION

The prevalent feeding system in sheep production in the municipality of Capulhuac was grazing (87.1 %); the flocks graze on native vegetation and crop residues in the region, with lack of technical management. In 9.7 % of the production units, the feed is mixed (grazing and feed supplement); the sheep graze on native vegetation and crop residues of the region, however, sheep can graze on improved forage such as white clover (*Trifolium repens* associated with rye grass (*Lolium perenne*), and at night they are complemented with concentrate. Better management can be applied to sheep according their physiological stage; 3.2 % of the production units keep their sheep in total confinement and feed them concentrate and ground corn stover; in addition, they carry out zootechnical control according to the productive phase. In the latter two systems, hair breeds (Dorper, Kathadin and Blackbelly) and wool breeds (Hampshire and Suffolk) are bred to obtain sheep for breeding and for meat production, mainly for barbacoa production.

Table 1. Mean values for calculating marketing margins of raw and cooked (barbacoa) sheep meat.

Concept	Measuring unit	Value
Initial body weight lamb (IBWL)	kg	26.2±3.4
Cost of lamb at the start of feeding (CLSF)	USD/kg	1.9±0.1
Cost of lamb during feeding (CLDF) = IBWL * CLSF	USD/sheep	50
Feeding period (FP)	days	64
Daily weight gain (DWG)	kg/day	0.32±0.04
Total weight gain (TWG) = (FP * DWG)	Total kg	20.2
As fed feed intake (AFFI)	kg/sheep/day	1.69±0.2
Total as fed feed intake (TAFFI) = FP * AFFI	kg/period	108.2
Feed cost (FC)	USD/kg	0.3
Total feed cost (TFC) = TAFFI * FC	USD/period	28.9
Amount of meat used in the calculations (K)	kg	1
Final body weight lamb sale at farm (FBWL) = IBWL + TWG	kg	46.4±3.8
Live weight at slaughter or market sale (LWS)	kg	42.6±3.4
Loss of final live weight at slaughter or market (LFLW) = FBWL - LWS	kg	3.9
Yield at *in situ* slaughter (YISS)= LWS *100/ FBWL	%	91.6
Feed conversion (FC)= AFFI / DWG	kg	5.3
Production cost calculated by the producer (PCCP)	USD/kg meat	1.7
Production cost calculated by the producer (PCCP1)	USD/sheep	34.3±4.7
Live sale price at municipal plaza (LKSPP)	USD/kg	2.1±0.1
Live sale price at municipal plaza (LSSPP)= LKSPP * LWS	USD/sheep	90.9
Live price at *in situ* slaughter (LPS)	USD/kg	2.1
Live price at *in situ* slaughter (LPS)= LKSPP * LWS	USD/sheep	90.9
Kilograms carcass price (KCP)	USD/kg	5±0.2
Price of raw meat to consumer (PRMC)	USD/kg	5±0.2
Price of non fried or raw viscera (PRV)	USD/kg	5±0.2
Sheep *in situ* slaughter (SCISS)	USD/sheep	3.8
Price of barbacoa meat (PBM)	USD/kg	17.5±0.8
Price of pancita (PP)	USD/kg	17.5±0.8
Price of barbacoa or pancita taco (PBPT)	USD/taco	0.9±0.04
Price of consommé (PC)	USD/plate	0.9±0.04
Price of sheep leather (PSL)	USD/sheep	2.7

The most common marketing channels in sheep meat production chain in Capulhuac, State of Mexico, are producer-stocker-barbacoa sellers, and final consumer. A description is made of the marketing channels found in the present investigation; 58.3 % of the producers participate with 89.2 % of the final price when they sell their live animals to a marketing agent (stocker) (Figure 2); this is done only when there is economic urgency or when the lambs have reached the age and live weight for slaughter (45-50 kg); sometimes, the animals are overweight for marketing, and are consequently sanctioned in the price due to the accumulation of fat in the carcass, which causes a reduction in the yield of meat and barbacoa. Sex, live weight and age were the factors of highest relevance in determining the price of purchase by stocker. The stocker finalizes the lambs until they reach 45-50 kg weight, or sells them directly to another marketing and transformation agent (barbacoa seller) with 10.7 % increase with respect to the price received by producer. Carrera (2008); López *et al.* (2008); Nuncio *et al.* (2001); Vázquez *et al.* (2009); and Góngora *et al.*

(2010) mentioned that the sheep are sold to small and large intermediaries who later sell the animals to meat processers, which does not differ from what we found in this study.

The 32.3 % of the producers sell their live sheep directly to the barbacoa sellers; only 9.7 % of the producers close the marketing channel, they are producers that also prepare and sell the barbacoa directly to the final consumer; in this way, the participation of the producer is 100 % in the final price. In this marketing channel there is certain investment mainly by enterprising producers. One of the most important characteristic of this marketing channel is that the principal actor is open to adopt new technologies, technical advice in raising and feeding sheep, and possesses knowledge in preparing and marketing barbacoa (Figure 2).

The above information was used to calculate the relative and equivalent values of uncooked (Table 4) and cooked carcass meat (Table 5).

Table 2. Additional mean values for calculating marketing margins of raw and cooked (barbacoa) sheep meat, pancita and consommé.

Concept	Measuring unit	Value
Live kg farm sale (LKFS)	USD/kg	1.9±0.1
Live sheep farm sale (LSFS)= LKFS * FBWL	USD/sheep	88.6
Hot carcass yield (HCY)	kg	20.1±1.9
Cold carcass yield (CCY)	kg	19.4±2
Loss from hot to cold carcass (LHCC)= HCY – CCY	kg	0.7
Yield kg of head (YKH)	kg/sheep	1.9±0.3
Yield % of head (YPH)= YKH *100/ LWS	%/sheep	4.5
Yield kg of feet (YKF)	kg/sheep	1.4±0.2
Yield % of feet (YPF)= YKF *100/ LWS	%/sheep	3.2
Yield kg of sheep leather (YKL)	kg/sheep	6.8±1.4
Yield % of sheep leather (YPL)= YKL *100/ LWS	%/sheep	16
Yield kg of blood (YKB)	kg/sheep	1.5±0.3
Yield % of blood (YPB)= YKB *100/ LWS	%/sheep	3.5
Yield kg of testicles (YKT)	kg/sheep	0.3±0.1
Yield % of testicles (YPT)= YKT *100/ LWS	%/sheep	0.8
Yield kg of red viscera (lungs, heart, liver) (YKRV)	kg/sheep	2.1±0.3
Yield % of red viscera (YPRV)= YKRV *100/ LWS	%/sheep	5.0
Weight kg of full green viscera (stomach, intestines) (WKGV)	kg/sheep	8.4±0.6
Yield kg of green viscera (rumen, reticle, psalterium, abomasum and intestines) (YKGV)	kg/sheep	3.8±0.6
Content kg of intestines (feces) (CKI)= WKGV – YKGV	kg/sheep	4.6
Yield % of intestines (feces) (YPIF)= CKI *100/ LWS	%/sheep	10.7
Yield % of intestines (YPI)= YPIF *100/ LWS	%/sheep	9.0
Total yield kg of green and red viscera and testicles (TKYGRV)= YKRV + YKGV+ YKT	kg	6.3
Total yield % of green and red viscera, and testicles (TYPGRV)= TKYGRV*100/ LWS	%/sheep	14.8
Cost of kg preparation and sale of barbacoa (CKPSB)	USD/kg	2.7±0.1
Cost of sheep preparation and sale of barbacoa (CSPSB)= CKPSB * CCY	USD/sheep	51.6±2.7
Carcass kg yield as barbacoa (CKYB)	kg/sheep	13.0±0.2
Carcass kg loss as barbacoa (CKLB)= HCY-CKYB	kg/sheep	7.1
Carcass % loss as barbacoa (CPLB)= CKLB *100/ HCY	%/sheep	35.4
Average % carcass yield as barbacoa/live animal at slaughter (APCYB/AS)= CKYB *100/ HCY	%/sheep	64.7
Average kg yield of viscera as pancita (AKYVP)	kg/sheep	3.7±0.1
Loss of kg viscera as pancita (LKVP)= TKYGRV- AKYVP	kg/sheep	2.6
Loss % of viscera as pancita (LPVP)= LKVP *100/ TKYGRV	%/sheep	42.0
Yield % of viscera as pancita (YPVP)= AKYVP *100/ TKYGRV	%/sheep	58.1
Amount of consommé per plate (ACPP)	ml/plate	308.0±33.8
Yield of consommé (YC)	Plates/sheep	49.3±6.2
Average yield of consommé (AYC)	Liters/sheep	15.6±0.7
Weight of barbacoa meat per taco (WBPT)	g/taco	42.2±13
Yield of barbacoa meat in tacos (YBT)	Tacos/kg	26.5±10.2
Yield of barbacoa from carcass in tacos (YBCT)=CKYB*YBT	Tacos/sheep	344.5
Yield kg of pancita in tacos (YKPT)	Tacos/kg	26.5±10.2
Yield of sheep pancita in tacos (YSPT)= AKYVP*YKPT	Tacos/sheep	96.7

Most of the barbacoa sellers carryout *in situ* slaughter (at their home), similar to what was reported by Abbott (1987) and FAO (2009), who indicated that after the purchase of the livestock, it is slaughtered *in situ* in urban or rural regions; some researchers have indicated that the slaughter of pigs is also carried out *in situ* by retailers (González *et al.*, 2010). Later, the carcass is air cooled for 24 hours and is then cut without a definite pattern. The most complete as possible pieces are obtained, with a size that permits them to be place in stainless steel pots for cooking. There is no difference with respect the type of cut and the price at which it can be sold in the market; the entire carcass is sold as barbacoa, at the same price. This dish is mainly sold as tacos in stands placed on the street, municipal markets and in restaurants of the State of Mexico and the Federal District.

Table 3. Prices for comparing marketing levels of raw and cooked (barbacoa) sheep meat and pancita.

Concept	Measuring unit	Value
Sheep price at farm price to producer (SPF)	USD/kg	1.9
Sheep price at *in situ* slaughter (SPS)	USD/kg	2.1
Exit price of sheep carcass from *in situ* slaughter (EPSCISS)	USD/kg	5.0
Price of sheep as kg barbacoa (PSKB)	USD/kg	17.5
Price of kg sheep uncooked viscera (PKSUV)	USD/kg	5.0
Price of kg sheep cooked viscera (PKSCV)	USD/kg	17.5
Weight of sheep bought from producer (WSP)	kg/sheep	46.4
Loss transport of sheep at entrance to *in situ* slaughter (LTSISS)	%	8.4
Weight of sheep entering at *in situ* slaughter (WSEISS)	kg	42.6
Yield coefficient of carcass from *in situ* slaughter (YCCISS)	%	47.3
Weight loss of carcass after *in situ* slaughter (WLCISS)	%	3.5
Yield coefficient of cooked carcass (barbacoa) (YCCC)	%	30.6
Yield coefficient of viscera from carcass (YCC)	%	52.7
Yield viscera byproducts (pancita) coefficient (YVC)	%	6.9
Total yield coefficient in transformation process (TYCTP)	%	37.5

Table 4. Relative and equivalent values of uncooked sheep meat in Capulhuac, State of Mexico.

Concept	Measuring unit	Value
Kilograms sheep *in situ* slaughter to obtain one kg of meat to consumer (KSISS)=1/YCCISS*100	kg	2.1
Kilograms sheep at farm to obtain one kg of meat to consumer (KSF)= KSISS/(1-LTSISS)*100	kg	2.3
Participation of meat value in total value		
By Processing KSISS the carcass meat (principal product) is obtained (K)= KSISS*YCCISS/100	kg	1.0
Kilograms of byproduct (viscera to prepare pancita) obtained (KB) = KSISS * YCC/100 I	kg	1.1
Meat value		
Value of carcass meat (principal product) (VCM) = K*EPSCISS	USD	5.0
Byproduct value		
Byproduct value (BV) =KB*PKSUV	USD	5.6
Relative value of uncooking sheep meat (RVUCSM)		
RVUCSM=(VCM/(VCM+BV)*100	%	47.0
Equivalent values		
Equivalent value to the producer (EVP)=(KSF) (KSISS) (RVUCSM)/100	USD/kg	2.0
Equivalent value at entrance to *in situ* slaughter (EVEISS)=(KSISS) (SPS) (RVUCSM)/100	USD/kg	2.1
Equivalent value at exit from *in situ* slaughter (EVEISS)=(KSISS) (YCCISS) (EPSCISS)/100	USD/kg	5.0

Studies carried out in goats (Rebollar *et al.*, 2007) and pigs (González *et al.,* 2010) in the south of the State of Mexico indicated that the marketing channel was producer, regional stocker, retailer and final consumer, which is similar to what was found in this investigation. On the other hand, D'Aubeterre *et al.* (2007) found four sheep meat marketing channels in Venezuela: 1) producer and consumer; 2) producer, butcher shops/supermarket and consumer; 3) producer, stocker, carrier, retailer and consumer; 4) producer, stocker-carrier, slaughterhouse, butcher shop/supermarket, restaurants and consumer. Bravo *et al.* (2002) indicated that the beef meat marketing channel is producer, stocker, introducer, municipal slaughterhouse, meat carcass whole seller, retailer and final consumer. However, when the distance between the points of production and consumption is short, the marketing channel is simple, that is, the butchers buy live animals from the producers at the production unit or at local market, they slaughter and prepare the animals in a local slaughterhouse and sell the meat in a market stand or in a retail establishment (Abbott, 1987). Pittet *et al.* (1994) mentioned that the sheep meat marketing channel in U.S.A. is very long with important degrees of inefficiency and with a tendency to shorten it to improve the profitability of the business.

Table 5. Relative values and equivalents of cooked meat (barbacoa) in Capulhuac, State of Mexico.

Concept	Measuring Unit	Value
Kilograms sheep at *in situ* slaughter to obtain one kg of meat to the consumer(KSISS)= K/TYCTP *100	kg	2.7
Kilograms sheep at farm needed to obtain one kg of meat to the consumer (KSF)= KSISS/1-LTSISS/100	kg	2.9
Kilograms of carcass meat at exit from *in situ* slaughter needed to obtain one kg of meat to the consumer (KCISS)=(KSISS) (YCCISS)/100	kg	1.3
Participation of the value of the barbacoa meat in total value		
By processing KSISS barbacoa meat (principal product) obtained by K= KSISS*YCCC/100	kg	0.8
Kilograms of byproducts (viscera to prepare pancita) obtained (KB)= (KSISS) (YVC)/100	kg	0.2
Amount of principal product (APP)= (KSISS) (YCCC)/100	kg	0.8
Value of meat		
Value of principal product (VPP) = (PSKB) (APP)	USD	14.3
Value of byproducts (viscera to prepare pancita) (VB) = (KB) (PKSCV)	USD	3.2
Relative value of barbacoa meat (RVB)		
RVB = (VPP/(VPP + VB) * 100	%	82.0
Equivalent values		
1. Equivalent value at entrance to *in situ* slaughter (EVISS) = (KSF) (SPF) (RVB)/100	USD/kg	4.5
2. Equivalent value at entrance to *in situ* slaughter (EVISS) = (KSISS) (SPS) (RVB)/100	USD/kg	4.6
3. Equivalent value at exit from *in situ* slaughter (EVEISS) = (KCISS) (EPSCISS)	USD/kg	6.4

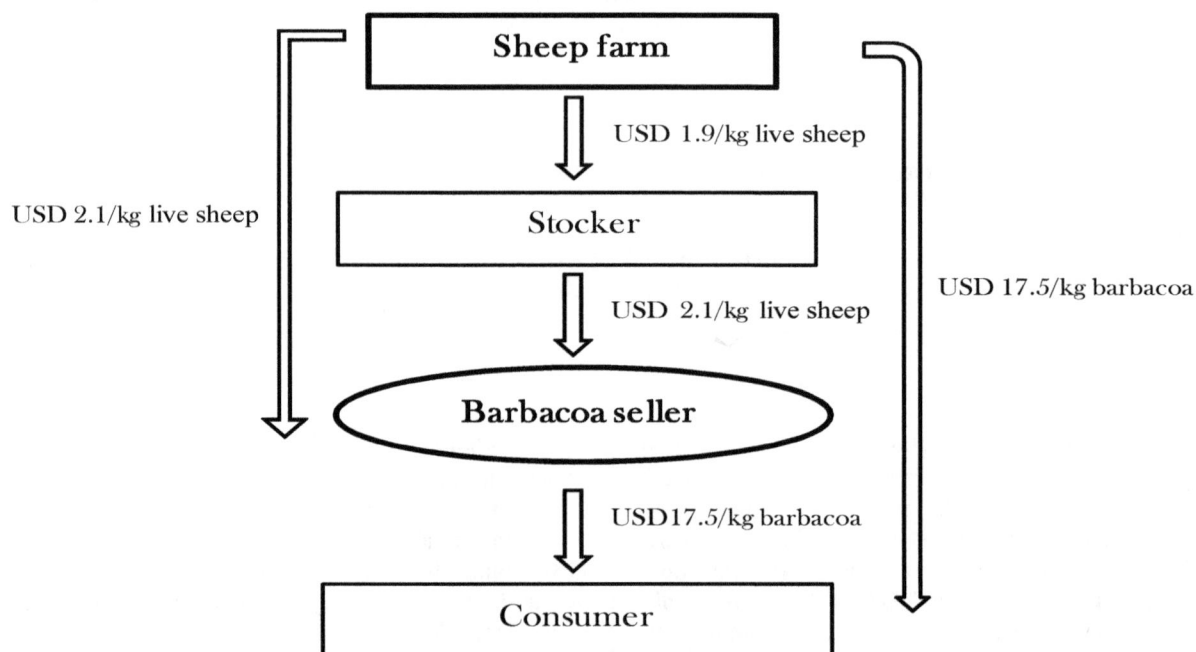

Figure 2. Sheep meat marketing channels in Capulhuac, State of Mexico.

Table 6. Calculation of the net economic profit per live sheep obtained by the farm.

Tipe of sale	Total cost, USD	Total income, USD	Profit, USD[a]	B/C[b]
If the sale is as:				
Live sheep in corral	84.2[1]	88.6[6]	4.4	1.05
Live sheep in plaza	85[2]	90.9[7]	5.9	1.06
Carcass meat	88.8[3]	101.4[8]	12.6	1.14
Barbacoa and pancita	51.6[4]	291.2[9]	239.6	5.64
Tacos (barbacoa, pancita and consome)	51.6[5]	430.6[10]	379	8.34

[1]Cost of production per sheep (feed, health plus 15 % per kg of sheep produced from labor, operation costs, depreciation of installations and equipment, financial costs, light, water, plus the cost of sheep at the start of fattening) (Lara, 2008).
[2]Sale price of live sheep in corral plus the cost of transport to plaza or *in situ* slaughter (0.8 USD/sheep).
[3]Sale price of live sheep in plaza or *in situ* sacrifice plus the cost of *in situ* slaughter.
[4]Cost of preparation, sale of barbacoa, pancita and consommé per sheep.
[5]Cost of preparation of barbacoa, pancita and consome per sheep.
[6]Cost of sheep at start of fattening multiplied by the final live weight farm sale.
[7]Sale price of live sheep in plaza or *in situ* slaughter multiplied by live weight of sheep placed in plaza or *in situ* slaughter.
[8]Carcass yield of sheep multiplied by carcass sale price.
[9]Carcass yield in barbacoa plus yield of pancita multiplied by the sale price for each kg of pancita.
[10]Yield of tacos of barbacoa per sheep plus yield of tacos as pancita/animal plus the yield of consommé per sheep multiplied by the sale price of barbacoa as taco, pancita as taco and consommé per plate.
[a]Profit (total income minus total cost).
[b]Ratio benefit (B)/cost (C) (total income divided by total cost).

Table 7. Calculation of net economic profit per live sheep obtained by the barbacoa seller or stocker.

Tipe of sale	Total cost USD	Total income USD	Profit USD[a]	B/C[b]
If the sale is in:				
Live sheep in plaza	90.9[1]	90.9	0	1
Carcass meat	94.8[2]	101.4[5]	6.6	1.06
Barbacoa and pancita	146.3[3]	291.2[6]	144.9	1.99
Tacos (barbacoa, pancita and consommé)	146.3[4]	430.6[7]	284.3	2.94

[1]Sale price of sheep placed in plaza or *in situ* slaughter multiplied by live weight of sheep placed in market or *in situ* slaughter.
[2]Total price of live sheep at plaza or *in situ* slaughter plus the cost of *in situ* slaughter.
[3]Total price of sheep carcass plus cost of preparation, sale of barbacoa and pancita.
[4]Total price of sheep carcass plus cost of preparation, sale of barbacoa, pancita and consommé.
[5]Yield of sheep carcass multiplied by the carcass price per kg.
[6]Yield of carcass as barbacoa plus the yield of pancita multiplied by the sale price per kg of barbacoa and pancita.
[7]Yield of barbacoa tacos per sheep plus yield of pancita tacos per sheep plus yield of consommé per animal multiplied by the sale price of barbacoa per taco, pancita per taco and consommé per plate.
[a]Profit (total income minus total cost).
[b]Ratio benefit (B)/cost (C) (total income divided by total cost).

When a comparison was made of the net profit obtained by producer at farm and the stocker or barbacoa processor at the plaza (Tables 6 and 7), it was found that the total income of the producer was lower (2.5 %) with respect the stocker or barbecue processor. To this respect, in the study carried out by Rebollar *et al.* (2007), a lower income (19.1 %) was obtained by the goat breeder with respect to the stocker or processor of goat meat as birria (typical dish).

The absolute margin of marketing of the raw and cooked meat (barbacoa) was 3 USD and 13 USD/kg

(76.9 % higher margin for the cooked meat) (Table 8). In relative terms, in the raw meat, the producer obtained 40 % of the price paid by the final consumer per kg of carcass meat and the intermediaries obtained 60 %. With respect the cooked meat, the producer obtained 25.7 % and the intermediaries 74.3 % of the price paid by the consumer per kg of barbacoa. The absolute and relative margins of the sheep stocking presented the lowest values.

In general, the study reflected lower marketing margin with cooked and raw meat for the producer with respect to the intermediaries; this was similar to what was found in the marketing of meat of other animal species; for example, in pork meat the producer obtained the lowest relative margin (40.4 vs 59.7 %) with respect to intermediaries, (Sierra *et al.*, 2005), 26.2 vs 73.7 % (González *et al.*, 2010); in beef meat, 43 vs 57 % (Iturrioz and Iglesia, 2009; in goat meat, 4.2 vs 52.7 %; in goat meat as birria, 20.4 vs 79.6 % (Rebollar *et al.*, 2007). In contrast, Abbott (1987) found in sheep meat, 64 vs 36 %; in beef meat 66 vs 34 %; in pork meat, 75 vs 25 % for producers and intermediaries. These results were attributed to the fact that the productive chains of sheep, beef and pork

meat were organized or structured for the marketing of the final product.

In this investigation, the barbacoa seller obtained the greatest part of the total marketing margin, followed by the stocker and the producer (Table 9). The barbacoa seller obtained higher benefit - cost ratios, followed by the stocker and producer, which coincides with Rebollar *et al.* (2007), who reported that goat producer and birria processor obtained the highest benefit - cost ratios, followed by the regional stocker.

Table 8. Mean values of sheep meat marketing margins of cooked (barbacoa) and raw meat in Capulhuac, State of Mexico.

Raw material/Agent	Values	
	Absolute (USD/kg)	Relative (%)
Cooked meat (barbacoa)		
Equivalent value to producer	4.5	25.7
Equivalent value of entrance to *in situ* slaughter	4.6	26.6
Equivalent meat value exit from *in situ* slaughter	6.4	36.3
Gross margin of sheep stocking (2-1)	0.1	0.7
Gross margin of transformation of carcass meat (3-2)	1.7	9.8
Total gross margin of marketing of cooked meat (7-1)	13	74.3
Price paid by the final consumer of cooked meat	17.5	100
Raw meat		
Equivalent value to producer	2	40
Equivalent value entrance at *in situ* slaughter	2.1	42.4
Equivalent value of live sheep to carcass meat	5	100
Gross margin of sheep stocking (2-1)	0.1	1.1
Gross margin of transformation of carcass meat (3-2)	2.9	57.6
Gross margin of marketing of the raw meat (7-1)	3	60
Price paid by the final consumer of raw meat	5	100

Table 9. Structure of costs and profits of the marketing margins by each agent of the sheep meat chain in Capulhuac, State of Mexico.

Agent	Margin (USD/kg)	Costs (USD/kg)	Costs[a] (%)	Profit (USD/kg)	Profit[b] (%)	B/C[c]
Producer	1.9	1.8[3]	94.7	0.1[7]	5.3	1.05
Stocker	2.1	2.0[4]	95.2	0.1[8]	4.8	1.05
Barbacoa seller[1]	5.0	4.4[5]	88.0	0.6[9]	12.0	1.13
Barbacoa seller[2]	17.5	3.1[6]	17.7	14.4[10]	82.3	5.64

[1]Agent that transforms the sheep at *in situ* slaughter to carcass meat.
[2]Agent that transforms carcass meat to barbacoa and pancita.
[3]Sale price of live sheep in corral divided by final live weight of sheep in corral.
[4]Total cost of sheep at plaza or *in situ* slaughter divided by live weight at plaza or *in situ* slaughter.
[5]Total cost of carcass meat divided by the yield of carcass meat.
[6]Total cost of barbacoa and pancita divided by the live sheep (46.4 kg) divided by the yield of the carcass meat to barbacoa and pancita.
[7]Total profit of sale of live sheep in corral divided by final live weight of live sheep in corral.
[8]Total profit of sale of sheep at plaza or *in situ* slaughter divided by live weight at plaza or *in situ* slaughter.
[9]Total profit of sale of sheep meat carcass divided by yield of sheep carcass.
[10]Net profit of sale of barbacoa and pancita of live sheep (46.4 kg) divided by yield of carcass meat to barbacoa and pancita.
[a]Costs, USD/kg multiplied by 100 divided by margin.
[b]Profit, USD/kg multiplied by 100 divided by margin.
[c]Benefit (B)/cost (C) ratio (margin divided by costs, USD/kg). Technical

A relevant aspect that was observed in this study and that is one of the factors that can cause the producers to obtain a lower marketing margin, is that they do not know the characteristics of the product demanded by the stoker or barbacoa seller. This would be minimized if the sheep production and marketing systems adhere to the recommendations of the Mexican norm PROY-NMX-FF-106-SCFI-2006, where the excesses of fat or the poor musculature of the carcass are sanctioned in the price earned by the producer. That's why, it is necessary that different actors form part of the productive chain (producers, marketers, processors, barbacoa sellers, consumers), as well as investigators, technicians and government sectors to recognize the need of integration of a more efficient chain to achieve strategies that improve production (articulated work according to demand). The primary purpose should be to respond to the needs of the market by means of a shared vision of cooperation, communication, and coordination, which would make it possible to identify alternatives and strategies of action that could benefit all of the actors that participate in each one of the production links of sheep meat, thus making it more competitive and equitable at regional or national level.

CONCLUSIONS

According to the present investigation, it is concluded that three sheep marketing channels were identified; the first was producer-stoker-barbacoa selller-consumer. This channel is considered long and with degrees of inefficiency, where the producer receives a low profit with respect to the other agents, which does not motivate him to continue to carry out this activity. A second channel was producer-barbacoa seller-consumer; in this channel, it was observed that the stoker does not participate, which implies a better remuneration to producer. However, this channel is weak because there is no agreement for the sale of the product to the barbacoa seller. Consequently, the stoker may intervene at any time. In any case, it is an alternative channel of higher remuneration for the low investment producers, when there is an agreement of purchase-sale between these actors. A third channel was producer-barbacoa seller (complete cycle), where it was observed that 100 % of the profits were obtained by the producer. In this channel, a certain amount of investment is required, along with organization and knowledge of the elaboration and marketing of the product to the consumer; this producer is open to adopting new technologies for the success of the activity. Finally, with respect to raw and cooked (barbacoa) sheep meat, the producer obtained the lowest marketing margin with respect to the intermediary from the price paid by the final consumer. The producer obtains the highest benefit cost when he carries out the sale of the sheep carcass and barbacoa tacos, pancita and consommé. The profit

of this agent is low if the sale is as live animals in plaza or street market.

ACKNOWLEDGEMENTS

The authors would like to appreciate to sheep producers of Nthezee (C.C.S.O.), sheep introducers: Pablo Conde García, Alejandro Conde Gutiérrez and Oscar Torres Vega; Fine Cuts workshop "Maya": Agustín Zamora Aguilar; Fine Cuts workshop "El Rey", the Union de Barbacoyeros from Capulhuac A.C., State of Mexico, MVZ. Raúl Torres Ramos, Arturo Arellano García, Dr Fabio Montossi-Instituto Nacional de Investigación Agropecuaria (INIA)-Uruguay, the Consejo Nacional de Ciencia y Tecnología, to Facultad de Medicina Veterinaria y Zootecnia-Universidad Autónoma del Estado de México.

REFERENCES

Abbot, J.C. 1987. Mejora del mercadeo en el mundo en desarrollo. Colección FAO: Desarrollo económico y social, N° 37.

Aguilar, D.A. 2007. Evaluación alianza para el campo 2006. SAGARPA y Gobierno del Estado de México. p. 16-17.

Bravo, P.F.J., García, M.R., García, D.G., López, L.E. 2002. Márgenes de comercialización de la carne de res proveniente de la cuenca del Papaloapan, en el mercado de la ciudad de México. Agrociencia. 36, 255-266.

Bustamante, P.W. 2001. Apuntes de Mercadotecnia para la Microempresa Rural. Santiago de Chile. pp. 77-82.

Caldentey, A.P. 1979. Comercialización de productos Agropecuarios. Editorial Agrícola Española. Capítulo VII. pp. 108.

Carrera, Ch.B. 2008. Situación de la ovinocultura en México: En Ganadería y Desarrollo Rural en Tiempo de Crisis. Universidad Autónoma Chapingo, México. pp. 275-283.

Cochran, W.G. 1984. Técnicas de muestreo. Ed. C.E.C.S.A. México, D.F. pp. 513.

D'Aubeterre, R., Delgado, A., Armas, W.J., Rueda, M. 2007. Canales de mercadeo y comercialización del producto cárnico ovino (*Ovis aries*) en el Estado Lara, Venezuela. Zootecnia Tropical. 25, 205-209.

FAO. 2009. Carne y productos cárnicos. http://www.fao.org. Consulta 3 de diciembre de 2010.

FAO. 2010. La situación de los recursos zoogenéticos mundiales para la alimentación y la agricultura, editado por Barbara Rischkowsky y Dafydd Pilling. http://www. fao.org/docrep/011/a1250s/a1250s00.htm. Consulta 30 septiembre del 2010.

García, M.R., García, D., Montero, H.R. 1990. Notas sobre mercados y comercialización de productos agrícolas. Centro de economía. Colegio de Postgraduados. Montecillo, Edo. de México. pp. 120-200.

Góngora, P.R.D., Góngora, G.F. y Magaña, M.M.A. 2010. Caracterización y socioeconómica de la producción ovina en el Estado de Yucatán, México. Agronomía Mesoamericana. 21, 131-144.

González, R.F.J., Hernández, M.J., Rebollar, R.S., Rojo, R.R. 2010. Production and Marketing characteristics of pig production in the south of the State of Mexico. Tropical and Subtropical Agroecosystems. 12, 167-174.

Instituto Nacional de Estadística y Geografía (INEGI). 2009. Mujeres y hombres en México. Decimotercera edición. México. pp. 1-133.

Iturrioz, G.M., Iglesias, D.H. 2006. Los márgenes brutos de comercialización en la cadena de la carne bovina de la provincia de Pampa. Cuadernos de CEAGRO. 8, 51-56.

Lara, P.S.J. 2008. Engorda de corderos con dietas a base de granos, altas en energía. Fortalecimiento del sistema producto ovino. México. pp. 26-31.

López, S.Y., Soriano, R.R., Arias, M.L. 2008. Mujeres y ganadería ovina de Santa Catarina Tayata, Oaxaca: En Ganadería y Desarrollo Rural en Tiempo de Crisis. Universidad Autónoma Chapingo, México. pp. 285-293.

Lovelock, C.H., Reynoso, J., D'Andrea, G., Huete, L.M., Sánchez, C.M.A. 2004. Administración de Servicios Estrategias de Marqueting Operaciones y Recursos Humanos. Ed. Prentice Hall. México. pp. 54-72.

Montossi, F. 2002. Investigación aplicada a la cadena agroindustrial cárnica: avances obtenidos: carne ovina de calidad (1998-2001). INIA-Uruguay. Serie técnica. 126, 5-7.

NAFIN. 1998. Diplomado en el ciclo de vida de los proyectos de inversión. Ed. NAFIN-OEA: México. p. 35.

Norma Mexicana PROY-NMX-FF-106-SCFI-2006 de Clasificación de Carne de Ovino en Canal. http://200.77.231.100/work/normas/nmx/2006/proy-nmx-ff-106-scfi-2006.pdf. Consulta 3 de enero de 2010.

Nuncio-Ochoa, G., Nahed, T.J., Díaz, H.B., Escobedo, A.F. y Salvatierra, I.E.B. 2001. Caracterización de los sistemas de producción ovina en el Estado de Tabasco. Agrociencia. 35, 469-467.

Rebollar-Rebollar, S., Hernández-Martínez, J., García-Salazar, J.A. García-Mata, R., Torres-Hernández, G. Bórquez-Gastélum, J.L., Mejía-Hernández, P. 2007. Canales y márgenes de comercialización de caprinos en Tejupilco y Amatepec, México. Agrociencia. 41, 363-370.

Pittet, D.J., Maino, M.M., Pérez, M.P., Morales, S.M.S. 1994. Caracterización del mercado de la carne ovina en los Estados Unidos de Norteamérica. Avances en Medicina Veterinaria. 9.

Samaniego, G.J.A. 2000. Limitantes para el desarrollo y transferencia de tecnología agrícola en la Región Lagunera. Revista Mexicana de Agronegocios. 4, 486-497.

Sistema de Administración Tributaria.sat.gob.mx http://www./sitio_internet/asistencia_contribuyente/informacion_frecuente/tipo_cambio/42_19677.html

Schwentesius, R.R., Gómez, C.M.A. 2004. Márgenes y costos de comercialización: Aspectos conceptuales. Reporte de Investigación. UACH. CIESTAAM. pp. 9-15

Servicio de Información Agroalimentaria y Pesquera (SIAP). Servicio de Información Agroalimentaria y Pesquera. www.siap.gob.mx. Consulta el 25 de enero de 2010.

Sierra, M.L.D., Ortiza de la Rosa, B., Sierra, V.A.C., Rivera, L.A., Sanginés, G.J.R., Magaña, M.M.A. 2005. Estructura del mercado en Yucatán 1990-2003. Técnica Pecuaria en México. México. 43, 347-360.

Suárez, D.H., Sagarnaga, V.M. 2000. Efecto de la globalización de mercados sobre la ovinocultura. En: Memorias del V Curso: Bases de la cría ovina. Agosto 23-24; Texcoco (Edo. de México) México (DF): Asociación Mexicana de Técnicos

Especialistas en Ovinocultura, A.C. p. 178-190.

Trejo, G.E. 2008. La producción ovina ¿Negocio que se nos va de las manos? Termómetro Financiero. FIRA. p. 24.

Tomillo, E. Z. 2001. La investigación zootécnica española. Las razones de un fracaso Archivos de Zootecnia. 50, 441-463.

Vázquez, M.I., Vargas, L.S., Zaragoza, R.J.L., Bustamante, G.A., Calderón, S.F. Rojas, A.J. y Casiano, V.M.A. 2009. Tipología de explotaciones ovinas en la sierra norte del estado de Puebla. Técnica Pecuaria en México. 47, 357-369.

Wohlken, E. 1991. Einfuhrung in die land wirtschaftliche Marktlehre. Ed. Ulmer, Stuttgart. Alemania. p. 42.

12

The Soil Biota: Importance in Agroforestry and Agricultural Systems

Hortensia Brito-Vega[1*], David Espinosa-Victoria[2],
José Manuel Salaya-Domínguez[1] and Edmundo Gómez-Méndez[1]

[1]*División Académica de Ciencias Agropecuarias, Universidad Juárez Autónoma de Tabasco, Km 25.5 Carretera VHA-Teapa, Rancheria 3er. Sección, Vhsa, Tabasco, Mexico. C.P. 86040. E-mail: hortensia.brito@ujat.mx*

[2]*Colegio de Postgraduados, Campus Montecillo, Carretera México-Texcoco Km 36.5, Montecillo, Texcoco, C.P. 56230 Estado de Mexico*
Corresponding author

SUMMARY

The biological component of soil is important for the maintenance and functioning of ecosystems. Currently there have been some studies on the diversity of soil biota and their role in key soil processes. Microorganisms are critical to the functioning of biological systems and the maintenance of life on the planet, since they participate in metabolic, ecological and biotechnological processes on which we depend for survival and for facing future challenges. Soil organisms maintain soil processes such as carbon capture and storage, nutrient cycling, nitrogen fixation, water infiltration, aeration, and organic matter degradation. The effects of these ecosystem services are not yet fully explored, especially soil microorganisms (bacteria and fungi), and macro-organisms (earthworms).

Key words: Processes; carbon capture and storage; organic matter; microorganisms.

INTRODUCTION

The intense perturbation on soils caused by population increases has been recorded in many parts of the tropics, can lead to lower crop yields as well as promote the invasion of difficult to control weeds. One of the options for stopping this process is land use through agroforestry systems (Jenkins *et al.*, 2004). In almost all traditional farming systems, including livestock systems, trees are interspersed with crops or managed zonally by alternating trees and crops and/or pasture. That is, they are agroforestry systems and even with the modernization of the agricultural region, agricultural landscapes still contain a high number of trees. These trees fulfill many purposes such as production (timber, firewood, fodder, fruits and medicines) as well as services (shade for crops and/or animals, protection in the case of windbreaks). In addition, trees increase the biodiversity of the agroecosystems by creating homes for other organisms in their branches, roots and litter (Reyes and Valery, 2007). Agroforestry techniques are used in regions with diverse ecological, economic and social conditions, as well as in regions with fertile soils where agroforestry systems can be very productive and sustainable. Equally, these practices have a high potential to maintain and improve productivity in areas that present problems of low fertility and excess or shortage of soil moisture (Musálem, 2001).

Agroforestry can also play an important role in conserving biodiversity in deforested and fragmented landscapes, providing habitats and resources for plant and animal species, maintaining the connection of the landscape (and, thereby, facilitating the movement of animals, seeds and pollen), making landscape living conditions less harsh for the inhabitants of the forest, reducing the frequency and intensity of fires, potentially reducing effects on remaining adjacent fragments, and providing buffer zones to protected areas (Jenkins et al., 2004). The objective of this report is about some services microorganisms that contribute to different ecosystems.

The ecosystem-service

An ecosystem-service is defined as goods, functions and processes of an ecosystem that provide benefits to the human population (Boyd and Banzhaf, 2005; Balvanera and Cotler, 2007). The agro-forestry-grazing system is a production system (production services) for beneficial services: soil formation, nutrient cycling and primary production. Soils are also involved in regulating services (climate regulation on greenhouse gas fluxes and C capture, flood control, detoxification, protection of plants against parasites). This influences the dynamics of organic matter and the wide effects on soil physical properties (Table 1) (Jenkins et al., 2004).

The microbial component of soil is important for health, maintenance, function and quality of ecosystems (Olalde and Aguilera, 1998). The role of the organisms is the transformation and availability of organic matter for plants, an activity without which the world would be a huge dump. Others have played a significant role in relation to man, and productivity, participating in agriculture, food processing and medicine production (van Eekeren, 2010).

Table 1. Ecosystem services of the soil biota.

	Goods/services	Ecosystem process	Soil biota contribution
Production	Pest and disease regulation	Interaction plants-organisms	Reduction and control of pathogenic organisms
	Increase food	Biofertilizers Biopesticides Phytostimulators o Bioremediators	Carbon and nitrogen storage
	Nutrient cycles	CO_2 and N fixation Organic matter degradation Carbon capture	
Environmental	Climate regulation	Ecosystem conservation	Reduced global warming
	Air quality regulation		
Contamination	Water cycle	Maintenance of plantations	Water quality and conservation
	Biodegradation		Transformation of harmful compounds into substances with lesser impact

Biological involvement in the process for formation of soil aggregates is important for their activity, mesofauna excrements, trapping of particles by roots and glues produced by fungi, mainly hyphae structures of arbuscular mycorrhizal fungi that trap and bind primary particles for the development of aggregates, and bacteria. Aggregates that form it are generally stable in water (González-Chávez et al., 2004, van Eekeren, 2010). The importance of aggregates in the ecosystem is for soil stability and structure, participating in processes such as infiltration, aeration, water holding capacity, less encrusting of the soil surface and greater resistance to erosion (González-Chávez et al., 2004).

The concept of ecosystem services is a newly formed, so there is a need to develop methodologies to identify, quantify and rank (if possible) the provided services. It is essential to identify the relationships between services and ecosystem processes that regulate them, to know the perception of the players, to model scenarios of loss of services and the potential population affected, coupled with an economic valuation study (Almeida-Leñero et al., 2007). Only this will generate solid management proposals to conserve and restore ecosystems without forgetting that the main goal should be human welfare (Almeida-Leñero et al., 2007).

Agricultural production

The relationships established by soil microorganisms may benefit to the plants when they occur in the area close to their roots (rhizosphere) characterized by having a large number of organic compounds that are exuded by the plants (Kloepper, 1994). Beneficial soil biota around the root, set and accelerates biochemical processes affecting growth and development of vegetation, and it is related to an increase in available nutrients and the production of growth substances while inhibiting parasitic organisms and pathogens (Lara and Echeverría, 2007).

Reyes and Valery (2007) studied the phosphate solubilizing, which varied with soil physical and chemical conditions and plant species. Where microbial population density showed a low rhizosphere/soil ratio opposed to one with high physical and chemical degradation caused by mining and by the corn shade house trial, the Azotobacter strain MF1b significantly increased the dry weight in the NK chemical treatments and phosphate rock with K, giving an agricultural service as a strain with potential use in soil fertility.

Ecosystems agricultural same ecological property, the cycling of nutrients in the soil, is considered as an ecosystem service. In agricultural fields, food production depends on the cycling of nutrients in the soil (fertility), but also of various human interventions such as spraying of agricultural chemicals, raw material processing and food distribution. Developing appropriate indicators may provide a better understanding and quantifying of the link between the benefits provided by ecosystems and their ecological properties (Quétier et al., 2007).

Earthworms provide production services to the ecosystem by digesting organic matter through their mutual interaction with the microflora registred into their digestive tract. The effects caused by the worms can be seen at different time and space scales. In short time, such as digestion for the worm, organic residues is fractionated, and some nutrients (nitrogen, phosphorus and potassium) can be assimilated by plants are released to the soil (Lavelle et al., 1997).

Thus earthworms affect the nitrogen cycle directly by consuming and assimilating inorganic and organic-N. The worms can process large amounts of organic matter, with the consumption rate of Aporrectodea tuberculata ranging from 5 to 13 mg g^{-1} organic matter g^{-1} per worm per day^{-1}, while for Lumbricus terrestris goes from 14 to 2.7 mg g^{-1} per day^{-1}. The efficiency of assimilated nitrogen (^{15}N) ranges from 10 to 26% in A. tuberculata and 25-30% in L. terrestris. Brussaard (1999) estimated that consumption by earthworms may be 11.8 to 17.1 Mg organic matter ha^{-1} per yr^{-1}; this can amount to 19-24% of organic waste matter yields. Geophagus earthworms can ingest huge amounts of land. The population of Pontoscolex corethrurus in grasslands of "Plan de las Hayas" Veracruz, can ingest of 400 Mg per ha per year, which means a worm of this kind can ingest from 1 to 3 times its own weight per day (Lavelle et al., 1997).

Bacteria

Kloepper (1994) used the term plant growth promoting rhizobacteria to refer to rhizobacteria capable in plants. This is a crucial feature for microbial inoculum selection for use in agriculture and food, mainly as biofertilizers, biopesticides, phytostimulators or bioremediators (Lugtenber et al., 2001; Peña and Reyes, 2007).

The use of rhizosphere microorganisms in biotechnology for the human community has been implemented by inoculating seeds and plants (Peña and Reyes, 2007). Exploited and abandoned soils from a phosphate rock mine, where populations were 29% for fungi and 13% for cultivable calcium phosphate dissolving bacteria for plants under field conditions (Reyes et al., 2006).

Biological nitrogen fixation is carried out exclusively by prokaryotes, which have the ability to reduce atmospheric nitrogen (N_2) to ammonium (NH_4) that can be used by plants, contributing to, and improving productivity of crops (Zehr *et al.*, 2003, Philippot and Germon 2005). Several studies have demonstrated the importance and effect of the presence of nitrogen-fixing bacteria in agricultural and ecosystems processes, as well as sustainable land use (Mantilla-Paredes *et al.*, 2009).

Arbuscular mycorrhizal fungi (AMF)

Mycorrhizae type of symbiosis between the fungi and plant roots; however, not all fungi are mycorrhizal, some are also saprophytic (Barker, 2009). With the establishment of arbuscular mycorrhizal fungi in the root system (Figure 1), the morphological and physiological integration of the symbiosis is satisfied, and will determine mutuality and nutrient exchange in both symbionts (Jones and Smith, 2004).

The benefit provided by AMF to the human community is the colonization of roots for absorption of some essential plant nutrients such as phosphorus and potassium. Also absorption of low mobility nutrients in the soil such as Cu and Zn, increasing plant tolerance to abiotic soil stress conditions and protection from pathogens (Smith and Read, 1997). The mutualistic symbiosis between fungi and plants is seen in the forage crops, ornamentals, fruits, vegetables and maple and other forests and pine (van Den Heijden., 1998).

The use of AMF in agriculture contributes to improve plant nutrition; however, the monoculture condition of agroecosystems, may be causing a decrease in the diversity of AMF and as a result (Figure 2), these microorganisms could be providing a beneficial effect even though limited to the hosts (Alarcon, 2007).

Another role for which AMF in ecosystems provide to plant is adaptability, establishment and growth. As to the soil stability, hyphae allow the aggregation of soil particles, which prevent soil erosion (Abbott and Gazey, 1994). Moreover, the activity of AMF allows the presence of microbial populations to be modified, participating as regulatory agents of beneficial and pathogenic microbiota, thus, influencing the dynamics of organic carbon and soil fertility (Alarcón, 2007).

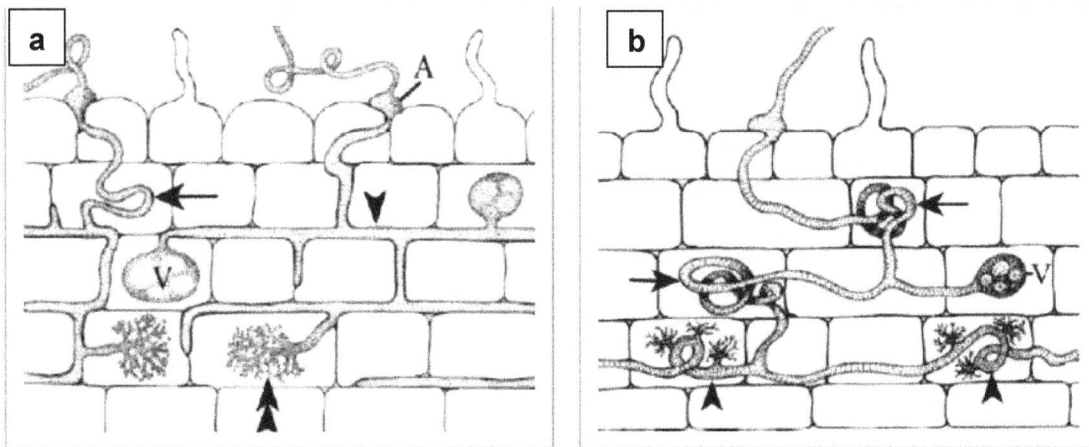

Figure 1. Types of arbuscular mycorrhizal association. a) Type Arum, characterized by hyphae that grow in an intercellular form (arrow head). b) Type Paris, characterized by hyphae with intracellular growth (arrow head). Vesicles (V), Appressorium (A), hyphae (arrows) y arbuscules (two arrowheads). Modified from Barrer (2009).

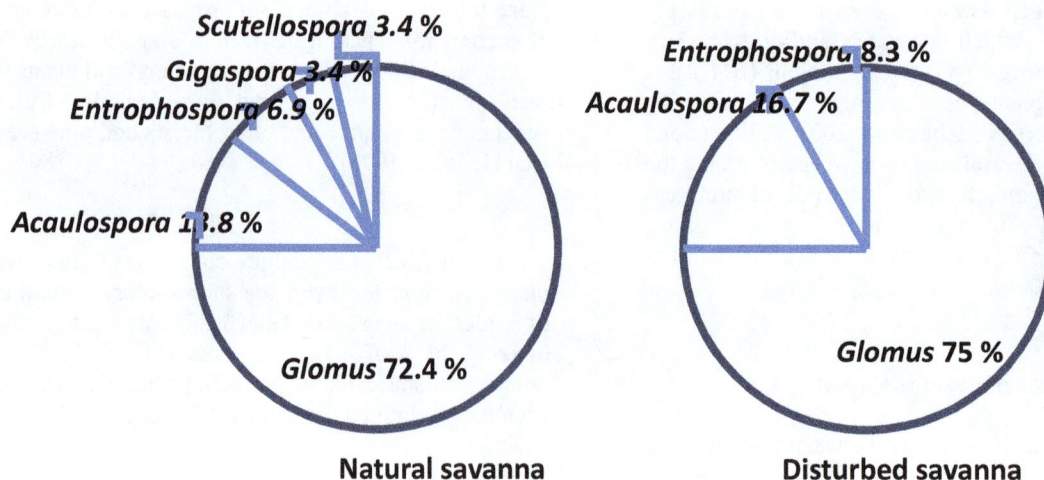

Figure 2. Percentage of the genus of AMF in natural and disturbed savanna (Lovera and Cuenca, 2007).

Antagonistic fungi, the case of *Trichoderma* spp

Trichoderma spp. has been widely used as antagonistic fungal agents against several pests as well as plant growth enhancers. Faster metabolic rates, antimicrobial metabolites, and physiological conformation are key factors which chiefly contribute to antagonism of these fungi. Mycoparasitism, spatial and nutrient competition, antibiosis by enzymes and secondary metabolites, and induction of plant defence system are typical biocontrol actions of these fungi. On the other hand, *Trichoderma* spp. has also been used in a wide range of commercial enzyme productions, namely, cellulases, hemicellulases, proteases, and -1,3-glucanase. The classification of the genus, *Trichoderma*, antagonic mechanisms and role in plant growth promotion has been well documented. However, fast paced current research in this field should be carefully updated for the fool-proof commercialization of the fungi (Verma *et al.*, 2007).

Previously, Cook (1993) classified the mycoparasitic interactions as: (i) replacement (unilateral antagonism), (ii) deadlock (mutual antagonism), and (iii) intermingling (no antagonism), with lack of explanation at microscopic level. However, more recently the understanding of mycoparasitism has considerably improved (Yedidia *et al.*, 1999). Interestingly, the studies were carried out at genetic

(Zeilinger *et al.*, 1999; Brunner *et al.*, 2003) and microscopic (Benhamou *et al.*, 1999) levels. However, a broader concept concerning living plants (with the exception of preservation of wood, where *Trichoderma* spp. alone kill plant pathogenic fungi; discussed later) would be that after being treated with mycoparasites, plants induce defense mechanisms on their own. Further, this phenomenon leads to production of fungal inhibitory compounds by plants in addition to *Trichoderma* spp., thereby, facilitating mycoparasitism. Moreover, the mycoparasitism shown by *Trichoderma* sp. was a host specific (*Pythium oligandrum*) (Benhamou *et al.*, 1999). The complex group of extracellular enzymes has been reported to be a key factor in pathogen cell wall lysis during mycoparasitism (Verma *et al.*, 2007).

The microorganisms in biodegradation of toxic compounds

Torres (2003) indicates that bioremediation practices consist mainly of the use of different organisms (plants, yeasts, fungi, and bacteria) to neutralize toxic substances, or transforming them into less toxic substances or making them harmless to the environment and human health. One of the most used bio-correctional measures is the use of microorganisms for soil decontamination (ecosystem service) and is based on the absorption of organic substances by these microorganisms, which they use as the carbon source needed for growth and energy for their metabolic functions.

The bacteria were found to be a group of bacteria mainly from the genus *Pseudomonas* which degrade the hydrocarbon 3-phenoxybenzoic in soils (Halden *et al.*, 1999), and *Sphingomonas wittichii* RW1 which under anaerobic conditions is capable of transforming the metabolite 2,7-dichlorobenzene producing the metabolites 4-chloroethanol and 1,2,3,4-tetrachlorodibenzo.

The best known bacterial genus in terms of efficiency in the degradation of petroleum fuels are *Pseudomonas*, *Acinetobacter*, *Agrobacterium*, *Flavobacterium*, *Arthrobacter* (Torres, 2003), *Aeromonas*, *Corynebacterium* and *Bacillus*. Of these, *Pseudomonas*, *Agrobacterium*, and *Bacillus* are also characterized for having the ability to fix atmospheric N (Lugtenberg *et al.*, 2001).

The AMF increased tolerance to metals in most of their host plants (van Der Heijden, 1998). *Glomus mosseae* BEG-132 captured between 470 and 680 μg g^{-1} of Cu (based on dry weight) and increased tolerance to As and Cu (Sánchez *et al.*, 2004).

Hernández *et al.* (2003) evaluated the behavior of free living atmospheric nitrogen fixing bacteria populations (ANFB) and hydrocarbonoclastic ANFB (HC's) in soils contaminated by fuels such as kerosene. The presence of kerosene did not drastically affect populations of ANFB and HC's-ANFB, and after evaluating atmospheric N fixation, it was found that the bacteria could perform this function. The largest populations were: for ANFB $(410 \times 10^4$ CFU g^{-1} root) in the rhizoplane in a concentration of 2500 mg kg^{-1} kerosene and HC's-ANFB- $(299 \times 10^4$ CFU g^{-1} soil) in the rhizosphere in a concentration 500 mg kg^{-1} kerosene [CFU-colony forming units]. These results showed that HC's-ANFB can be an option for bioremediation of kerosene contaminated soil and its application will induce the incorporation of N into these soils.

The microbiota in soil formation

The earthworm participates in ecosystem services such as soil formation, regulation and supply of water by means of mechanical activity within the soil due to its ability to move, creating structures that affect aeration and water infiltration (galleries), degradation of organic material and its incorporation into soil. Part of this degradation is due to the digestion that takes place in their digestive tract and with the production of manure (excretas) as part of the biogenic structure, for which they are also called

ecosystem engineers (Brown *et al.*, 2001, Lavelle *et al.*, 2006).

There are three functional groups of earthworms which based on their feeding style and habitat contribute in one way or another to ecosystem services: the epigeous (litter inhabitants, consumers of decaying organic matter, dorsoventrally flattened and pigmented); the endogenous (inhabitants of ground, not pigmented, consuming land and subdivided into poly-, meso- and oligo-humic (Lavelle and Spain, 2001); and anecic (ground dwellers and consumers of leaves) forming galleries (Brown *et al.*, 2001; Brito-Vega *et al.*, 2006). These behaviors usually lead to a stimulation of plant growth (Whalen and Parmelee, 1999; Hallaire *et al.*, 2000; Eriksen-Hamel and Whalen, 2007). For example, Eriksen-Hamel and Whalen (2007) analyzed two species of earthworms, the endogenous *Aporrectodea caliginosa* which was dominant in the field and the anecic *Lumbricus terrestris* which barely survived field conditions in the crop, and found that the N-total in soybeans increased 25% with the increase of the earthworm population (100 to 500 individuals m^{-2}).

Organic matter

Soil organic matter (**SOM**) is constituted of many diverse components with different states of decomposition, which vary depending on the quality of material, during mineralization (sugars, amino acids, hemicellulose, cellulose, and lignin) and humification. During the process of decomposition, organic matter is enhanced by the participation of soil organisms that ingest and transform a mixture of organic substrates and inorganic soil (micro- and macro-organisms). At the end of this process, part of the final product is absorbed by plant roots or other organisms. The organisms are involved at different levels of the trophic system, including earthworms that participate favorably in changes of N from the topsoil; some feed on microbes (microbivores), organic matter in decomposition (detritivores), or a mixture of microbivores - detritivores (Domínguez *et al.*, 2009). It is still uncertain as to what trophic level earthworms should be located, given that they may use different feeding strategies, from selective and nonselective mechanisms, plus they have the ability to obtain energy from both living and dead carbon sources (Rodríguez-Echeverria, 2009).

The activity of earthworms in the organic matter

The decomposition of organic matter occurs in two distinct phases in relation to the activity of earthworms: 1) active phase, where the worms process organic matter, alter the physical and chemical properties, and the microbial composition (Lores *et al.*, 2006); and 2) the maturation phase, where microorganisms take over in the decomposition of organic material previously processed by earthworms (Domínguez *et al.*, 2009). Earthworms are involved in the decomposition, and transformation organic matter through processes that occur in their digestive system. The processes include modification of microbial and diversity, the modification of microfauna populations, the homogenization of the substrate and the intrinsic processes of digestion and assimilation, including also the production of mucus, which are a source of easily assimilated nutrients for microorganisms (Domínguez *et al.*, 2009). These microorganisms produce extracellular enzymes that degrade cellulose and certain phenolic compounds, increasing the degradation of ingested material (Salzman, 2005).

The mineralization of nutrients is performed by the metabolic activity of bacteria and fungi. But this metabolic activity is influenced by soil fauna that lives in interaction with microorganisms, and also by various interactions that determine the transfer of nutrients through the system. In this sense, the excreta of earthworms play an important role in decomposition, because they contain nutrients and microorganisms that are different from those contained in the organic material before ingestion (Domínguez *et al.*, 2009).

CONCLUSION

The concept of ecosystem services enables the analysis of the link between ecosystem functioning and human welfare. The activity of soil biota may decrease or increase the productivity of ecosystems. Negative effects can cause a considerable decrease in plant productivity (pests) or increase the positive effects (beneficial organisms.) Each organism may have a different influence on soil processes and plant production, and the abundance or biomass can reach thresholds, both positive and negative on the ecosystems services provided to humanity.

REFERENCES

Abbott, L. K., Gazey C. 1994. An ecological view of the formation of VA mycorrhizas. Plant Soil. 159:69-78.

Alarcón, A. 2007. Micorriza arbuscular. In: Microbiología agrícola. (Ids): Ferrera-Cerrato, R., Alarcón, A. Trillas, México. pp. 568.

Almeida-Leñero L., Nava M., Ramos A., Ordoñez M. de J., Jujnovsky J. 2007. Servicios ecosistématicos en la cuenca del río Magdalena, Distrito Federal, México. Gaceta Ecológica. Número especial 84-85: 53-64.

Balvanera, P., Cotler H. 2007. Acercamiento al estudio de los servicios ecosistématicos. Gaceta Ecologica. 84-85:8-15.

Barrer, S. E. 2009. El uso de hongos micorrizicos arbusculares como una alternativa para la agricultura. Facultad de Ciencias Agropecuarias. 7:124-132.

Benhamou, N., Rey P., Picard K., Tirilly Y. 1999. Ultrastructural and cytochemical aspects of the interaction between the mycoparasite Pythium oligandrum, and soilborne plant pathogens, Phytopathology 89:506–517.

Boyd, J. W., H. S. Banzhaf. 2005. Ecosystem services and government: the need for a new way of judging nature's value. Resources. 158:16-19.

Brito-Vega, H. Espinosa-Victoria D., Figueroa-Sandoval B., Patrón-Ibarra J.C. 2006. Diversidad de lombrices de tierra con labranza de conservación y convencional. Terra latinoamericana. 24: 99-108.

Brown, G.G., Fragoso C., Barois I., Rojas P., Patrón C.J., Bueno J., Moreno G.A., Lavelle P., Ordaz V., Rodríguez C. 2001. Diversidad y rol funcional de la macrofauna edáfica en los ecosistemas tropicales mexicanos. Acta Zoologica Mexicana. 1:79-110.

Brunner K., Montero M., Mach R.L., Peterbauer C.K., Kubicek C.P. 2003. Expression of the ech42 (endochitinase) gene of Trichoderma atroviride under carbon starvation is antagonized via a BrlA-like cis-acting element. FEMS Microbiology Letters. 218:259–264.

Brussaard, L. 1998. Soil fauna, guilds, functional groups and ecosystem processes. Applied Soil Ecology 9:123-135.

Brussaard, L. 1999. On the mechanisms of interactions between earthworms and plants. Pedobiologia 43:880-885.

Cook, R.J. 1993. Making greater use of introduced microorganisms for biological control of plant pathogens. Annual Review of Phytopathology 31: 53–80.

Domínguez, J., Aire, M., Gómez-Brandón, M. 2009. El papel de las lombrices de tierra en la descomposición de la materia orgánica y el ciclo de nutrientes. Ecosistemas. 18:20-31.

González-Chávez, M. C. A., Gutiérrez-Castorena M. C., Wright, S. 2004. Hongos micorrízicos arbusculares en la agregación del suelo y su estabilidad. Terra Latinoamericana. 22:507-514.

Hernández, A. E.,Ferrera-Cerrato, R., Rodríguez V. R. 2003. Bacterias de vida libre fijadoras de nitrógeno atmosférico en rizósfera de frijol contaminada con queroseno. TERRA Latinoamericana. 21: 81-89.

Jenkins, M., Scherr, S. J., Inbar, M. 2004. Markets for biodiversity services: potential roles and challenges. Environment. 46: 32-42.

Jenkins, M., Scherr, S. J., Inbar, M. 2004. Markets for biodiversity services: potential roles and challenges. Environment 46: 32-42.

Jones, M., Smith, S. 2004. Exploring functional defitions mycorrhizas: are mycorrhizas always mutualism? Canandia Journal of Botany. 82:1089-1109.

Kloepper, J. 1994. Plant growth-promotiong rhizobacteria (other systems). In Okon J (Id.) *Azospirillum*. CRC. Boca Raton, FL. EEUU. pp 137-167.

Lara, A., Echeverría, C. 2007. Conclusiones del Congreso Internacional de los servicios Ecosistémicos en los Neotrópicos: Estado del arte y desafíos futuros. Bosques. 28:10-12.

Lavelle, P., A. Spain. 2001. Soil ecology. Kluwer. Amsterdam, The Netherlands.

Lavelle, P., D. Bignell, M. Lepage. 1997. Soil function in a changing world: the role of invertebrate ecosystem engineers. European Journal of Soil Biology. 33: 159-193.

Lavelle, P., Decaëns, T., Aubert, M., Barot, S., Blouin, M., Bureau, F., Margerie, F., Mora, P., Rossi, J. P. 2006. Soil invertebrates and ecosystem services. European Journal of Soil Biology. 42: 3-15.

Lavelle, P., E. Blanchart, A. Martin, A.V. Spain, S. Martin. 1992. Impact of soil fauna on the properties of soils in the humid tropics. pp. 157-185. *In:* R. Lal, P.A. Sanchez (eds.).

Myths and science of soils in the tropics Special Publication 29. Soil Science Society of America. Madison, WI, USA.

Lovera, M., Cuenca, G. 2007. Diversity of arbuscular mycorrhizal fungi (AMF) and soil mycorrhizal potential of natural savanna savannah disturbed the great plains, Venezuela. Interciencia. 32: 108-114.

Lugtenberg, B., Dekkers L., Bloemberg G. 2001. Molecular determinants of rhizosphere colonization by *Pseudomonas*. Annual Review of Phytopathology. 39:461-90.

Mantilla-Paredes, A. J. Cardona, G. I., Peña-Venegas, C. P., Murcia, U., Rodríguez, M., Zambrano, M. M. 2009. Distribución de bacterias potencialmente fijadoras de nitrógeno y su relación con parámetros fisicoquímicos en suelos con tres coberturas vegetales en el sur de la Amazonia colombiana. Revista de Biologia Tropical. 57: 915-927.

Olalde P. V., Aguilera, G. L. I. 1998. Microorganismos y biodiversidad. Terra Latinoamericana. 3: 289-292.

Pankhurst, C.E., Doube, B. M., Gupta, V.V.S.R., Grace, P.R. 1997. Soil Biota: Management in sustainable farming systems. CSIRO. East Melbourne. pp. 262.

Peña, H. B., Reyes, I. 2007. Aislamiento y evaluación de bacterias fijadoras de nitrógeno y disolventes de fosfatos en la promoción de crecimiento de la lechuga (*Lactuca sativa* L). Interciencia. 32:560-565.

Philippot, L., Germon, J.C.. 2005. Contribution of bacterial to initial input and cycling of nitrogen in soils. *In* Buscot, F., and Varma, A. (Ids.). Microorganisms in soils: roles in genesis and functions. Springer, Nueva York, EEUU. pp. 159-176.

Quétier, F. Tapella, E., Conti, G., Cáceres, D., Díaz, S. 2007. Servicios ecosistématicos y actores sociales. Aspectos conceptuales y metodológicos para un estudio interdisciplinario. Gaceta ecológica. 84-85:17-26.

Reyes, I., Valery, A. 2007. Efecto de la fertilidad del suelo sobre la microbiota y la promoción del crecimiento del maíz (*Zea mays* L.) con *Azotobacter* spp. Bioagro. 3:117-126.

Reyes, I., Valery, A., Valduz, Z. 2006. Phosphate-solubilization and colonization of maize rhizosphere by wild and genetically modified strains of *Penicillium rugulosum*. Microbiology Ecology. 44:39-48.

Rodríguez-Echeverria, S. 2009. Organismos del suelo: la dimensión invisible de las invasiones por plantas no nativas. *Ecosistemas* 18:32-43.

Salzman, J. 2005. The promise and perils of payments for ecosystem services. International Journal of Innovation and Sustainable Development 1:5-20.

Sánchez, V. G., Carrillo, G. R., Martínez, G. A., González, Ch. Ma.C. 2004. Tolerancia adaptativa de hongos micorrízicos arbusculares al crecer en sustratos contaminados con As y Cu. Revista internacional de Contaminación Ambiental. 20:147-158.

Smith, S. Read, D. 1997. Mycorrhizal Symbiosis. 2 ed. Londres: Academic Press Limited, p.605.

Torres, R. D. 2003. El papel de los microorganismos en la biodegración de compuestos tóxicos. Ecosistemas. 2:1-5.

van Der Heijden, M. 1998. Different arbuscular mycorrhizal fungal species are potential determinants of plant community structure. Ecology. 79: 2082-2091.

Van Eekeren, N. 2010. Grassland management soil biota and ecosystem. Wageningen University, Wageningen. English.Thesis. p. 264.

Yedidia, I., Benhamou, N., Chet, I. 1999. Induction of defense responses in cucumber plants (*Cucumis sativus* L.) by the biocontrol agent *Trichoderma* harzianum, Applied Environmental Microbiology. 65:1061–1070.

Zehr, J.P., Jenkins, B.D., Short, S.M., Steward, G.F. 2003. Nitrogenase gene diversity and microbial community structure: a cross-system comparison. Environental. Microbiology. 5: 539-554.

Zeilinger, S., Galhaup, C., Payer, K., Woo, S.L., Mach, R.L., Fekete, C., Lorito, M., Kubicek, C.P. 1999. Chitinase gene expression during mycoparasitic interaction of *Trichoderma harzianum* with its host, Fungal Genet. Fungal Genetics of Biology. 26:131–140.

13

Presence of BCMV and BCMNV in Five Dry Bean-Producing States in Mexico

D. Lepe-Soltero[1], B.M. Sánchez-García[2], Y. Jiménez-Hernández[2],
R.A. Salinas-Perez[4], M.A. García-Neria[1], D. González de León,
N.E. Becerra-Leor[3], J.A. Acosta-Gallegos[2*] and L. Silva-Rosales[1*]

[1] *Laboratorio de Interacciones Planta-Virus del Depto. de Ing. Genética. Cinvestav
Unidad Irapuato. Km. 9.6. Lib. Nte. Carr. Irapuato-León. CP 36821.*

[2] *Programa de Frijol CEBAJ-INIFAP Km 6.5 Carretera Celaya a San Miguel de
Allende Celaya, Gto. Mexico. CP 38110.*

[3] *Programa de Frijol CECOT-INIFAP Km 34.5 Carretera Federal Veracruz-
Córdoba, Medellín de Bravo CP 94270.*

[4] *Programa de Frijol CEVAF-INIFAP Km 1609 Carretera Internacional Mexico-
Nogales, CP 81100 Juan Jose Rios, Los Mochis, Sin.*

*dlepe@ira.cinvestav.mx ; bmsgsma@yahoo.com.mx; yajiher_1013@yahoo.com.mx
salinas.rafael@inifap.gob.mx; margarcia@ira.cinvestav.mx; gdeleon09@gmail.com
becerra.noe@inifap.gob.mx; jamk@prodigy.net.mx; lsilva@ira.cinvestav.mx
[*] Corresponding authors*

SUMMARY

A survey was conducted to assess the frequency of BCMV and BCMNV in five of the main dry bean producing states in Mexico during the spring-summer 2009 and fall-winter growing seasons 2009-2010. States included in the survey were Nayarit, Sinaloa and Sonora in the pacific west coast, Veracruz in the gulf coast and Guanajuato in central Mexico. A total of 338 samples were collected and analyzed by RT-PCR with specific primers for each viral species. Forty-four samples (13%) gave positive reaction for BCMV, 70 (21%) for BCMNV and 30 (9%) were positive for both viral species, 164 (48%) were negative for both viruses and 30 (9%) could not yet be defined. As for cultivars, Azufrado Higuera (Nueva Granada race) grown at Sinaloa showed the highest frequency (33%) of BCMV, whereas Negro Jamapa (Mesoamericana race) from Nayarit displayed highest frequency (50%) of BCMNV. In these two states the percentage of positive samples for either viral species was 80%. In addition, in cultivar Negro Jamapa mixed infections of both viruses were detected. Results point out a high risk of viral infection with seed movement across states, particularly since both viral species are seed transmitted and in the states at the pacific west coast, large seed lots are produced during the fall-winter season.

Key words: BCMV; BCMNV; bean plants viruses

INTRODUCTION

In Mexico common bean grown during the spring-summer season is located at the semiarid and central highlands, 96% of 1.4 million Ha under rainfed conditions (SIAP, 2012). At the Bajio subregion and central highlands, BCMV can reduce seed yield and infect the seed that might be used in subsequent plantings. During the fall-winter season dry beans are grown in the humid and dry tropic regions. In the first region, mostly grown on residual moisture and in the second, under irrigation. In the first region that includes the southeast states of Veracruz and Chiapas, BCMNV is more prevalent as well as in Nayarit in the dry tropics. In Sinaloa, BCMV along with the 'calico viral disease' can damage the bean crop. Losses due to viral diseases can fluctuate between 20 to 100%. In Mexico BCMV has been reported to damage production ranging from 30 to 80% (Chew *et al.*, 2010), however reported losses due to BCMNV are scarce.

BCMV and BCMNV are two strongly related pathogens infecting bean plants. Up until 1992, they were considered as two serotypes, A and B, but now they are considered as two separate species from the same Genus (*Potyvirus*) and Family (*Potyviridae*) (Morales (1998). The separation into species initially based on the serological reactivity was further confirmed by different properties of the virus (Berger *et al.*, 1997) such as the molecular weight and peptidic profile of the capsid protein (McKern, 1992, Huguenot *et al.*, 1994), as well as the cytological effects (Vetten *et al.*, 1992), produced on the infected tissue and general responses on the infected plant (Kelly, 1997). Nucleotide sequence backup for this taxonomic demarcation has also been reported by Saiz *et al.* (1994). Both viral species are seed transmitted (Hall, 1991), making their control more difficult with the use of residual seed from previous harvests and due to the exchange of seed between users from different localities. Both practices have an implication on the presence and diversity of BCMV and BCMNV. Symptoms produced by both viral species are very similar and they include: mosaic, stunt, chlorosis and leaf deformation. At higher temperatures systemic necrosis may be observed, depending on one of the seven pathogroups affecting the crop. BCMV and BCMNV are categorized as pathogroups according to a classical study by Drijfhout (1978) and Kelly (1997). This division is based on the response of different bean cultivars to viral isolates due to the genetic composition of the set of cultivars used. One important response, systemic necrosis, is a hypersensitive reaction to the necrotic viral strains as studied by Collmer *et al.* (1996). It starts with either pinpoint lesions or veinal necrosis appearing as "cross road" or pinpoint spots, on certain genetic composition of the host plant; i.e, the presence of the dominant I gene with combination of the recessive bc genes.

In Mexico, initial reports only make reference to the presence of BCMV in the federal states of Puebla (Diaz-Plaza *et al.*, 1992), Guanajuato (Montes-Rivera and Arévalo-Valenzuela, 1985), Veracruz (López-Salinas *et al.*, 1994) and Sonora (Jiménez-García and Nelson, 1994). A subsequent work showed the prevalence of BCMV over BCMNV in Mexico mostly at the central states, whereas BCMNV proliferated toward the eastern tropical states (Flores-Estévez and Silva-Rosales, 2000; Flores-Estévez *et al.*, 2003). In this report, the frequency of both species is reported in terms of the different cultivars sampled in five different major dry-bean producing federal states in Mexico. Also, the implications of germplasm movement as a factor shaping the distribution of both viral species are discussed. Some possible explanations are given to understand the presence of both viral species in black seeded cultivars.

MATERIALS AND METHODS

Sample collection

Field samples were obtained from commercial and experimental bean fields collected in five federal states in Mexico representing four agricultural systems: irrigated and rainfed crop during winter-spring and spring-summer seasons (Guanajuato); irrigated crop during fall-winter season (Sinaloa and Sonora); residual moisture crop at the fall-winter season (Nayarit) and rainfed and residual moisture crops during summer and fall-winter seasons at the humid tropics (Veracruz). In order to obtain uniformly collected samples and recorded data, a registration manual was implemented specifically for this project, containing main visual criteria guidelines on virus symptoms (Figure 1), and main strategies for sample collection such as the gathering of the widest diversity of symptomatic varieties in as many cultivated fields as possible, in order to increase viral diversity and collection of bean plants in a defined perimeter around a source of infection. A total of nine samples per infection foci, within a hectare were initially spotted, eight of which were symptomatic and one asymptomatic plant. Symptomatic plants were collected at equidistant plows. A total number of 338 samples were collected (Table 1), in about 24 localities and from about 35 bean cultivars and photographed with a registration number to generate a database for the frequency account of both viral species in the generated collection.

Table 1. Cultivars sampled by state.

Cultivar/States	Guanajuato	Nayarit	Sinaloa	Sonora	Veracruz
Aluyori			8		
Azufrado Criollo		9			
Azufrado Higuera 1				18	
Azufrado Higuera 2			54		
Azufrado Noroeste			17		
Azufrado Peruano 87				5	
Azufrado Regional 87				5	
Azufrado Reg. Criollo			18		
Bayo Berrendo		9			
Bayo Blanco				5	
Bayo Madero	1				
CIAT 103-25		8			
Criollo Negro					1
Criollo Vaina Blanca y Morada					2
Criollo Vaina Morada					9
ELS 15-55		7			
FJB 08046	1				
Flor de Junio	5				
Flor de Junio Marcela	18	10			
Flor de Mayo	4				
Flor de Mayo Anita	12				
Flor de Mayo Bajío	1				
Flor de Mayo Dolores	2				
Negro 8025	3				
Negro Chapingo		9			
Negro Guanajuato		7			
Negro Huasteco 81					5
Negro Jamapa	7	46			6
Negro Papaloapan					3
Negro San Luis Criollo	1				
Pinto Durango	6				
PTB 08005	1				
Rosa de Castilla 62	6				
Zac. 524/8025/vax-4-2	9				
	77	105	97	33	26

cDNA synthesis and RT-PCR analysis

Total leaf RNA was extracted with TRIzol™, according to the manufacturer instructions. After quality verification with "GelRed" staining in a 1% non-denaturing agarose gel, RNA was quantified and stored at -80 °C until cDNA synthesis.

Total RNA was used as a template for RT-PCR (reverse transcription followed by polymerase chain reactions) using specific primers directed toward the coat protein cistron to obtain products of 890 bp and 740 bp for BCMV and BCMNV respectively as previously described by Flores-Estévez *et al.* (2003). For each sample 1 ng up to 5 μg total RNA was used plus 1 μL of 10 μM oligo (dT)$_{18}$ and 1 μL of dNTP Mix (10 mM each) in a volume of 12.5 μL with sterile, distilled water. Reaction tubes were incubated at 65 °C for 5 minutes and quickly chilled on ice. Then 4 μL of 5x First-Strand Buffer, 2 μL of 0.1 M DTT and 1 μL de RNaseOUT^{TM} (40 U/μL) were added to each tube. The tubes were shaken gently and incubated at 42 °C for 2 minutes. Finally 0.5 μL of SuperScript^{TM} RT (200 U/μL) were added per reaction tube and mixed with the aid of a micropipet. The tubes were incubated at 42 °C for 50 minutes and then heated at 70 °C for 15 minutes to inactivate the reverse transcriptase.

PCR amplification was performed using between 0.5 and 1 μL of cDNA in 20 μL reaction volume containing 2 μL of 10x PCR buffer without Mg^{2+}, 0.4 μL of 10 mM dNTP mixture, 0.6 μL of 50 mM $MgCl_2$, 1 μL of primer mix (10 μM of each nucleotide) and 0.2 μL of *Taq* DNA Polymerase (5 U/μL). The contents of the tubes were mixed with the aid of a pipet. cDNA synthesis of the coat protein cistron of BCMV and BCMNV was carried out with a initial denaturation at 94 °C for 1 minute followed by 30 cycles each of: 15 seconds of denaturation at 94 °C, 30 seconds of annealing at 63 °C and 40 seconds of elongation at 72 °C. A final elongation extension was done at 72 °C for 10 minutes in a thermal cycler (Applied Biosystems). The PCR fragments were verified in a 1 % non-denaturing agarose gel after "GelRed" staining. They were then cloned and sequenced by capillary electrophoresis in Cinvestav, Sede Irapuato, at LANGEBIO Unit. The sequence analysis and alignment were performed using the ClustalW option from the Geneious ^{TM} software package.

RESULTS

A total of 338 samples were collected at the federal states of Guanajuato, Nayarit, Sinaloa, Sonora and Veracruz (Table 1). The database from the collection is available at (www.frijol.inifap.gob.mx). RT-PCR reactions were carried out for all the samples and the presence or absence of the corresponding 740 and 890 bp bands was indicative of the presence of BCMNV and BCMV, respectively (Figure 2). Some of the bands (9% of them) were difficult to interpret since the product was not clearly visible, possibly due to poor RNA quality coming from a leaf tissue on suboptimal conditions and were therefore registered as not defined in this report, and left out for further hybridization analyses in a different study. Almost half of the samples were scored as negatives (48%). Less than half of all samples (34%) were either positive for BCMV (13%) or BCMNV (21%), or for both viral species (9%). There were more positive samples having BCMNV (21%) than BCMV (13%). Besides single infections, by any of BCMV or BCMNV, mixed infections were detected in 30 out of 144 positive samples equivalent to a 20%, this is a higher percentage than the 12% obtained by Flores-Estévez *et al.* (2003). Neither on that occasion, nor at this time, mixed infections were found in Veracruz and Guanajuato (Table 1 in reference 15) .In this study, mixed infections were found on Sinaloa and Nayarit (Table 2). Interestingly, as opposed to the 2003 study where mixed infections were found in light seed-colored cultivars, in this work, mixed infections were found in black seeded cultivars like Negro Jamapa from Nayarit, however samples of the same cultivar from Veracruz did not have BCMV. Sinaloa and Nayarit had the highest percentage of mixed infections. Sinaloa had the highest frequency of BCMV whereas Nayarit the highest of BCMNV (Table 2).

Table 2. Occurrence of BCMV and BCMNV by Federal State.

Federal States	BCMV positives	BCMNV positives	BCMV and BCMNV positives	Negatives	Not defined	Total
Guanajuato	7 (9%)	3 (4%)*	-	43 (56%)	24 (31%)	77
Nayarit	-	43 (41%)	12 (11%)	46 (44%)	4 (4%)	105
Sinaloa	34 (35%)	7 (7%)	18 (19%)	36 (37%)	2 (2%)	97
Sonora	3 (9%)	3 (9%)	-	27 (82%)	-	33
Veracruz	-**	14 (54%)	-	12 (46%)	-	26
Total	44 (13%)	70 (21%)	30 (9%)	164 (48%)	30 (9%)	338

*Percentage of incidence was expressed as the number of incidence divided by the total number of samples for a particular State.
**- no positives were found for the searched virus.

Table 3. BCMV and BCMNV in single or mixed infections (in different bean cultivars) in Sonora and Nayarit

Cultivar	Race	Frequency					Total samples
		BCMV positive	BCMNV positive	BCMV and BCMNV positive	Negatives	Not defined	
Azufrado Higuera 2	Nueva granada	18 (33%)*	3 (6%)	11 (20%)	22 (41%)	-	54
Negro Jamapa	Mesoamericana	-	23 (50%)	10 (22%)	13 (28%)	-	46
Azufrado Reg. Criollo	Mesoamericana	14 (78%)	-	-	4 (22%)	-	18
Azufrado Noroeste	Nueva Granada	1 (6%)	3 (18%)	7 (41%)	6 (35%)	-	17
Flor de Junio Marcela	Jalisco	-	2 (20%)	-	8 (80%)	-	10
Azufrado Criollo	Mesoamericana	-	4 (44%)	1 (12%)	4 (44%)	-	9
Bayo Berrendo	Mesoamericana	-	7 (76%)	1 (12%)	-	1 (12%)	9
Negro Chapingo	Mesoamericana	-	6 (67%)	-	3 (33%)	-	9
Aluyori	Nueva Granada	1 (12%)	1 (12%)	-	4 (50%)	2 (26%)	8
CIAT 103-25	Mesoamericana	-	-	-	8 (100%)	-	8
ELS 15-55	Mesoamericana	-	-	-	5 (71%)	2 (29%)	7
Negro Guanajuato	Mesoamericana	-	1 (14%)	-	5 (72%)	1 (14%)	7

*Percentage of incidence was expressed as the number of incidence divided by the total number of samples for a particular variety.
Reg. stands for Regional

Almost half of the collected samples did not contain any of the viral species reflecting the sampling of plants without any symptoms (healthy), as part of the survey plan; or plants with virus-like symptoms (mosaics, choloris or leaf deformations), probably due to toxemias caused by insects feeding from phloem sap or else, by fungi and bacteria infected plants. These pathogens may cause similar infection symptoms as those caused by BCMV and BCMNV making it difficult to differentiate in the field, which one is causing the infection. This is why incidence and severity are not reliable parameters in this study and were not an addressed issue here.

Thirty-four bean lines or cultivars were collected in the five sampled federal states mentioned before. Lines from the different states from which samples were collected are shown in Table 3. Guanajuato was the state where more cultivars were collected due to the presence of INIFAP experimental station where periodic evaluation of national germplasm normally takes place. Eight different cultivars were collected in Nayarit, six in Veracruz, and four in Sinaloa and Sonora. The highest sample number was from Negro Jamapa and Azufrado Higuera 2 cultivars, followed by Flor de Junio Marcela, Azufrado Higuera 1, Azufrado Regional Criollo, Azufrado Noroeste, etc. (Table 3).

As mixed infections were found in Sinaloa and Nayarit, a closer examination of the viral species present per cultivar was done for these two states. Five cultivars had both viruses present in mixed infections, these were: Azufrado Higuera 2 (11 out of 54); Negro Jamapa (10 out of 46); Azufrado Noroeste (7 out of 17); Azufrado Criollo and Bayo Berrendo (both with 1 out of 9). As single infections is concerned, nine cultivars were positive for BCMNV: Azufrado Higuera 2 (3 out of 54 samples); Negro Jamapa (23 out of 46); Azufrado Noroeste (3 out of 17); Flor de Junio Marcela (2 out of 10); Azufrado Criollo (4 out of 9); Bayo Berrendo (7 out of 9); Negro Chapingo (6 out of 9); Aluyori (1 out of 8) and Negro Guanajuato (1 out of 7). Lastly, four were positive for BCMV; Azufrado Higuera 2 (18 out of 54); Azufrado Regional Criollo (14 our of 18) Azufrado Noroeste (1 out of 17) and finally, Aluyori, with only one out of 8.

Figure 1. Two examples of the 338 collected samples from the different states under survey in this work (Guanajuato, Sonora, Sinaloa, Nayarit and Veracruz). One sample (left), shows typical symptoms of BCMV with dark green thick areas circumventing primary and secondary veins. BCMNV symptoms are seen as areas where the minor veins become necrotic resulting in an apparent net or cross-road appearance (right).

In short, in Nayarit and Sinaloa, where most of Azufrados and Negros come from, mixed infections were detected. BCMNV was found in all sampled states in single infections whereas BCMV was found in Guanajuato, Sinaloa and Sonora.

DISCUSSION

The presence of the species of the bean common mosaic virus was analyzed, namely BCMV and BCMNV, within 338 samples collected in the states of Guanajuato, Nayarit, Sinaloa, Sonora and Veracruz. Both species were detected in a targeted survey on symptomatic plants. This is why at least one of the viral species was detected in most of the 24 localities sampled. On a previous work, the presence of BCMV and BCMNV was monitored in Mexico (Flores-Estévez et al. (2003) in 2003. At that time, sixteen federal states were surveyed but no samples from Sonora, Sinaloa and Nayarit were included. In the present survey, in these three States, BCMNV was present but BCMV was absent in the states of Nayarit and Veracruz. The absence of BCMV in Veracruz was also observed in the survey of 2003. It is in this state that black seed samples from the

Mesoamericana race (Singh et al. (1991) were permissive for BCMNV. One salient difference with that first survey is that in this work, mixed infections in black bean cultivars such as Negro Jamapa were found, although only in Nayarit. Such an event was not recorded before; mixed infections were only detected in light seed-colored materials.

The question remains as to how this black seeded cultivar, Negro Jamapa, has acquired BCMV. It is possible that the necrotic species are more abundant in dry and humid tropical climates due to the climate per se and to alternate hosts within the year. Other possibility would be that in the breeding process, while developing this cultivar, the presence of different combinations of the bc type recessive resistance genes was allowed along with the I resistance gene providing resistance for BCMV but not to BCMNV in mixed infections. In fact, this cultivar is a multiline cultivar made up by the intermixing of eleven lines (Rosales-Serna et al. (2004). Interstate movement of this cultivar might have allowed the presence of the BCMV species on seeds where high viral pressure occurs (places where this species are prevalent).

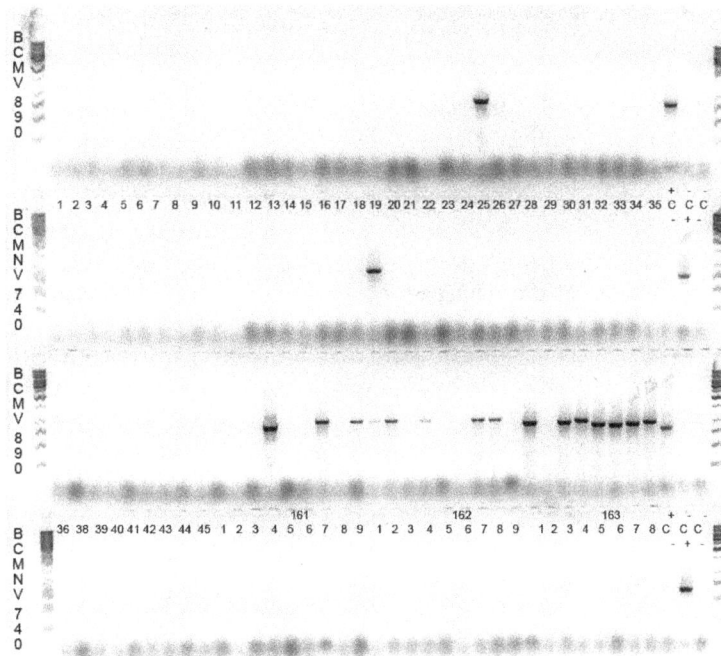

Figure 2. Representative photograph of a 1% agarose gels with the amplified RT-PCR products for the coat protein (CP) for either BCMV or BCMNV as shown on the left side of each gel indicating the expected band size for each viral species. Positive and negative controls as represented by + and – symbols respectively.

Another possibility is that the presence of the necrotic species facilitates the presence (replication) of the non-necrotic species, an equivalent simile to synergism. However, replication rates would need to be measured before the proper use of this term in this system (black seeded cultivar with mixed infections). The last possibility would be that the presence of the *I* gene in the cultivar Negro Jamapa can cause a permissive state for the presence of BCMV at the average temperatures in Nayarit providing that a pressure of this virus prevails in this state as compared to Veracruz.

However, more studies would be needed to understand the presence of both species in black seeded germplasm putatively having the BCMV-resistance *I* gene. Further studies are being conducted to characterize the pathogroups of BCMV and BCMNV present in Mexico and also to try to understand the permissiveness of both species in black seeded cultivars. Due to its geographical position, the Bajío region is a potential region with the high risk of occurrence of necrotic strains if there is an indiscriminate introduction of contaminated seed from Nayarit and Sinaloa, mostly during the winter-spring season. Also, at the Bajío region due to high temperatures during the winter-spring irrigated crop, temperature-dependent necrotic strains might have resulted with the hypersensive reaction in plants without the I gene.

The prevalence of BCMV found among the Azufrados, improved cultivars (Nueva Granada race) and landraces (Mesoamerican race) at Sinaloa, might indicate the presence of effective genes against BCMNV but less towards BCMV. Another possibility is that suboptimal temperatures during the winter crop cycle at this northern state is not favorable to the presence of BCMNV.

Unfortunately, since a high percentage of farmers buy certified seed at each planting season, the high prevalence of BCMV found in most samples of Azufrado cultivars in Sinaloa suggest that seed production system does not comply with all the requirements for the production of clean-disease-free seed and/or the re-use of contaminated grain as seed. This result also indicates that effective resistant genes against BCMV need to be incorporated by breeding into the popular cultivars grown at these two states.

CONCLUSIONS

The high presence of negative samples found in this research indicates that around half of the visual scores given on the symptoms of BCMV in the field, are wrong, therefore their presence must be defined by other means such as the RT-PCR technique used here. In Mexico, both BCMV and BCMNV are present in some of the main bean growing areas. At the pacific coastal areas at Sinaloa and Sonora BCMV is prevalent whereas at Nayarit and at the lowlands of Veracruz, BCMNV is more abundant.

A high prevalence of BCMNV was found at the state of Nayarit, mostly on black seeded cultivar Negro Jamapa and of BCMV at Sinaloa on Azufrado type cultivars.

Mixed infection with both BCMV and BCMNV were found in the black seeded cultivar Negro Jamapa grown at Nayarit during the fall-winter season.

Since both viral species BCMV and BCMNV, are seed transmitted, there is a high risk of epidemics at the states of Sinaloa and Nayarit, respectively, and into other states through the movement of seed across states.

ACKNOWLEDGMENTS

This work was carried out with financial support from Sectorial Project Funds SAGARPA-CONACYT 2009-C01-109621. Special thanks are given to José Luis Hernández and Alicia Rangel for RNA extractions and RT-PCR reactions.

REFERENCES

Berger, P.H., Wyatt, S.D., Shiel, P.J., Silbernagel, M.J., Druffel, K., Mink, G.I., 1997. Phylogenetic analysis of the Potyviridae with emphasis on legume-infecting potyviruses. Arch Virol 142: 1979-1999.

Chew, M.Y.I., Velásquez, V.R., Mena, C.J., Gaytán, M.A., 2010. Virus de frijol en la Comarca Lagunera y Zacatecas. Folleto Técnico Campo Experimental Zacatecas CIRNOC-INIFAP, Zacatecaz, p. 41.

Collmer, C.W., Marston, M.F., Albert, S.M., Bajaj, S., Maville, H.A., Ruuska, S.E., Vesely, E.J., Kyle, M.M., 1996. The nucleotide sequence of the coat protein gene and 3' untranslated region of azuki bean mosaic potyvirus, a member of the bean common mosaic virus subgroup. Mol Plant Microbe Interact 9: 758-761

Diaz-Plaza, R., Téliz, O.D., Muñoz-Orozco, A., 1992. Efecto de la enfermedades en frijol de temporal en la Mixteca Poblana. Revista Mexicana de Fitopatología 9: 21-30.

Drijfhout, E., 1978. Genetic interaction between Phaseolus vulgaris and bean common mosaic virus with implications for strain identification and breeding for resistance., Agricultural Research Reports 872. Centre for Agricultural Publishing and Documentation, Wageningen, p. 98.

Flores-Estévez, N., Acosta-Gallegos, J.A., Silva-Rosales, L., 2003. Bean common mosaic virus and Bean common mosaic necrotic virus in México. Plant Disease 87: 21-25.

Flores-Estévez, N., Silva-Rosales, L., 2000. First report of bean common mosaic necrotic potyvirus infecting bean plants in Aguascalientes and Veracruz. Plant Disease 84: 923.

Hall, R., 1991. Compendium of bean diseases. APS, p. 73.

Huguenot, C., Furneaux, M.T., Hamilton, R.I., 1994. Capsid protein properties of cowpea aphid-borne mosaic virus and blackeye cowpea mosaic virus confirm the existence of two major subgroups of aphid-transmitted, legume-infecting potyviruses. J Gen Virol 75 (Pt 12): 3555-3560

Jiménez-García, E., Nelson, M.R., 1994. Los virus del frijol en las áreas agrícolas de Sonora. Instituto Nacional de Investigaciones Forestales y Agropecuarias, Celaya, Mexico.

Kelly, J.D., 1997. A review of varietal response to bean common mosaic potyvirus in Phaseolus vulgaris. . Plant varieties and seeds 10: 1-6.

López-Salinas, E., Durán-Prado, A., Becerra-Leor, E.N., 1994. Manual de producción de frijol en el estado de Veracruz. Instituto Nacional de Investigaciones Forestales y Agropecuarias., Veracruz, Mexico.

McKern, N.M., 1992. . Isolates of bean common mosaic virus comprising two distinct potyviruses. Phytopathology 82: 923-929.

Montes-Rivera, R., Arévalo-Valenzuela, A., 1985. Guía para cultivar frijol de riego en el centro y sur de Guanajuato. Instituto Nacional de Investigaciones Forestales y Agropecuarias, Celaya, Mexico.

Morales, F.J., 1998, Present status of controling Bean Common Mosaic virus. APS Press: , St. Paul Minnesota.

Rosales-Serna, R., Acosta-Gallegos, J.A., Muruaga-Martínez, J.S., Hernández-Casillas, J.M., Esqueda-Esquivel, V.A., Pérez-Herrera, P., 2004. Las variedades mejoradas de fijol del Instituto Nacional de Investigaciones Forestales Agrícolas y Pecuarias. Libro Técnico 6: 148

Saiz, M., Dopazo, J., Castro, S., Romero, J., 1994. Evolutionary relationships among bean common mosaic virus strains and closely related potyviruses. Virus Res 31: 39-48.

SIAP, 2012. PRODUCCION AGRICOLA. Ciclo: Ciclicos y Perennes 2010. Modalidad: Riego + Temporal. Servicio de Información Agroalimentaria y Pesquera. SAGARPA.

Singh, P., Gepts, P., Debouck, D.G., 1991. Races of common bean (Phaseolus vulgaris, Fabaceae). . Econ. Bot. 45: 379-396.

Vetten, H.J., Lesemann, D.E., Maiss, E., 1992. Serotype A and B strains of bean common mosaic virus are two distinct potyviruses. Arch Virol Suppl 5: 415-431.

14

Occurrence in the Soil and Dispersal of *Lecanicillium lecanii*, a Fungal Pathogen of the Green Coffee Scale (*Coccus viridis*) and Coffee Rust (*Hemileia vastatrix*)

Doug Jackson[1*], Kate Zemenick[1] and Graciela Huerta[2]

[1]*University of Michigan, Department of Ecology and Evolutionary Biology, 2077 Kraus Natural Science, 830 North University, Ann Arbor, Michigan, U.S.A., 48109.*

[2]*El Colegio de La Frontera Sur, Departamento de Entomología Tropical, Carretera Antiguo Aeropuerto km 2.5, Tapachula, Chiapas, Mexico*

E-mail: dougjack@umich.edu
* Corresponding author

SUMMARY

The fungus *Lecanicillium lecanii* attacks the green scale (*Coccus viridis*), a pest of coffee, and is also a hyperparasite of coffee rust (*Hemileia vastatrix*). Knowledge of the epizootiology of this fungus is potentially important for conservation biological control in coffee agroecosystems. The presence of viable propagules of *L. lecanii* in the soil, a possible environmental reservoir, was assessed using two baiting methods: the standard *Galleria mellonella* bait method and a *C. viridis* bait method. Infectious propagules of *L. lecanii* were detected in soil samples taken from a 45 ha study plot, both nearby and far from recent epizootics of *L. lecanii*. To test the potential for the transmission of *L. lecanii* conidia from the soil via rain splash or wind, coffee seedlings with populations of *C. viridis* were placed near *L. lecanii*-inoculated soil and then subjected to artificial rain and wind treatments. Rain splash was shown to be a potential transmission mechanism. Dispersal of *L. lecanii* conidia by the ant *Azteca instabilis* was tested using field and laboratory ant-exclusion experiments. *Azteca instabilis* was shown to transport conidia of *L. lecanii*; however, dispersal by *A. instabilis* may not be important under field conditions.

Key words: *Lecanicillium lecanii*; *Azteca instabilis*; *Coccus viridis*; dispersal; fungal entomopathogen; epizootiology

INTRODUCTION

Conservation biological control, based on management practices that promote the survival and effectiveness of natural enemies of potential pest species, has attracted considerable attention for sustainable crop production (Barbosa, 1998; Gurr et al., 2000; Bale et al., 2008; Cullen et al., 2008; Fiedler et al., 2008; Jackson et al., 2009). Fungi are promising candidates for conservation biological control programs, as they are known to attack a variety of pest organisms (Butt et al., 2001), including arthropods (Shah and Pell, 2003; Cruz et al., 2006), plants (Hasan and Ayres, 1990; Te Beest et al., 1992; Charudattan and Dinoor, 2000; Sauerborn et al., 2007), and plant pathogens (Kiss, 2003; Fravel, 2005). However, effective conservation biological control using fungal pathogens requires a thorough knowledge of their ecology (Pell et al., 2010), which is still lacking, particularly in semi-natural habitats such as complex agroecosystems (Hesketh et al., 2010).

The fungal entomopathogen and mycoparasite Lecanicillium lecanii (Zimmerman) Zare and Gams is a promising candidate for use in conservation biological control in our study system – an organic, shade coffee agroecosystem in Chiapas, Mexico. Lecanicillium lecanii has been shown to be an important natural enemy of the green scale, Coccus viridis Green (Hemiptera: Coccidae) in coffee (Easwaramoorthy, 1978; Reddy and Bhat, 1989; Uno, 2007; Jackson et al., 2009). It also is known to attack the coffee rust, Hemileia vastatrix Berkeley and Broome (Shaw, 1988; Eskes, 1989; González et al., 1995; Vandermeer et al., 2009), and may suppress this potentially devastating coffee disease (McCook 2006, Suffert et al. 2009).

In addition to its direct, negative effects on potential coffee pests, L. lecanii may have an important influence on a keystone mutualism between an arboreal-nesting ant, Azteca instabilis F. Smith (Hymenoptera: Formicidae), and C. viridis. Azteca instabilis tends C. viridis in a typical ant-hemipteran mutualism, wherein the ants protect the scale insects, which are sedentary as adults, from predators and parasitoids. In exchange, the scales excrete a carbohydrate-rich honeydew that the ants consume. Recent studies have shown that this mutualism may play a key role in maintaining multiple natural pest control agents in this agroecosystem. Because the ants also inadvertently protect the larvae of the coccinellid scale predator Azya orbigera Mulsant (Coleoptera: Coccinellidae), the A. instabilis-C. viridis mutualism provides enemy-free space and high prey density for this important biological control agent (Liere and Perfecto, 2008). This mutualism also

contributes to the management of the coffee berry borer, Hypothenemus hampei Ferrari (Coleoptera: Scolytidae) through the deterrent effect of A. instabilis foragers (Perfecto and Vandermeer, 2006).

Lecanicillium lecanii may strongly influence the location and abundance of A. instabilis colonies, and hence may determine the extent of the aforementioned biological control effects of the ant-hemipteran mutualism. In this system, L. lecanii often becomes a local epizootic, killing nearly all of the C. viridis on a single coffee plant or a small group of neighboring plants. Therefore, L. lecanii reduces the amount of carbohydrate food available to an ant colony, which may have an indirect negative effect on colony survival. The potential for L. lecanii to cause the ant nest density-dependent mortality of A. instabilis colonies — one of the fundamental processes underlying the spatial self organization that generates the low-density, clustered spatial distribution of ant nests in this farm — has recently been demonstrated through a combination of field studies and computer modeling (Jackson et al., 2009).

Although a substantial amount of research has been done on the systematics (Zare et al., 2000; Gams and Zare, 2001; Sung et al., 2001; Zare and Gams, 2001; Zare et al., 2001; Kouvelis et al., 2008) and production (Feng et al., 2000; Kamp and Bidochka, 2002; Gao et al., 2007; Gao et al., 2009, Shi et al., 2009) of L. lecanii, much less is known about its basic ecology and natural history, including in the context of coffee agroecosystems.

In the current study, we investigated mechanisms contributing to the development of local epizootics of L. lecanii. Epizootics in this system are strongly influenced by the pronounced seasonality in this region, which is characterized by a wet season and a dry season. During the dry season, scale populations, and hence the prevalence of L. lecanii, are drastically reduced. Lecanicillium lecanii is re-established every wet season following the resurgence of the scale populations. Therefore, the initiation and progression of epizootics depend on one or more initial infection events following the onset of the wet season (primary dispersal) and the subsequent spread of infection from infected C. viridis individuals to susceptible individuals (secondary dispersal).

Three fundamental questions follow from the basic epizootiology of this system: 1) where do the propagules of L. lecanii persist during the dry season, 2) what are the mechanisms of primary dispersal, i.e., how are propagules initially dispersed onto the coffee plants and the scale insects during the wet season, and 3) what are the mechanisms of secondary dispersal within and between coffee plants following an initial infection? In this study, we investigate a subset of the mechanisms that may be operative in this system. We hypothesize that the soil provides an environmental reservoir for *L. lecanii*, and that propagules are transmitted from the soil to susceptible scale populations via rain splash or wind dispersal. We also explore the possibility that *A. instabilis* itself is primarily responsible for the dispersal of *L. lecanii* conidia within and between coffee plants.

MATERIALS AND METHODS

The study was performed in a 45 ha plot located at Finca Irlanda, an approximately 300 ha, organic coffee farm in the Soconusco region of Chiapas, Mexico (15° 11' N, 92° 20' W). The farm is a shade coffee plantation, with coffee plants growing beneath trees that have been planted in an approximately uniform distribution. The locations of every shade tree in the 45-hectare plot were obtained from biannual censuses; the locations of *A. instabilis* colonies, which nest in the shade trees, were also recorded during each census. All experiments were performed in the months of July and August, during the wet season (typically early May through November), which is within the peak season for the growth and spread of *L. lecanii* (unpublished data).

Soil sample baiting

Two independent soil sample baiting methods were performed to detect the presence of viable propagules of *L. lecanii* in soil samples. The first employed larvae of the wax moth *Galleria mellonella* L. (Lepidoptera: Pyralidae), which is a standard method for detecting entomopathogenic fungi in soil (Zimmermann, 1986). As an alternative method, we used populations of *C. viridis* on coffee leaves to detect the presence of *L. lecanii* propagules.

We obtained soil samples from a total of 40 locations: 10 locations far from *A. instabilis* nests, and therefore far from where epizootics of *L. lecanii* had occurred the previous year; 15 locations near the center of a previous epizootic, site A; and 15 locations near the center of another epizootic, site B (sites and locations indicated in Figures 1 and 2). The first 10 locations were chosen to determine the potential for *L. lecanii* propagules to persist in the soil even without a recent influx of propagules from a nearby epizootic. The other 30 locations were chosen to determine if the prevalence of propagules in the soil decreases with distance from the center of recent epizootics.

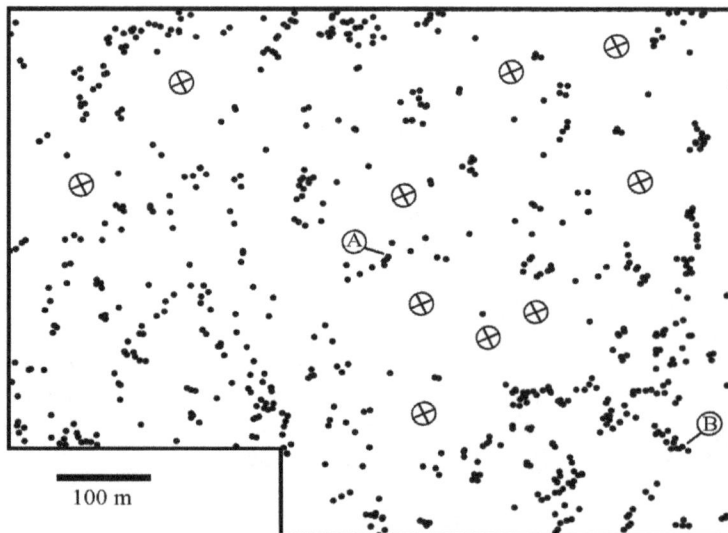

Figure 1. Location of *A. instabilis* ant nests (solid circles) in 45 ha plot: soil sample locations far from *A. instabilis* nests, and therefore far from recent epizootics of *L. lecanii* (circles with crosses); Site A; and Site B.

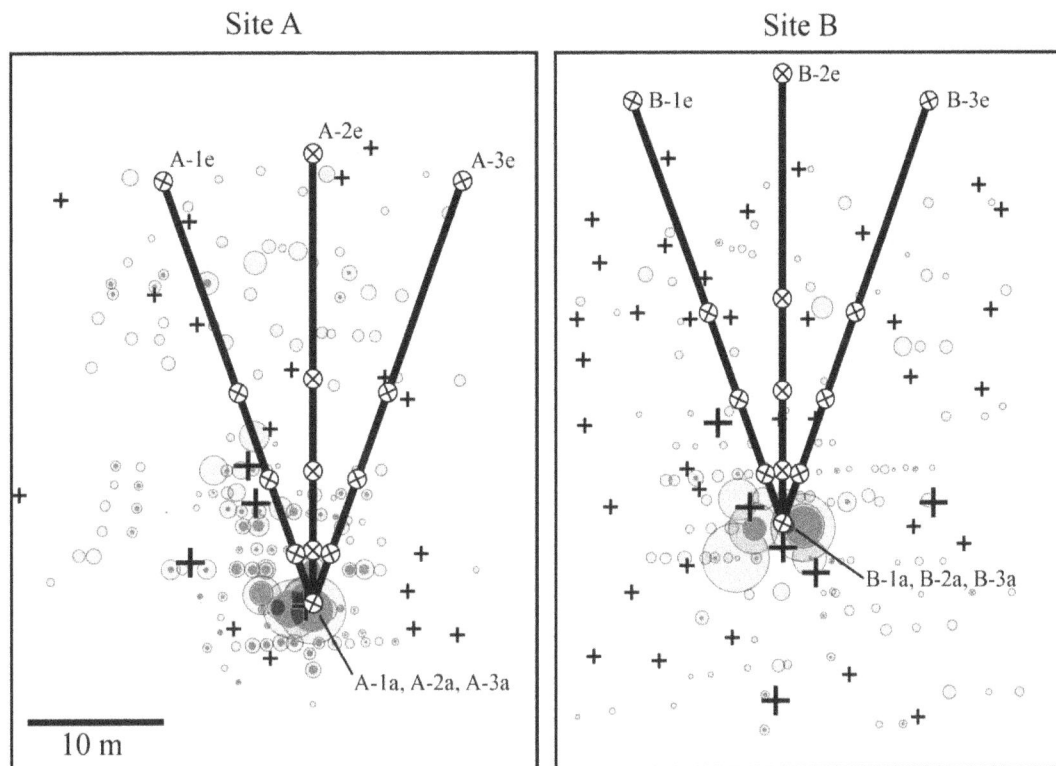

Figure 2. Locations of soil samples on transects leading away from foci of two *L. lecanii* epizootics. Small crosses indicate locations of shade trees. Large crosses indicate shade trees occupied by *A. instabilis* colonies. Light gray circles are proportional to the number of healthy *C. viridis* on individual coffee plants in the previous year, and dark gray circles are proportional to the number of *C. viridis* infected with *L. lecanii*. Circles with crosses show the locations of soil samples. Survey data are adapted from Jackson *et al.* (2009).

Soil samples were taken to a depth of 10 cm using a 2 cm-diameter, manual core sampler. The litter layer, when present, was included in the samples. At each location, 10 samples from a 40 cm X 80 cm rectangular area were taken. The core sampler was thoroughly cleaned and rinsed with 100% ethyl alcohol between samples. The 10 samples from each location were combined in separate polyethylene bags. After collection, the soil was spread on paper under aseptic conditions to dry for 24 hours at ambient temperature in the dark. The soil was then homogenized by rolling and passing through a sieve (Niblack and Hussey, 1987).

After the soil was allowed to dry, 90 cc (approximately 80 g) from each sample were placed in a plastic container and moistened evenly with 20 mL of distilled water. Laboratory-reared *G. mellonella* larvae were prepared by placing them in 56 °C water for 7 seconds in order to reduce their activity and discourage them from producing silk webbing in the soil. Each sample was baited with 10 larvae. The plastic containers were then sealed with

perforated lids and incubated at room temperature (26-28 °C) for 2 weeks. The larvae were inspected daily, and dead larvae were removed and placed in humidity chambers for later evaluation. In lieu of the usual step of inverting the containers to ensure that the larvae penetrate the soil evenly, the soil was thoroughly mixed during the daily inspection process. At the end of the incubation period, larvae exhibiting fungal growth were inspected using a stereomicroscope at 400x magnification to identify the fungi morphologically.

For the second soil sample baiting, we collected branches with uninfected *C. viridis* populations from three adjacent coffee plants located within the 45 ha plot; there were no scale insects with any visible signs of infection by *L. lecanii* on any of these three plants or the adjacent coffee plants. The average number of large (greater than approximately 0.7 mm in width) scales was 35.8 per leaf (s.d. = 14.3). We then divided the branches into sections of three leaves, selecting one section at random for each soil sample. Ten grams of soil from each sample were suspended in 10 mL of distilled water. The suspension was applied, using a small paintbrush, to inoculate the scale insects on a leaf. This procedure was immediately replicated for the other two leaves assigned to the soil sample, i.e., a separate suspension was prepared for each leaf. As a control, 10 groups of leaves with scale insects (30 leaves) were treated with distilled water. The leaves were placed in humidity chambers at 100% relative humidity and incubated for 2 weeks. Fungal infections were identified morphologically using a stereomicroscope (400x magnification).

Rain splash and wind dispersal

The potential for rain splash and wind dispersal of conidia from the soil was tested using coffee seedlings containing susceptible scale insect populations placed in four treatments: rain, rain-wind, wind, and control. The average number of scale insects per seedling was 112.6 (s.d. = 92.7). For this and all other experiments, only adult scales larger than approximately 0.7 mm in width were counted. The seedlings used in this and all other experiments were obtained from the farm's nursery, where they were planted and reared in 10 x 20 cm black polyethylene bags. Four seedlings were randomly assigned to each treatment, for a total of 16 seedlings. The seedlings were placed in the four corners of white 60 × 60 × 60 cm insect rearing tents (BugDorm-2, MegaView Science Co., Ltd., Taiwan). A plastic tray (26.5 × 17.5 × 6.0 cm) with soil that had been inoculated with an aqueous suspension of *L. lecanii* conidia was placed in the center of each group of four seedlings.

The inoculum was an aqueous suspension of *L. lecanii* conidia cultured from spores and hyphae acquired from an infected *C. viridis* obtained within the 45 ha plot. The *L. lecanii* isolate originated in a single *C. viridis* individual from a population affected by a severe epizootic, with nearly 100% prevalence of *L. lecanii*, and therefore was likely of average, or possibly above average (for our study site), pathogenicity to *C. viridis*. Following isolation of *L. lecanii* from the scale insect, conidia were mass-produced via solid-state fermentation using cooked rice as a substrate. We then suspended the resultant conidia in 5 L of 0.1% Tween 80 solution. Approximately 0.45 mL of suspension was added per cubic centimeter of soil at the start of the experiment. The conidial concentration, approximately 1.9×10^5 conidia/mL, was determined using a hemacytometer.

Seedlings in the rain and rain-wind treatments were removed from their tents once every 24 hours during the two week experiment to be exposed to artificial rain splash. During the rain treatment, the seedlings were placed around their respective plastic trays, with one seedling on each edge. Two minutes of simulated rain were created using a 2.5-gallon plastic bladder connected to a hose with a spray nozzle and filled with room-temperature tap water. Prior to the experiment, the volume and intensity of the simulated rain was compared and adjusted to qualitatively match rainfall typical of the study site, approximately 20 mm rain/day during the wet season (Richter, 2000). The simulated rain was focused on the center of the plastic tray such that the rain impinged primarily on the soil but also fell on the seedlings. After one minute, the plants were moved in a clockwise manner to an adjacent edge of the tray to account for the rectangular shape of the tray, i.e., so that each plant was exposed to equivalent rain splash intensity. The bottoms of the plastic trays were perforated to allow the water from the simulated rain to drain. To prevent any potential loss of conidia from the inoculated soil, we placed the rain-wind treatment tray underneath the rain treatment tray while the simulated rainfall was performed on the rain treatment, and vice versa. To balance the net washout of conidia, we alternated the order of the simulated rain treatment, i.e., every other day the same treatment was rained on first. The plants from all of the treatments were taken out of their cages and left outside while the simulated rain was being applied so that each plant spent the same amount of time outside of the tents. The seedlings were always returned to the same corners of the tents in order to avoid cross contamination between plants.

After all plants were returned to their tents, the wind and rain-wind treatments were exposed to simulated wind that was created by small electric fans (one fan per tent). The fans were run for 30 minutes at maximum speed, which is qualitatively similar to the typical maximum daily wind speed at the study site (3-4 on the Beaufort Scale). The orientation of each fan was changed daily by rotating the fan 90 degrees clockwise; this was done to vary the direction of the airflow impinging on the plants.

Seedlings were inspected daily for scale individuals exhibiting the white halo of mycelia characteristic of infection by *L. lecanii*. A final count of infected and healthy *C. viridis* adults was performed after two weeks, at the conclusion of the experiment.

Ant exclusion

Two ant exclusion experiments were performed: a laboratory experiment, in which most potential conidia dispersal mechanisms were eliminated, and a field experiment, which included the full complement of potential conidia dispersal pathways (e.g., wind, rain splash, arthropods, and other animals).

For the laboratory ant exclusion experiment, eighteen small coffee seedlings inhabited by populations of *C. viridis* were obtained from the farm's nursery. The *C. viridis* populations on six of the seedlings showed signs of being infected with *L. lecanii*, with some of the scales surrounded by the white halo of mycelia indicative of *L. lecanii* infection. The scales on the other 12 seedlings showed no signs of infection. The average number of scales on these 12 seedlings was 99.8 per plant (s.d. = 38.5). The six seedlings harboring infected scales were set aside as sources of fungal conidia, and the 12 infection-free seedlings were designated for use in the treatments.

For each replicate, three plastic flower pots were attached in a line to a wooden board, with approximately five cm separating the pots. An infected seedling was planted in the center pot and then covered with an enclosure of clear plastic in order to prevent transmission of fungal conidia by air currents or flying insects. The top of the plastic enclosure was rolled up and sealed with metal clips to allow for periodic access to the seedling. A small opening covered with mosquito netting was included on one side at the top of the enclosure as a vent to prevent condensation from accumulating inside. Two fungus-free seedlings were then planted in the two adjacent pots. These seedlings were also covered with plastic enclosures, with the vents on both of these enclosures facing in the opposite direction from the infected seedling's vent. To allow the passage of ants from the center seedling to the ant inclusion treatment seedling, an approximately 2.5 cm-diameter clear plastic tube penetrating the enclosures was routed between the two seedlings. An identical tube was routed between the center seedling and the ant exclusion treatment, with the exception that one end of the tube was covered with mosquito netting to prevent ants from entering the tube. Hot glue was used to thoroughly seal the enclosures to ensure that ants could not escape and that other arthropods could not enter the enclosures. Six identical replicates were constructed.

At the beginning of the experiment, a single coffee leaf with scales heavily infected by *L. lecanii* was tied to the base of each infected seedling in order to increase the amount of conidia available for the ants to spread. The coffee leaves were collected from the site of a severe epizootic, with nearly 100% prevalence of *L. lecanii*, and therefore it is probable that the pathogenicity of the strain(s) of *L. lecanii* used as inoculum were of at least average (for our study site) pathogenicity to *C. viridis*. Approximately 150 *A. instabilis* ants were then placed in the enclosures with the seedlings and leaves harboring infected scales. After three weeks, the scales on the

seedlings were counted and the number of scales showing signs of infection by *L. lecanii* was noted.

For the field ant exclusion experiment, twenty coffee seedlings inhabited by *C. viridis* populations, with a mean of 202.1 scales per plant (s.d. = 136.9), were placed in plastic pots and arranged in a circle around a shade tree containing an active *A. instabilis* colony. Since the purpose of the *A. instabilis* colony was simply to provide a source of ant foragers, all of the field ant exclusion replicates were located near a single, vigorous colony. The plants were placed two meters from the base of the shade tree, with 20 cm separating each pot. To encourage discovery of the seedlings by the ants, bridges of plastic twine were tied between the shade tree and the bases of the seedlings.

The seedlings were assigned in an alternating manner to either the ant exclusion treatment or the ant inclusion treatment, i.e., 10 seedlings were assigned to each treatment type. A piece of a coffee leaf covered with approximately 10 *C. viridis* that had been infected by *L. lecanii*, obtained from a site subject to a severe epizootic, was tied around the stem at the base of each coffee seedling to provide a source of conidia. An approximately eight centimeter wide strip of flagging tape was wrapped around the base of the seedlings, just beneath the infected coffee leaf; Tanglefoot® (Tanglefoot Co., Grand Rapids, Michigan, USA) was applied to the flagging tape on the ant exclusion seedlings. Surrounding vegetation was cleared to ensure that no bridges were available whereby the ants could access the seedlings from neighboring vegetation. All ants were removed from the ant exclusion seedlings by hand, using a small paintbrush, following the application of Tanglefoot®. The seedlings were left in the field from 15 July to 4 August. They were inspected daily to ensure that no ants had gained access to the ant exclusion seedlings. To encourage a more typical number of ants to discover and tend the scale insects on the ant inclusion seedlings, small pieces of tuna were placed at the bottom of all seedlings on 18 July. The leaves with fungus that had been tied to the base of the seedlings were beginning to show signs of decomposition by 27 July, so a single coffee berry with approximately five fungus-infected scales from the location of a major epizootic was attached with a wire-tie to the base of each seedling to provide a fresh source of inoculum. Following the experiment, prevalence of *L. lecanii* was assessed.

Statistical analyses were performed following the resampling, or bootstrapping with permutation, method described in Liere and Perfecto (2008). In this method, synthetic treatment and control populations are created by resampling without replacement from the original observations. The relevant statistical measure (e.g., the mean number of infections) is then calculated for these synthetic treatment and control populations. Next, the difference between the values of this metric for the two synthetic populations is compared to the difference between the values of this metric for the actual treatment and control populations. This procedure is repeated many times, and a *P*-value is calculated based on the proportion of repeats for which the difference between synthetic populations is as extreme or more extreme than the difference between the actual populations. The result is an estimate of the probability that the treatment and control populations could be as different as they are by chance alone. Data were resampled 10,000 times. The rain splash and wind dispersal data were resampled using a custom script in Matlab, while the ant exclusion data were analyzed using the Resampling Stats Excel add-in version 3.2 (Resampling, 2006).

RESULTS

Soil sample baiting

Of the 400 larvae used in the *G. mellonella* larvae baiting (10 larvae/sample X 40 samples), 202 were infected by one or more entomopathogenic fungi. Of these, six were infected with *L. lecanii,* based on morphological identification using the characteristic conidia and diagnostic phialides (Zare and Gams, 2001). Two of the *L. lecanii*-infected larvae were from samples taken from the points nearest to the focus of the *L. lecanii* epizootic at Site B (B-1a and B-2a); one was from a sample taken at one of the fourth-furthest transect points at Site A (A-2d); and the other three larvae were from samples taken far from *A. instabilis* nests (Table 1). In no case was there more than one larva per soil sample infected by *L. lecanii.*

The *C. viridis* baiting method yielded eight positive identifications of *L. lecanii* from the 40 soil samples, at the following locations: the fourth-furthest point at Site A (A-1d); all five distances at Site B (B-2a through B-2e); and two locations far removed from *A. instabilis* nests (Table 1). All of the positive samples from Site B were taken from the middle transect. All three replicates from the third-furthest point at Site B were positive, and two of the replicates from the fifth-furthest point were positive, meaning that a total of 11 of the 120 assays (3 replicates per sample X 40 soil samples) were positive. None of the scale insects on the control leaves were infected.

Of the 14 sampling locations that tested positive for the presence of *L. lecanii,* only one location – a point nearest to the center of the epizootic at Site A – tested

positive using both methods. That is, a total of 13 of the 40 sampling locations tested positive for *L. lecanii* using one or the other of the two methods.

Table 1. Locations of positive *G. mellonella* and *C. viridis* baiting results. Three locations near the centers of previous *L. lecanii* epizootics yielded positive results with the *G. mellonella* method; six were positive with the *C. viridis* method. Three positives were obtained from locations far from previous epizootics using *G. mellonella,* while the *C. viridis* method yielded two. See Figures 1 and 2 for location information.

Location	G. mellonella	C. viridis
A-1a		
A-1b		
A-1c		
A-1d		X
A-1e		
A-2a		
A-2b		
A-2c		
A-2d	X	
A-2e		
A-3a		
A-3b		
A-3c		
A-3d		
A-3e		
B-1a	X	
B-1b		
B-1c		
B-1d		
B-1e		
B-2a	X	X
B-2b		X
B-2c		X
B-2d		X
B-2e		X
B-3a		
B-3b		
B-3c		
B-3d		
B-3e		
Far from nests	3	2

Rain splash and wind dispersal

At the end of the experiment, three of the four rain treatment seedlings had scales infected by *L. lecanii;* one of the rain-wind treatment seedlings had infected scales; and none of the wind or control treatment seedlings had infected scales. The mean percentages of scales infected with *L. lecanii* were 0.0%, 3.2 ±

2.6% (SE), 0.0%, and 0.3 ± 0.3% for the control, rain, wind, and rain-wind treatments, respectively. The difference in the number of scales infected in the rain treatments compared to the control, wind, and rain-wind treatments was significantly greater than the random expectation ($P < 0.0001$, $P < 0.0001$, and $P < 0.01$, respectively). The difference in the number of scales infected in the rain-wind and the control treatments, however, was not greater than expected by chance ($P = 0.27$). There was no significant linear relationship between the number of scales per plant and the rate of infection ($P = 0.28$).

Ant exclusion

In the laboratory ant exclusion experiment, scales on five of the six ant inclusion seedlings exhibited the white mycelial mat characteristic of *L. lecanii* infection, while only one scale on the ant exclusion seedlings showed signs of possibly being infected. On the ant inclusion seedlings with *L. lecanii*-infected scales, the percentage of infected scales ranged from 1.8% to 12.5%. The mean percentage of scales killed by *L. lecanii* was significantly greater for the ant inclusion seedlings than for the ant exclusion seedlings (0.1 ± 0.2% [SE] without ants, 4.3 ± 1.8% with ants, $P < 0.01$).

In the field ant exclusion experiment, the percentage of infected scales on the ant exclusion seedlings ranged from 3.0% to 46.5%, while on the ant inclusion seedlings the range was 3.6% to 42.2%. The mean percentages of scales killed by *L. lecanii* with or without ants were not significantly different (17.4 ± 4.6% [SE] without ants, 18.2 ± 4.3% with ants, $P = 0.44$).

There was no significant linear relationship between the average number of scales per plant and the rate of infection in either the lab experiment ($P = 0.80$) or the field experiment ($P = 0.84$)

DISCUSSION

These results suggest the following scenario for the development of epizootics in this coffee agroecosystem. During the dry season, the populations of *C. viridis* are markedly smaller than during the wet season. Therefore, individual populations of scale insects are below the epizootic threshold density, and *L. lecanii* persists primarily in the environmental reservoir provided by the soil. As the scale populations increase following the onset of the wet season, they are exposed to *L. lecanii* propagules splashed up from the soil, which provide the inocula necessary to initiate epizootics. Further development of an epizootic almost certainly requires transmission of conidia between individuals in the scale population, which can be effected by *A.*

instabilis and other, as yet unknown, vectors. These processes lead to a rapid increase in the prevalence of *L. lecanii* shortly after the start of the wet season, which has been observed in our study site (unpublished data) and others (Reimer and Beardsley, 1992).

The baiting results demonstrate that viable propagules of *L. lecanii* can be found in locations that are as far removed as possible in this system (up to approximately 50 m) from recent *L. lecanii* epizootics. This suggests that either 1) *L. lecanii* can persist in the soil for multiple seasons or 2) *L. lecanii* is not dispersal limited in this system.

The fact that the soil can act as an environmental reservoir for *L. lecanii* in this system has important implications for the epizootiology of this fungus. The temporal dynamics of diseases have been shown to be strongly influenced by the presence of a pathogen reservoir: Hochberg (1989) showed that intermediate levels of translocation of a pathogen from a reservoir result in damped oscillations and relative stability of an otherwise oscillatory system. While the results of the rain splash experiment demonstrate that translocation of *L. lecanii* from the soil is possible, further study will be necessary to determine the actual level of translocation under field conditions. In particular, the concentration of *L. lecanii* in the soil in the field, and how this concentration varies spatially and temporally, are unknowns that could significantly affect the realized translocation rate.

The spatial dynamics of this system will also be strongly affected by the apparent ubiquity of infectious propagules in the soil. Transmission of *L. lecanii* upwards from infected soil widely distributed within the farm would likely result in much more rapid and widespread infection at the onset of the wet season compared to transmission from multiple point sources, e.g., from isolated cadavers left over from epizootics that occurred in the previous wet season. The potential for *C. viridis* to escape foci of previous epizootics by dispersing is also likely to be greatly reduced by the widespread occurrence of *L. lecanii* propagules in the soil.

The results of the two soil sample baitings are also interesting from a methodological perspective. In none of the samples were multiple replicates of the *G. mellonella* larvae infected by *L. lecanii*, which suggests that there is a large element of chance with this method, i.e., the presence of infectious material in a sample will not necessarily result in infection of the larvae. This may be due to the larvae failing to come into contact with the infectious material, possibly due to a very low density of infectious material in the sample; resistance of the larvae to infection; mortality due to other causes that occurs before the larvae can become infected; or *L. lecanii* being outcompeted within a single larva by another entomopathogenic fungus. Negative results of this method, therefore, should be treated with caution. Results from the *C. viridis* baiting were similarly subject to chance. However, the issue of other entomopathogenic fungi outcompeting *L. lecanii* was not a concern with this method, as *C. viridis* did not become infected by any fungi other than *L. lecanii*, perhaps because it is not susceptible to the broad range of entomopathogenic fungi that infected the *G. mellonella* larvae.

Another consideration raised by our study is that using a bait species known to be a target of the entomopathogen of interest may be a more powerful detection strategy than using a non-target bait species. Although there was not a significant difference in the total number of positive samples obtained using the two bait species employed in our study, Klingen *et al.* (2002) reported that using a pathogen-specific host species as a bait yielded significantly more positives than using *G. mellonella*. Therefore, when considering the apparent rarity of *L. lecanii* in our study system (15% and 20% positive samples with the *G. mellonella* and *C. viridis* methods, respectively) and other agroecosystems [e.g., 0.4-2.6% in a study by Meyling and Eilenberg (2007)], the potential influence of the sensitivity of the bait species should be kept in mind. An understanding of the role of the soil as an environmental reservoir for fungal entomopathogens in a given system would likely benefit from a combination of standard baiting methods (e.g., the *G. mellonella* bait method), baiting methods that are specifically tailored to the system (e.g., the *C. viridis* method used here), and molecular approaches (Enkerli and Widmer, 2010), including those that allow for quantitative assessments. A quantitative assessment of the abundance of *L. lecanii* propagules may reveal a dispersal kernel dependent on the distance from recent epizootics, which we were unable to detect using our experimental methods.

In the rain splash and wind dispersal experiment, the lower infection rate in the rain-wind treatment relative to the rain treatment suggests that there may be an important interaction between rain splash and wind in this agroecosystem. Wind increases the rate of evaporation of rain splash from the surface of the scale insects, and therefore may decrease infection rates due to desiccation of conidia. Airflow may also remove rain splash-dispersed conidia from the scale insects before they are able to germinate. This potential interplay between rain splash and wind may have important implications for management of shade levels in coffee agroecosystems. As shade level increases, the intensity of rain splash and wind will both decrease, which may serve to simultaneously decrease dispersal of conidia from the soil while increasing the probability of success of the conidia that are dispersed. Therefore, prevalence of *L. lecanii* may be maximized at an intermediate shade level. To our knowledge, although the effect of shade on prevalence following artificial inoculation has been studied (Easwaramoorthy and Jayaraj, 1977), the effects of shade level on the occurrence of natural epizootics of *L. lecanii* has not been investigated.

Rain splash dispersal of fungal entomopathogens has not been studied extensively, but has been previously noted by other researchers, including dispersal of *Beauveria bassiana* from the soil onto leaves of corn plants (Bruck and Lewis, 2002) and of the mealybug pathogen *Hirsutella cryptosclerotium* (Fernandez-Garcia and Fitt, 1993). Fitt *et al.* (1989) identify characteristics of fungi that tend to be rain splash dispersed, such as mucilaginous conidia; Heale (1988) notes that *Verticillium lecanii* conidia are produced in mucilaginous heads and dispersed by water splash or insects. There is also a substantial literature on rain splash dispersal of fungal pathogens of plants (for example, Madden, 1997; Geagea *et al.*, 2000; Ahimera *et al.*, 2004; Huber *et al.*, 2006).

The results from the laboratory ant exclusion experiment suggest that *A. instabilis* is capable of transporting conidia of *L. lecanii*, and hence may play a role in dispersing the fungus throughout populations of *C. viridis*. This would seem to indicate that transmission of conidia via ants between branches in a coffee plant, or perhaps between coffee plants themselves, is possible. However, the proportion of scale insects infected by the fungus was very low in the laboratory experiment relative to the field experiment, so the ants appear to be relatively poor dispersal agents. It is important to consider, however, that differences in pathogenicity of the inocula used in the two experiments could be partially responsible for the disparity in infection rates.

These results are consistent with a previous study that showed that the common black ant, *Lasius niger* (Hymenoptera: Formicidae), was capable of retaining conidia of an entomopathogenic fungus previously grouped in the *V. lecanii* species complex (Sitch and Jackson, 1997; Bird *et al.*, 2004) and that by transporting conidia to tended aphids, it can serve as a vector. Bird *et al.* (2004) demonstrated that *L. niger* workers artificially inoculated with *Lecanicillium longisporum* (Zimmerman) Zare and Gams [*Verticillium lecanii* (Zimmerman) Viégas] conidia could infect aphid populations, causing significant mortality under laboratory, semi-field, and field conditions. Aphid mortality due to *L. longisporum* was greatest under laboratory conditions and least under field conditions, which contrasts with the observations of this study. However, relative mortality under laboratory and field conditions depends heavily on the specific attributes of the methodologies and the laboratory and field environments (e.g., microclimate, presence of other potential vectors, etc.), so it is not possible to draw any general conclusions from this discrepancy.

The coffee seedlings used in the field ant exclusion experiment were most representative of smaller coffee plants and the lowest branches of larger plants. Based on the results from this study, other dispersal mechanisms besides *A. instabilis*-vectored dispersal from one scale insect to another dominate in these locations. There are a number of dispersal agents that could disperse *L. lecanii* conidia, such as rain splash from the soil or between *C. viridis* individuals, or any of the sundry flying and crawling arthropods that visit the coffee plants.

Roditakis *et al.* (2000) showed that aphids are capable of transporting conidia of *L. lecanii*, so it is likely that other arthropods in this system are also capable of spreading conidia of *L. lecanii*. Sitch and Jackson (1997) demonstrated that resistant arthropods from a variety of orders are capable of retaining *Verticillium lecanii* conidia, albeit at lower rates than target aphid species. A particularly intriguing possibility is that the predatory beetle *A. orbigera,* a key predator of scale insects in this system that is positively associated with the presence of the *A. instabilis-C. viridis* mutualism (Liere and Perfecto, 2008), may be a primary vector of *L. lecanii*. Such a phenomenon would not be unprecedented, as the coccinellid aphid predator *Coccinella septempunctata* (Coleoptera: Coccinellidae) has been shown to be a potential vector of an entomopathogenic fungus when artificially inoculated, causing significant aphid mortality due to fungal infection (Roy *et al.*, 2001). Whatever the dominant dispersal agents are, previous work showing a signal of dispersal-limited spread between coffee plants (Jackson *et al.*, 2009) suggests that these mechanisms are primarily transmitting the fungus between adjacent plants.

It is important to note that *A. instabilis* very likely plays a central role in the dynamics of *L. lecanii* infection of *C. viridis* even if it is not primarily responsible for dispersal of conidia. There appears to be a minimum abundance and density of *C. viridis* that are necessary for an outbreak of *L. lecanii* to occur, i.e., an epizootic threshold density (unpublished data). When such an outbreak occurs, the fungus kills the vast majority of scales on entire coffee plants. Without *A. instabilis* tending the scales and providing protection from predators and

parasitoids, the scale population is unlikely to reach a sufficient size for a fungal outbreak to occur (Reimer *et al.*, 1993; Uno, 2007). Therefore, *A. instabilis* is likely an important factor in determining the local prevalence of *L. lecanii*.

CONCLUSIONS

Our results suggest that a complete understanding of the epizootiology of *L. lecanii* will require knowledge of multiple phases of transmission and persistence: persistence in the soil, particularly during the dry season; translocation of propagules from the soil via rain splash; secondary dispersal between coffee plants, branches, and *C. viridis* individuals; and subsequent replenishment of the environmental reservoir in the soil. The spatial extent, phenology, and dynamics of epizootics in this system are all influenced by the details of these processes.

Understanding the development of *L. lecanii* epizootics in this system is crucial because of the role *L. lecanii* may play in the biological control of important coffee pests: directly, by attacking *C. viridis* and the coffee rust *H. vastatrix*, and indirectly, via its potential to influence the spatial distribution of the *A. instabilis-C. viridis* keystone mutualism. Consequently, enhanced understanding of the mechanisms controlling the occurrence of *L. lecanii* epizootics in this system, and appropriate management practices informed by this knowledge (e.g., coffee plant height and planting density, shade levels, etc.), appear to have an enormous potential benefit in terms of improved conservation biological control in this and other similar coffee agroecosystems.

ACKNOWLEDGEMENTS

We thank the Peters family for permission to work on their farm. Ricardo Alberto Toledo Hernández and Juan Cisneros Hernández at El Colegio de La Frontera Sur provided valuable assistance with the laboratory work. John Vandermeer and two anonymous reviewers provided helpful comments on earlier versions of this manuscript. We thank Heidi Liere for help with translation. This work was supported by the International Institute of the University of Michigan and NSF Grant DEB-0349388 to Ivette Perfecto and John Vandermeer

REFERENCES

Ahimera, N., Gisler, S., Morgan, D., Michailides, T. 2004. Effects of single-drop impactions and natural and simulated rains on the dispersal of *Botryosphaeria dothidea* conidia. Phytopathology 94:1189-1197.

Bale, J., Van Lenteren, J., Bigler, F. 2008. Biological control and sustainable food production. Philosophical Transactions B 363:761-776.

Barbosa, P. 1998. Conservation biological control. Academic Press, San Diego, California.

Bird, A.E., Hesketh, H., Cross, J.V., Copland, M. 2004. The common black ant, *Lasius niger* (Hymenoptera : Formicidae), as a vector of the entomopathogen *Lecanicillium longisporum* to rosy apple aphid, *Dysaphis plantaginea* (Homoptera : Aphididae). Biocontrol Science and Technology 14:757-767.

Bruck, D., Lewis, L. 2002. Rainfall and crop residue effects on soil dispersion and *Beauveria bassiana* spread to corn. Applied Soil Ecology 20:183-190.

Butt, T. M., Jackson, C., Magan, N. 2001. Fungi as biocontrol agents: progress, problems, and potential. CABI Publishing, Wallingford.

Charudattan, R., Dinoor, A. 2000. Biological control of weeds using plant pathogens: accomplishments and limitations. Crop Protection 19:691-695.

Cruz, L., Gaitan, A., Gongora, C. 2006. Exploiting the genetic diversity of *Beauveria bassiana* for improving the biological control of the coffee berry borer through the use of strain mixtures. Applied Microbiology and Biotechnology 71:918-926.

Cullen, R., Warner, K., Jonsson, M., Wratten, S. 2008. Economics and adoption of conservation biological control. Biological Control 45:272-280.

Easwaramoorthy, S. 1978. Effectiveness of the white halo fungus, *Cephalosporium lecanii*, against field populations of coffee green bug, *Coccus viridis*. Journal of Invertebrate Pathology 32:88-96.

Easwaramoorthy, S., Jayaraj, S. 1977. The effect of shade on the coffee green bug, *Coccus viridis* (Green) and its entomopathogenic fungus, *Cephalosporium lecanii* Zimm. Journal of Coffee Research 7:111-113.

Enkerli, J., Widmer, F. 2010. Molecular ecology of fungal entomopathogens: molecular genetic tools and their applications in population and

fate studies. Biocontrol 55:17-37.

Eskes, A. B. 1989. Natural enemies and biological control. In: Kushalappa, A.C., and Eskes, A.B. (eds.). Coffee Rust: Epidemiology, Resistance, and Management. CRC Press, Boca Raton, FL. pp. 162-168.

Feng, K., Liu, B., Tzeng, Y. 2000. *Verticillium lecanii* spore production in solid-state and liquid-state fermentations. Bioprocess and Biosystems Engineering 23:25-29.

Fernandez-Garcia, E., Fitt, B. 1993. Dispersal of the entomopathogen *Hirsutella cryptosclerotium* by simulated rain. Journal of Invertebrate Pathology 61:39-43.

Fiedler, A., Landis, D., Wratten, S. 2008. Maximizing ecosystem services from conservation biological control: the role of habitat management. Biological Control 45:254-271.

Fitt, B., McCartney, H., Walklate, P. 1989. The role of rain in dispersal of pathogen inoculum. Annual Review Phytopathology 27:241-270.

Fravel, D. 2005. Commercialization and implementation of biocontrol 1. Annual Review Phytopathology 43:337-359.

Gams, W., Zare, R. 2001. A revision of *Verticillium* sect. Prostrata. III. Generic classification. Nova Hedwigia 72:329-337.

Gao, L., Liu, X., Sun, M., Li, S., Wang, J. 2009. Use of a novel two-stage cultivation method to determine the effects of environmental factors on the growth and sporulation of several biocontrol fungi. Mycoscience 50:317-321.

Gao, L., Sun, M., Liu, X., Che, Y. 2007. Effects of carbon concentration and carbon to nitrogen ratio on the growth and sporulation of several biocontrol fungi. Mycological Research 111:87-92.

Geagea, L., Huber, L., Sache, I., Flura, D., McCartney, H., Fitt, B. 2000. Influence of simulated rain on dispersal of rust spores from infected wheat seedlings. Agricultural and Forest Meteorology 101:53-66.

González, E., Bravo, N., Carone, M. 1995. Caracterización de *Verticillium lecanii* (Zimm.) Viegas hiperparasitando *Hemileia vastatrix* Berk y Br y *Coccus viridis* Green. Revista de Protección Vegetal 10:169-171.

Gurr, G.M., Wratten, S.D., Barbosa, P. 2000. Success in conservation biological control of arthropods. In: Gurr, G.M. and Wratten, S.D. (eds.). Biological control: measures of success. Kluwer Academic Publishers, London, UK. pp. 105-132.

Hasan, S. and Ayres, P. 1990. The control of weeds through fungi: principles and prospects. New Phytologist 115:201-222.

Heale, J. B. 1988. The potential impact of fungal genetics and molecular biology on biological control, with particular reference to entomopathogens. In: M. N. Burge, (ed.). Fungi in biological control systems. Manchester University Press, Manchester, U.K. pp. 211-234.

Hesketh, H., Roy, H.E., Eilenberg, J., Pell, J.K., Hails, R.S. 2010. Challenges in modelling complexity of fungal entomopathogens in semi-natural populations of insects. Biocontrol 55:55-73.

Hochberg, M. 1989. The potential role of pathogens in biological control. Nature 337:262-265.

Huber, L., Madden, L., Fitt, B.D. 2006. Environmental biophysics applied to the dispersal of fungal spores by rain-splash. In: Cooke, B.M., Gareth Jones, D., and Kaye, B. (eds.). The Epidemiology of Plant Diseases. Springer, Netherlands. pp. 417-444.

Jackson, D., Vandermeer, J., Perfecto, I. 2009. Spatial and temporal dynamics of a fungal pathogen promote pattern formation in a tropical agroecosystem. The Open Ecology Journal 2:62-73.

Kamp, A., Bidochka, M. 2002. Conidium production by insect pathogenic fungi on commercially available agars. Letters in Applied Microbiology 35:74-77.

Kiss, L. 2003. A review of fungal antagonists of powdery mildews and their potential as biocontrol agents. Pest Management Science 59:475-483.

Klingen, I., Eilenberg, J., Meadow, R. 2002. Effects of farming system, field margins and bait insect on the occurrence of insect pathogenic fungi in soils. Agriculture, Ecosystems & Environment 91:191-198.

Kouvelis, V., Sialakouma, A., Typas, M. 2008. Mitochondrial gene sequences alone or combined with ITS region sequences provide firm molecular criteria for the classification of *Lecanicillium* species. Mycological Research 112:829-844.

Liere, H., Perfecto, I. 2008. Cheating on a mutualism: indirect benefits of ant attendance to a coccidophagous coccinellid. Environmental Entomology 37:143-149.

Madden, L. 1997. Effects of rain on splash dispersal of fungal pathogens. Canadian Journal of Plant Pathology 19:225-230.

McCook, S. 2006. Global rust belt: *Hemileia vastatrix* and the ecological integration of world coffee production since 1850. Journal of Global History 1:177-195.

Meyling, N., Eilenberg, J. 2007. Ecology of the entomopathogenic fungi *Beauveria bassiana* and *Metarhizium anisopliae* in temperate agroecosystems: Potential for conservation biological control. Biological Control 43:145-155.

Niblack, L.T., Hussey, S.R. 1987. Extracción de nematodes del suelo y de tejidos vegetales. In: Zuckerman, B.M., Mai, W.F., and Harrison, M.B. (eds.). Fitonematologia: manual de laboratorio. Centro Agronómico Tropical de Investigación y Enseñanza, Turrialba, Costa Rica. pp. 235-242.

Pell, J., Hannam, J., Steinkraus, D. 2010. Conservation biological control using fungal entomopathogens. Biocontrol 55:187-198.

Perfecto, I., Vandermeer, J. 2006. The effect of an ant-hemipteran mutualism on the coffee berry borer (*Hypothenemus hampei*) in southern Mexico. Agriculture Ecosystems & Environment 117:218-221.

Reddy, K.B., Bhat, P.K. 1989. Effect of relative humidity and temperature on the biotic agents of green scale *Coccus viridis* (Green). Journal of Coffee Research 19:82-87.

Reimer, N.J., Beardsley, J.W. 1992. Epizootic of white halo fungus, *Verticillium lecanii* (Zimmerman), and effectiveness of insecticides on *Coccus viridis* (Green) (Homoptera: Coccidae) on coffee at Kona, Hawaii. Proceedings, Hawaiian Entomological Society 31:73-81.

Reimer, N.J., Cope, M., Yasuda, G. 1993. Interference of *Pheidole megacephala* (Hymenoptera: Formicidae) with biological control of *Coccus viridis* (Homoptera: Coccidae) in coffee. Environmental Entomology 22:483-488.

Resampling. 2006. Resampling stats for Excel user's guide version 3. Resampling Stats.

Richter, M. 2000. The ecological crisis in Chiapas: a case study from Central America. BioOne 20:332-339.

Roditakis, E., Couzin, I., Balrow, K., Franks, N. 2000. Improving secondary pick up of insect fungal pathogen conidia by manipulating host behaviour. Annals of Applied Biology 137:329-335.

Roy, H.E., Pell, J.K., Alderson, P.G. 2001. Targeted dispersal of the aphid pathogenic fungus *Erynia neoaphidis* by the aphid predator *Coccinella septempunctata*. Biocontrol Science and Technology 11:99-110.

Sauerborn, J., Müller-Stöver, D., Hershenhorn, J. 2007. The role of biological control in managing parasitic weeds. Crop Protection 26:246-254.

Shah, P., Pell, J. 2003. Entomopathogenic fungi as biological control agents. Applied Microbiology and Biotechnology 61:413-423.

Shaw, D. E. 1988. *Verticillium lecanii* a hyperparasite on the coffee rust pathogen in Papua New Guinea. Australasian Plant Pathology 17:2-3.

Shi, Y., Xu, X., Zhu, Y. 2009. Optimization of *Verticillium lecanii* spore production in solid-state fermentation on sugarcane bagasse. Applied Microbiology and Biotechnology 82:921-927.

Sitch, J.C. and Jackson, C.W. 1997. Pre-penetration events affecting host specificity of *Verticillium lecanii*. Mycological Research 101:535-541.

Suffert, F., Latxague, É., Sache, I. 2009. Plant pathogens as agroterrorist weapons: assessment of the threat for European agriculture and forestry. Food Security 1:221-232.

Sung, G.H., Spatafora, J.W., Zare, R., Hodge, K.T., Gams, W. 2001. A revision of *Verticillium* sect. Prostrata. II. Phylogenetic analyses of SSU and LSU nuclear rDNA sequences from anamorphs and teleomorphs of the Clavicipitaceae. Nova Hedwigia 72:311-328.

Te Beest, D., Yang, X., Cisar, C. 1992. The status of biological control of weeds with fungal pathogens. Annual Review Phytopathology 30:637-657.

Uno, S. 2007. Effects of management intensification on coccids and parasitic hymenopterans in coffee agroecosystems in Mexico. University of Michigan, Ann Arbor, MI.

Vandermeer, J., Perfecto, I., Liere, H. 2009. Evidence for hyperparasitism of coffee rust (*Hemileia vastatrix*) by the entomogenous fungus, *Lecanicillium lecanii*, through a complex ecological web. Plant Pathology 58:636-641.

Zare, R., Gams, W. 2001. A revision of *Verticillium* section Prostrata. IV. The genera *Lecanicillium* and *Simplicillium* gen. nov. Nova Hedwigia 73:1-50.

Zare, R., Gams, W., Culham, A. 2000. A revision of *Verticillium* sect. Prostrata - I. Phylogenetic studies using ITS sequences. Nova Hedwigia 71:465-480.

Zare, R., Gams, W., Evans, H.C. 2001. A revision of *Verticillium* section Prostrata. V. The genus *Pochonia*, with notes on *Rotiferophthora*. Nova Hedwigia 73:51-86.

Zimmermann, G. 1986. The *Galleria* bait method for detection of entomopathogenic fungi in soil. Journal of Applied Entomology 102:213-215.

15

Postharvest Seed Treatments to Improve the Papaya Seed Germination and Seedlings Development

Guillermo M. Carrillo-Castañeda[1*], Francisco Bautista-Calles[1]
and Angel Villegas-Monter[2]

Recursos Genéticos y Productividad:[1]Genética and [2]Fruticultura.
Colegio de Postgraduados-Campus Montecillo. carrillo@colpos.mx
Km. 36.5 Carretera México-Texcoco. Montecillo, Texcoco, México 56230, México
[*]*Corresponding author*

SUMMARY

Practical technologies are required to preserve the viability of seeds particularly those known to be short-term viable species like *Carica papaya* (papaya). Papaya seeds were imbibed in water or chemical solutions (CaCl$_2$ 10^{-5} M, salicylic acid 10^{-4} M, and gibberellic acid 10^{-5} M) combined with inoculation of bacterial cell suspension to determine their effects on seed germination, plant growth, biomass production and chlorophyll accumulation. Seeds imbibed in water germinated 40 % more than control seeds and the time required to reaching 50 % seed germination was reduced two days in comparison to untreated seeds; however, the untreated seeds generated the largest (9.2 cm) and most vigorous seedlings. When seeds were imbibed in CaSG solution, a significant increase of the growth parameters such as fresh and dry biomass weight was observed. Seeds that were primed in gibberellic acid solution followed by inoculation with a mixture of *Azospirillum brasilense* cell suspension exhibited high seed germination (69 %), plant emergence (47 %) and seedling height (19 %), higher than the control. Differences in chlorophyll accumulation by seedlings were minimal.

Key words: *Carica papaya*; priming; inoculation; seed germination; development; chlorophyll.

INTRODUCTION

Plant development is a programmed process that starts from seed germination to maturity and fruiting. It is mainly modulated by a combination of dormancy, plant cell regulators (Richards *et al*., 2001, Olszewski *et al*., 2002, Peng and Harberd, 2002; Sun and Gubler, 2004; Smalle and Vierstra, 2004) and environment factors such as moist, temperature, oxygen, and light (Toh *et al*., 2008). Identification of triggers of seed germination and seedling growth promotion factors is crucial for the development of technologies to enhance stand establishment (Andrade-Rodriguez *et al*., 2008; De Mello *et al*., 2009; Venier *et al*., 2012). The germination inhibitors present in the papaya seed testa and sarcotesta control its germination (Chow and Lin, 1991; Paz and Vázquez, 1998) and to eliminate them, papaya growers have applied actions such as removing the sarcotesta from seeds, soaking and washing seeds in water (Mirafuentes, 1997) or sun drying (Jiménez 1996). Growth and development in plants is controlled by the selective removal of short-lived regulatory proteins. Degradation of repressor proteins by the ubiquitin-26S proteasome pathway is a central mechanism of gibberellic acid signal transduction in seed dormancy and germination (Smalle and Vierstra, 2004). ABA signaling, which is essential for seed development and seedling responses to the environment, is also mediated by protein degradation (Rodríguez-Gacio *et al.*, 2009). MicroRNAs (miRNAs) are involved in the repression of transcription factors at the mRNA level during seed germination and seedling growth (Nonogaki *et al.*, 2008). Emergence of the embryo from seed is repressed by surrounding tissues such as the endosperm and testa (Pinto *et al.*, 2007; Sung *et al.*, 2008) and this event provide the precise control of seed germination.

Seed deterioration is generally associated with loss of membrane integrity, biochemical changes, affections in important enzymatic activities, reduction in protein and nucleic acid synthesis, and lesions in DNA molecules (McDonald, 1999), result of adverse physical conditions of storage. The fast deterioration of *C. papaya* seeds prejudices its germination and emergence in the field, therefore growth regulators and enhancers commercially available have been routinely used to stimulate its germination (Bautista-Calles *et al.*, 2008).

Seed priming is a technique for improving both seed germination and vigor which involves the imbibition of seeds in water under controlled conditions to allow the initiation of early events of germination, followed by drying the seed back to its initial moisture condition (Jamieson, 2008; Varier *et al.*, 2010). Seeds can be imbibed in organic or inorganic solutions (chemopriming) (Nagao *et al.*, 1992; Parera and Cantliffe, 1995; Grzesik and Nowak, 1998) as well as inoculated with beneficial microorganisms (biopriming) during or after being primed (Warren and Bennett, 1997; Callan *et al.*, 1997). Microbial inoculants, which can promote plant growth and productivity, have internationally been accepted as an alternative source of N-fertilizers (Baset Mia *et al.*, 2010). The aim of this study was to evaluate the relevance of applying priming techniques to improve papaya seed germination, growth parameters and plant emergence.

MATERIALS AND METHODS

Three lots of certified papaya seed cv. Maradol with slight differences in germination percentage, were proportioned by Semillas del Caribe, S.A. de C.V. (Guadalajara city, Mexico). The bacterial species *Azospirillum brasilense* strains UAP154 and Sp59 provided by the Universidad Autónoma de Puebla. Calcium chloride ($CaCl_2$), salicylic acid (SA), gibberellic acid (AG_3), and the fungicide Captan® (50%) were commercially available.

Chemopriming procedure

Seeds were imbibed in a solution, such as CaS solution ($CaCl_2$ 10^{-3} M and SA 10^{-4} M), G solution (AG_3 10^{-3} M) or CaSG solution ($CaCl_2$ 10^{-3} M, SA 10^{-4} M and AG_3 10^{-3}M), following the procedure described by Bautista-Calles *et al.* (2008). All these solutions were prepared with distilled water and pH adjusted to 5.8 ± 0.1.

Biopriming procedure

Once the process of imbibition was concluded, dry seeds were mixed with a bacterial cell suspension (100 seeds were mixed with 1.25 mL of a cell suspension containing 10^{-9} cells mL^{-1}, approximately) and allowed to stand by 30 min at room temperature. Seeds were air dried during 30 min at 28 to 30 °C. Lots of 25 seeds were spread on a 22 x 24 cm folded paper towel, moistened with 7 mL distilled water. Paper towels were rolled up, placed in a plastic bag in vertical position and preserved at 28 to 30 °C during 10 days. At the end of this period, total germination was recorded (a seed was considered germinated when the radical protrution was approximately 1 mm). Germinated seeds were placed at one edge of a new paper towel moistened with 7 mL distilled water, rolled up and positioned in vertical position inside a plastic bag with the seeds on the top, to expose the seedlings to the light. Seedlings were allowed to develop during 14 days at 28 to 30 °C, and 16 h light. At the end of this period, stem and root length and weight of fresh and dried plantlets were determined.

Bacterial cell suspensions

Overnight cultures of *Azospirillum* were prepared in 3 mL of King's B medium (Vincent, 1970). Fresh medium (10 mL) was inoculated with 0.1 mL samples of the overnight culture and then incubated with agitation (150 oscillations min^{-1}) during 24 h at 28 to 30 °C. Cultures were harvested and their turbidity adjusted to 0.9 (660 nm, Colleman Junior II spectrophotometer) with sterile water. Samples of 5 mL each of UAP154 and Sp 59 cell suspensions were mixed to obtain the *Azospirillum* mix suspension.

Greenhouse experiment

The greenhouse located in Atoyac de Alvarez, Guerrero, Mexico (400 m, average annual temperature 29 °C) was built with translucent plastic to allow 50 % penetration of daylight. The floor was a mixture of fine gravel and sand in a layer. Black plastic bags (300 mL capacity with five perforations at the bottom) were filled with non-disinfected soil containing 33 % clay, 33 % sand and 33 % slime.

Bags were placed on benches 1 m above the ground. Seed was sowed 24 h after being treated. Five seeds were placed 1 cm depth in each pot. Irrigation was provided daily during the first week before and after sowing, and after the first week every 2 to 3 days, depending on climate conditions.

Chlorophyll determination

Samples of stem and leaf (0.5 g fresh tissue) from plants, after 15 days of emergence, were collected to determine, per triplicate, their chlorophyll content according to the procedure of Bruinsma (2009).

Biomass

Died biomass was determine in samples of 10 plantlets (after 15 days of emergence) per replication, after being dehydrated in an oven at 70 °C until constant weight.

Experimental design

The experimental design used in the combination of the hydro-priming, chemo-priming and bio-priming experiments, as well as treatments taken in the laboratory and in the field, was a completely randomized block design using three replications of 25 seeds per treatment. In the greenhouse, eight replications of 50 seeds each were used. All experiments were analyzed with the statistical package SAS (Statistical Analysis System) 2000 through the analysis of the variance (ANOVA) and average comparison of Tukey (P = 0.05).

RESULTS AND DISCUSSION

In a previous paper, we demonstrated a significant increase, greater than that of the untreated seed, for germination, speed of germination and seedling growth when the papaya seed was imbibed in water and in solutions of calcium, salicylic and gibberellic acids (Bautista-Calles et al., 2008). Results presented in this paper deal with the ability of germination and seedling growth parameters exhibited by papaya seed when it was previously exposed to a combined treatment: imbibition in gibberellic acid solution followed by inoculation with the *Azospirillum* mix suspension. Seeds exposed to this combined treatment expressed its highest capacity of germination (69.3 %) in comparison with the untreated seed (30.7 %) (Figure 1). The improvement of seed germination displayed by the treated seed is explained in part, by the presence of gibberellic acid, involved in the dormancy control of the seed, being in consequence an important germination promoter (Groot and Karssen, 1987). GA stimulates the production of the enzyme amylase by the aleurone layer, which breaks down starch into maltose, allowing it to diffuse into the embryo, where it is required to promote the growth of seedlings. The application of GA_3 to scarified seeds significantly promoted germination and decreased the number of days until germination (Nagano et al., 2010). In the papaya seed, the combined action of gibberellic acid and potassium nitrate has been found to be advantageous for improving both germination and emergence of seedlings (Nagao and Frutani, 1986; Frutani and Nagao, 1987). The primed seed, in addition, is brought to a stage where the metabolic processes are already initiated (protein synthesis from existing mRNA and DNA and repair of mitochondria and sub-cellular damage), giving it a starting point over the unprimed seed. Upon further imbibition, the primed seed can take off from where it has left completing the remaining steps of germination faster than the untreated seed (Brocklehurst and Dearman, 1983; Heydecker and Coolbear 1997; Derek, 1997; Varier et al., 2010). Treated seeds; however, gave rise to a less vigorous seedling as compared with those generated by the untreated seeds (Table 1).

Obtaining a significant increase in the capacity of germination of the papaya seed is an important achievement; however, vigor is another advantageous factor that may be present in any quality seed. The strategy followed to obtain improvements on vigor condition of the seedlings was to expose seeds to the mutual effect of $CaCl_2$, GA_3 and SA, and coupling this treatment with seed inoculation. Today is accepted the statement that Ca^{2+} is a central regulator of plant development and growth (Hepler, 2005). Ca^{2+} plays an important role in controlling membrane structure and function by binding to phospholipids and thus, stabilizing lipid bilayers and providing structural integrity to cellular membranes (Burstrom, 1968), which is particularly important in the germinating seed. Calcium modulates the activity of certain phosphatases and kinase enzymes that participate in the signal transduction during the germination process (Derek, 1997; Harper et al., 2004). Research during the last two decades has established that different stresses cause signal-specific changes in cellular Ca^{2+} level, which functions as a messenger in modulating diverse physiological processes that are important for stress adaptation (Kim et al., 2009; Redy et al., 2011). The process in which this ion participates is large and involves nearly all aspects of plant development (Harper et al., 2004; Bothwell and Ng, 2005). A growing body of evidence points to the importance of Ca and Calmodulin in the regulation of the transcriptional process during plant responses to endogenous and exogenous stimuli (Kim et al., 2009).

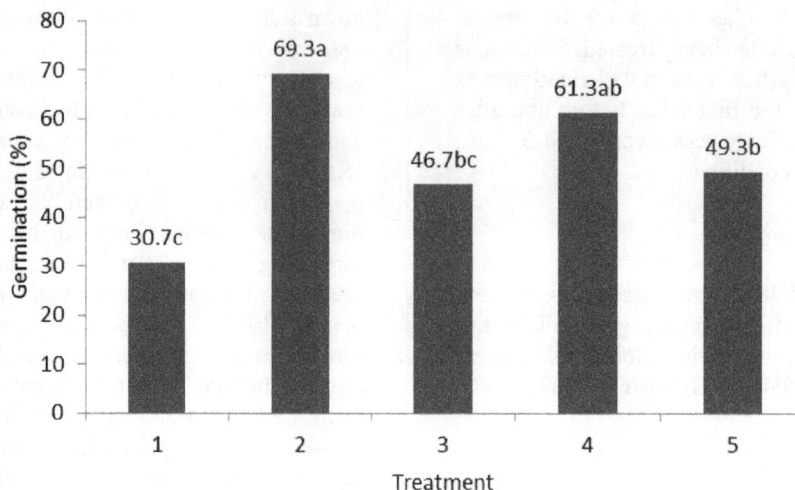

Figure 1. Germination. Papaya seeds were imbibed as indicated and inoculated with the *Azospirillum* mix suspension. 1) untreated; 2) imbibed in G solution and inoculated; 3) imbibed in CaS solution and inoculated; 4) imbibed in CaSG solution; 5) imbibed in CASG solution and inoculated. Different letters on a column represent different results (P ≤ 0.05).

Disease resistance in Arabidopsis is regulated by multiple signal transduction pathways in which salicylic acid function as key signaling molecule (Clarke *et al.*, 2000), acting in both local defense reactions at infection sites and the induction of systemic resistance; therefore, plants developed in the presence of salicylic acid may acquire the systemic resistance condition that might be beneficial to defend themselves, particularly at the stage of early development. Very little information exists on the establishment of defense mechanisms at the stage of seed germination (Rajjou *et al.*, 2006). In the root, salicylic acid acts increasing the content of certain enzymes, improving resistance of plant cells themselves, and in leaves accumulating some chloroplast proteins and enzymes capable of degrading the pathogen cell walls (Tarchevsky *et al.*, 2010).

Seeds that were imbibed in CASG solution germinated fast (T_{50} = 3 d) and generated seedlings with the highest dry biomass weight; however, when this treatment was coupled with inoculation, a reduction of growth parameters (seedling and root lengths as well as fresh and dry biomass production) and increase of one day in the T_{50} value was observed (Table 1). Thomas *et al.* (2007) demonstrated that papaya seeds inoculated with *Pantoea, Micobacterium,* or *Sphingomonas* spp., led to delayed germination or initial slow seedling growth; however, that slow seedling growth was overcome after 3 months and seedlings inoculated with *Pantoea, Microbacterium,* or *Sphingomonas* spp., displayed significantly better root and shoot growth.

Table 1. Seedlings 14 d of development. Seeds were imbibed as indicated and inoculated with a cell suspension of *Azospirillum*. 1) untreated; 2) imbibed in G solution and inoculated; 3) imbibed in CaS solution and inoculated; 4) imbibed in CaSG solution; 5) imbibed in CaSG solution and inoculated. Fresh and dry weight is an average of 10 seedlings

Treatment	Length (cm)			Fresh biomass weight (g)			Dry biomass weight (mg)			T_{50}
	Seedling	Stem	Root	Seedling	Stem	Root	Seedling	Stem	Root	(days)
1	9.2[a]	2.8[a]	6.3[a]	0.8[ab]	0.5[b]	0.3[a]	75.6[bc]	52.3[c]	23.3[a]	7
2	8.2[a]	2.2[a]	6.0[a]	0.6[bc]	0.4[b]	0.2[bc]	62.0[c]	46.0[c]	16.0[bc]	4
3	7.7[a]	2.4[a]	5.2[a]	0.8[ab]	0.5[ab]	0.3[ab]	84.0[ab]	65.3[ab]	18.6[ab]	3
4	7.6[a]	2.3[a]	5.3[a]	1.0[a]	0.6[a]	0.4[a]	94.6[a]	70.6[a]	24.0[a]	3
5	5.1[b]	2.1[a]	3.1[b]	0.6[c]	0.4[b]	0.2[bc]	67.6[bc]	55.6[bc]	12.0[c]	4
Significance	**	ns	**	**	**	**	**	**	**	
DMS	1.9	0.8	1.4	0.2	0.1	0.1	17.0	12.8	5.6	

Different letters in a column represent different results. ** Significance (P ≤ 0.05).

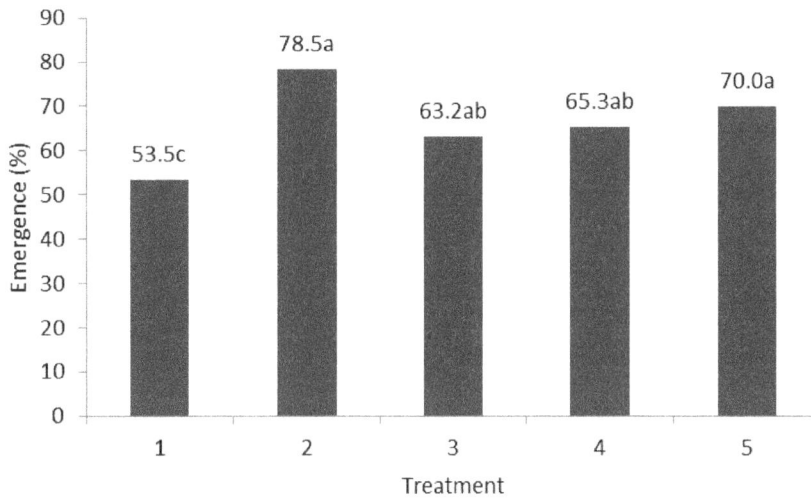

Figure 2. Seedling emergence in the greenhouse. Papayo seeds were imbibed as indicated and inoculated with a cell mixture of *Azospirillum*. 1) untreated; 2) imbibed in G solution and inoculated; 3) imbibed in CaS solution and inoculated; 4) imbibed in CaSG solution; 5) imbibed in CaSG solution and inoculated. Different letters on a column represent different results (P ≤ 0.05).

Seeds that were exposed to a combined treatment, imbibed in G solution and inoculated, exhibited the greatest emergence (Figure 2) seedling height (Figure 3) and accumulated the highest amount of chlorophyll as well (Figure 4). Gibberellic acid is essential for multiple processes of plant development, such as seed germination, stem elongation, and floral development (Richards *et al.*, 2001; Olszewski *et al.*, 2002; Peng and Harberd, 2002; Sun and Gubler, 2004; Cao *et al.*, 2006). Emergence of seeds that were imbibed in CASG solution and inoculated, were in second place followed by the seeds imbibed in CASG solution alone.

The presence of Gibberellic acid and the plant growth promoting microorganisms improved germination (Figure 1) and emergence (Figure 2). Some microorganisms are able to induce the systemic resistance in host plants to a broad variety of fungal, bacterial, and viral pathogens (Kaymak *et al.*, 2008; Aliye *et al.*, 2008; Jogaiah *et al.*, 2010); the presence of these plant growth promoting microorganisms seems to persist during plant lifetime, since they colonize active growth zones of roots and, for this reason, their beneficial effects will be effective until harvest.

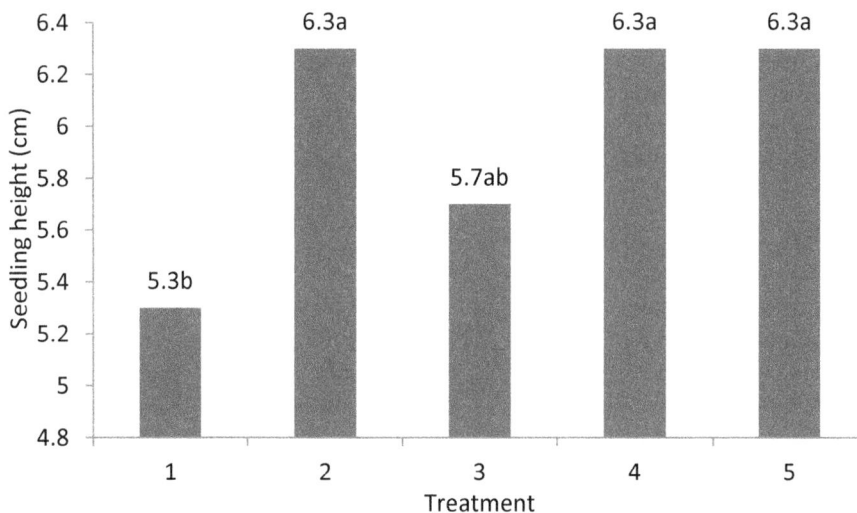

Figure 3. Seedling height. Papayo seeds were imbibed as indicated and inoculated with a cell mixture of *Azospirillum*. 1) untreated; 2) imbibed in G solution and inoculated; 3) imbibed in CaS solution and inoculated; 4) imbibed in CaSG solution; 5) imbibed in CaSG solution and inoculated. Different letters on a column represent different results (P ≤ 0.05).

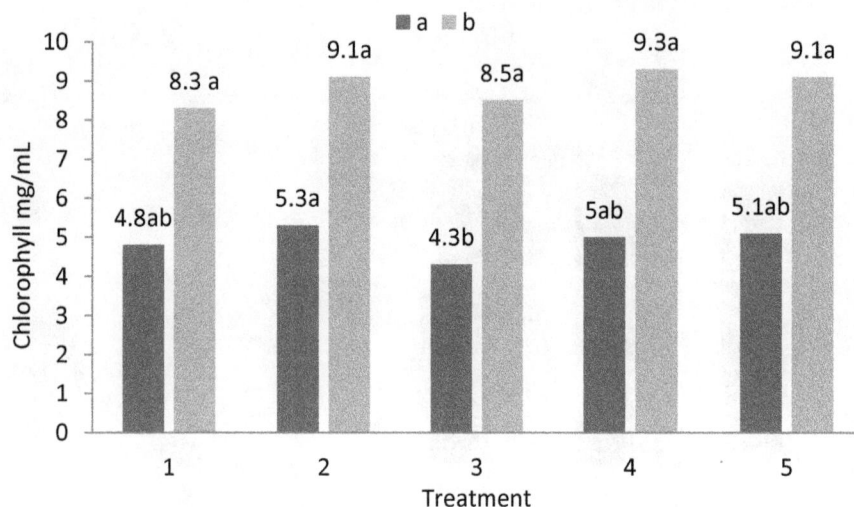

Figure 4. Chlorophyll determination after 15 days of emergence. Total (DMS = 1.71), chlorophyll a (DMS = 0.83) and chlorophyll b (DMS = 1.44). Seeds were imbibed as indicated and inoculated with a cell mixture of *Azospirillum*. 1) untreated, 2) imbibed in G solution and inoculated, 3) imbibed in CaS solution and inoculated, 4) imbibed in CaSG solution, 5) imbibed in CaSG solution and inoculated. Different letters on a column represent different results (P ≤ 0.05).

The use of microorganisms to improve crop development and yield is a technically simple process, though conceptually difficult to explain; however, results obtained when plants are inoculated with bacterial cells or mycorrhizal fungi are generous and this could become an everyday technique in the field. 'Maradol' papaya plants inoculated with arbuscular mycorrhizal fungi *Glomus mosseae* and *Entrophospora colombiana* exhibited increases in the number of fruits and yield by 41.9 and 105.2 % for *G. mosseae* and 22.1 and 44.1 % for *E. colombiana*, respectively, compared to control plants (Vázquez-Hernández *et al*., 2011). The gap between production potential of the crop and the current average production could be attributed to different factors, being diseases very significant. In consequence, seeds could be treated with chemical compounds and microorganisms to protect them and improve the development of papaya plants.

Seeds treated with Gibberellic acid solution followed by the inoculation with the cell mixture of *Azospirillum* exhibited the best response and therefore is the recommended procedure to achieve the best germination and emergence, and has great possibilities of being adopted by papaya growers.

CONCLUSIONS

Seeds that were exposed to a combined treatment, imbibed in G solution and inoculated a cell mixture of *Azospirillum* exhibited the maximum germination, the greatest emergence, seedling height, and accumulated

the highest chlorophyll amount as well. This treatment, which is economic and practical, can be attractive to enhance papaya seed germination capability.

REFERENCES

Aliye, N., Fininsa, C., Hiskias, Y. 2008. Evaluation of rhizosphere bacterial antagonists for their potential to bioprotect potato (*Solanum tuberosum*) against bacterial wilt (*Ralstonia solanacearum*). Biological Control. 47: 282-288.

Andrade-Rodriguez, M., Ayala-Hernandez, J. J., Alia-Tejacal, T., Rodriguez-Mendoza, H., Acosta-Duran, C. M., Lopez-Martinez, V. 2008. Effect of germination promoters and substrates in the development of papayo seedlings. Revista de la Facultad de Agronomia de la Universidad del Zulia. 25: 617-635.

Baset Mia, M. A., Shamsuddin, Z. H., Wahab, Z., Marziah, M. 2010. Effect of plant growth promoting rhizobacterial (PGPR) inoculation on growth and nitrogen incorporation of tissue-cultured *Musa* plantlets under nitrogen-free hydroponics condition. Australian Journal of Crop Science. 4: 85-90.

Bautista-Calles, F., Carrillo-Castañeda, G., Villegas-Monter, A. 2008. Recuperation of the high germinability of papaya seed through priming technology and biorregulators. Agrociencia. 42: 817-826.

Bothwell, J. H. F., Ng, C. K.-Y. 2005. The evolution of Ca^{2+} signalling in photosynthetic eukaryotes. New Phytologist. 166: 21-38.

Brocklehurst, P. A., Dearman, J. 1983. Interactions between seed priming treatments and nine seed lots of carrot, celery and onion. 1. Laboratory germination. Annals of Applied Biology. 12: 577-584.

Bruinsma, J. 2009. The quantitative analysis of chlorophylls a and b in plant extracts. Phytochemistry and Photobiology. 85:1-7.

Burstrom, H. G. 1968. Calcium and plant growth. Biological Review (Camb.) 43: 287-316.

Burton, M. G., Lauer, M. J., McDonald, M. B. 2000. Calcium effects on soybean seed production, elemental concentration, and seed quality. Crop Science. 40: 476-482.

Callan, N. W., Mathre, D. E., Miller, J. B., Vavrina, C. S. 1997. Biological seed treatments: factors involved in efficacy. Horticultural Science. 32: 179-183.

Cao, D. Cheng, H., Wu, W., Soo, H. M., Peng, J. 2006. Gibberellin mobilizes distinct DELLA-dependent transcriptomes to regulate seed germination and floral development in Arabidopsis. Plant Physiology. 142: 509-525.

Chow, Y. J., Lin, C. H. 1991. p-Hidroxy benzoic acid as the major phenolic germination inhibitor of papaya seed. Seed Science and Technology. 19: 167-174.

Clarke, J. D., Volko, S. M., Ledford, H., Ausubel, F. M., Dong, X. 2000. Roles of salicylic acid, jasmonic acid, and ethylene in cpr-induced resistance in Arabidopsis. Plant Cell. 12: 2175-2190.

De Mello, A. M., Streck, N. A., Blankenship, E. E., Paparozzi, E. T. 2009. Gibberellic acid promotes seed germination in penstemon digitalis cv. Husker red. HortScience. 44: 870-873.

Derek, J. B. 1997. Seed dormancy and germination. The Plant Cell. 9: 1055-1066.

Furutani, S. C., Nagao, M. A. 1987. Influence of temperature, KNO_3, GA_3 and seed drying on emergence of papaya seedlings. Scientia Horticulturae. 32: 67-72.

Groot, S. P. C., Karssen, C. M. 1987. Gibberellins regulate seed germination in tomato by endosperm weakening: a study with gibberellin-deficient mutants. Planta.171: 525-531.

Grzesik, M., Nowak, J. 1998. Effects of matriconditioning and hydropriming on Helichrysum bracteatum L. seed germination, seedling emergence and stress tolerance. Seed Science and Technology. 26: 363-376.

Harper, J. F., Breton, G., Harmon, A. 2004. Decoding Ca^{2+} signals through plant protein kinases. Annual Review of Plant Biology. 55: 263-288.

Hepler, P. K. 2005. Calcium: A central regulator of plant growth and development. The Plant Cell. 17: 2142-2155.

Heydecker, W., Coolbear, P. 1997. Seed treatments for improved performance–survey and attempted prognosis. Seed Science and Technology. 5: 353-424.

Jamieson, G. 2008. New perspectives on seed enhancement. Acta Horticulturae. 782: 143-150.

Jiménez, D. J. A. 1996. El Cultivo de la Papaya Hawaiana. Instituto para el Desarrollo de Sistemas de Producción del Trópico Húmedo de Tabasco. Serie Fruticultura Tropical. Gobierno del Estado de Tabasco, México. 111 p.

Jogaiah, S., Shivanna, R. K., Gnanaprakash, P. H., Hunthrike, S. S. 2010. Evaluation of plant growth-promoting rhizobacteria for their efficiency to promote growth and induce systemic resistance in pearl millet against downy mildew disease. Archives of Phytopathology and Plant Protection. 43: 368-378.

Kaymak, H. C., Yarali, F., Guvenc, I., Figen Donmez, M. 2008. The effect of inoculation with plant growth rhizobacteria (PGPR) on root formation of mint (Mentha piperita L.) cuttings. African Journal of Biotechnology. 7: 4479-4483.

Kim, M. C., Chung, W. S., Yun, D., Cho, M. J. 2009. Calcium and calmodulin-mediated regulation of gene expression in plants. Molecular Plant. 2: 13-21.

McDonald, M. B. 1999. Seed deterioration: Physiology, repair and assessment. Seed Science and Technology. 27: 177-237.

Mirafuentes, H. F. 1997. Manual para Producir Papaya en Tabasco. Instituto Nacional de Investigaciones Forestales y Agropecuarias. Centro de Investigación Regional del Golfo Centro. Campo Experimental Huimanguillo; Tabasco, México. 26 p.

Nagano, S., Mori, G., Oda, M. 2010. Promotion of seed germination in *Musa velutina* Wendl. & Drude by scarification and GA_3. Journal of Horticultural Science and Biotechnology. 85: 267-270.

Nagao, M. A., Furutani, S. C. 1986. Improving germination of papaya seed by density separation, potassium nitrate, and gibberellic acid. HortScience. 21: 1439-1440.

Nagao, M. A., Yoshimoto, J. M., Ho-A, E. B., Zee, F., Furutani, S. C. 1992. Assessment of KNO_3 preconditioning treatment on papaya seeds after extended storage. HortScience. 27: 490-700.

Nonogaki, H., Liu, P.-.P., Hewitt, J. R., Martin, R. C. 2008. Regulation of seed germination and stand establishment-importance of repression of developmental programs. Acta Horticulturae. 782: 51-58.

Olszewski, N., Sun, T. P., Gubler, F. 2002. Gibberellin signaling: biosynthesis, catabolism, and response pathways. Plant Cell (Suppl.). 14: S61-S80.

Parera, C. A., Cantliffe, D. J. 1995. Presowing seed priming. Horticultural Review. 16: 109-141.

Paz, L., Vázquez, Y. C. 1998. Comparative seed ecophysiology of wild and cultivated *Carica papaya* trees from a tropical rain forest region in Mexico. Tree Physiology. 18: 277-280.

Peng, J. R., Harberd, N. P. 2002. The role of GA-mediated signaling in the control of seed germination. Current Opinion on Plant Biology. 5: 376-381.

Pinto, L. V. A., Da Silva, E. A. A., Davide, A. C., Mendes De Jesus, V. A., Toorop, P. E., Hilhorst, H. W. M. 2007. Mechanism and controlof *Solanum lycopersicum* seed germination. Annals of Botany. 100: 1175-1158.

Rajjou, L., Belghazi, M., Huguet, R., Robin, C., Moreau, A., Job, C., Job, D. 2006. Proteomic investigation of the effect of salicylic acid on Arabidopsis seed germination and establishment of early defense mechanisms. Plant Physiology. 141: 910-923.

Reddy, S. N. A., Ali, S. G., Celesnik, H., Day, S. I. 2011. Coping with stresses: Roles of calcium- and calcium/calmodulin-regulated gene expression. Plant Cell. 23: 2010-2032.

Richards, D. E., King, K. E., Ait-Ali, T., Harberd, N. P. 2001. How gibberellins regulates plant growth and development: A molecular genetic analysis of gibberellin signaling. Annual Review of Plant Physiology and Plant Molecular Biology. 52: 67-88.

Rodríguez-Gacio, M. C., Matilla-Vázquez, M. A., Matilla, A. J. 2009. Seed dormancy and ABA signaling. The breakthrough goes on. Plant Signaling & Behavior. 4: 1035-1048.

Smalle J., Vierstra, R. D. 2004. The ubiquitin 26S proteasome proteolytic pathway. Annual Review of Plant Biology. 55: 555-590.

Sun, T.-P., Gubler, F. 2004. Molecular mechanism of gibberellin signaling in plants. Annual Review of Plant Biology. 55: 197-223.

Sung, Y., Cantliffe, D. J. R., Nagata, T., Nascimento, W. M. 2008. Structural changes in lettuce seed during germination at high temperature altered by genotype, seed maturation temperature, and seed priming. Journal of American Society of Horticultural Sciences. 133: 167-311.

Tarchevsky, I. A., Yakovleva, V. G., Egorova, A. M. 2010. Salicylate-induced modification of plant proteomes (review). Applied Biochemistry and Microbiology. 46: 241-252.

Thomas, P., Kumari, S., Swarna, G. K., Gowda, T. K. S. 2007. Papaya shoot tip associated endophytic bacteria isolated from *in vitro* cultures and host-endophyte interaction *in vitro* and *in vivo*. Canadian Journal of Microbiology. 53: 380-390.

Toh, S., Imamura, A., Watanabe, A., Nakabayashi, K., Okamoto, M., Jikumaru, Y., Hanada, A., Aso, Y., Ishiyama, K., Tamura, N., Iuchi, S., Kobayashi, M., Yamaguchi, S., Kamiya, Y., Nambara, E., Kawakami, N. 2008. High Temperature-induced abscisic acid biosynthesis and its role in the inhibition of gibberellin action in arabidopsis seeds. Plant Physiology. 146: 1368-1385.

Varier, A., Kuriakose, A. V., Dadlani, M. 2010. The subcellular basis of seed priming. Current Science. 99: 450-456.

Vázquez-Hernández, M. V., Arévalo-Galarza, L., Jaen-Contrerasa, D., Escamilla-García, J. L., Mora-Aguilera, A., Hernández-Castro, E., Cibrián-Tovara, J., Téliz-Ortiz, D. 2011. Effect of *Glomus mosseae* and *Entrophospora colombiana* on plant growth, production, and fruit quality of 'Maradol' papaya (*Carica papaya* L.). Scientia Horticulturae. 128: 255-260.

Venier, P., Funes, G., Garcia, C.C. 2012. Physical dormancy and histological features of seeds of five *Acacia* species (Fabaceae) from xerophytic forests in central Argentina. Flora. 207: 39-46

Vincent, J. M. 1970. A Manual for the Practical Study of Root Module Bacteria. International Biological Programme. Handbook No. 15. Blackwell Sci. Publ., Oxford.

Warren, J. E., Bennett, M. A. 1997. Bio-osmopriming tomato (*Lycopersicon esculentum* Mill.) seeds for improved stand establishment. Seed Science and Technology. 27: 488-499.

16

Weeds Suppression and Agronomic Characteristics of Maize Crop under Leguminous Crop Residues in No-Tillage System

Danielle Medina Rosa[1*], Lúcia Helena Pereira Nóbrega[1], Márcia Maria Mauli[1], Gislaine Piccolo de Lima[1] and Walter Boller[2]

[1] *Universidade Estadual do Oeste do Paraná (UNIOESTE), CCET – PGEAGRI, Caixa Postal, 711, (85819-110) Cascavel – PR – Brazil.*
Email: danimrosa@yahoo.com.br

[2] *FAMV/UPF/Passo Fundo – RS-Brazil.*
*Corresponding author

SUMMARY

This trial aimed at testing the leguminous *Mucuna deeringiana, Cajanus cajan and Stylosanthes capitata e macrocephala* before corn development and weeds incidence. The leguminous species were sown in October 2007 with a control treatment without legumes (fallow) in 4 x 5 m plots with five replications each, totaling 20 plots. At 90 days, plants were grazed and maize, 15 days after grazing, was planted on the wastes. The seedling emergence and plants growth were evaluated, besides the weeds incidence during culture development. At last, the experimental design was completely randomized and the means were compared by Scott-Knott at test 5 % of significance. The studied cover plants showed an efficient control over weeds and did not interfere in a negative way on the maize crop. Thus, it is an alternative to the integrated management of species concerning the green manure and crop rotation in no-tillage system for the Western region of Paraná.

Keywords: Green manure; *Mucuna deeringiana; Cajanus cajan; Stylosanthes capitata e macrocephala*; cover crop.

INTRODUCTION

Agriculture has undergone great changes and producers are always looking for monitoring trends and developments in technological and economic areas (Karam et al., 2008), with new and sustainable concepts as time passes by, in order to improve productivity and promote environmental performance. In this context, the no-tillage system associated with the practice of crop rotation and a capacity of producing wastes has been studied. Practices such as the addition of cover crops in this system provide benefits, which help in maintaining or improving the chemical, physical and biological soil conditions. As well as the control of weeds, diseases and pests, it is also an increase of nutrients to the next crop. The use of cover crops in soil has been a strategy to increase the agroecosystems sustainability, which also improves the economical crops, soil and environment. So, it comes as an economic, available and environmental sustainable alternative (Gamma-Rodrigués et al., 2007).

However, the interactions among cover crops and crops, especially when weed control is the main issue for both physical and allelopathic effects of cover crops, are poorly studied. This knowledge can assist in a suitable planning of crop rotation. Such practice may be important in controlling pests, diseases and weeds and as an alternative to manage soil fertility, due to its capacity to recycle nutrients from the topsoil and those which have percolated along horizons below it (Borkert et al., 2003).

The weed control is one of the main benefits of crop rotation, since some species occur and develop better in certain crops than others, due mainly to differences in cultural practices and vegetation cycles (Souza and Pires, 2002). Green manure is a renewable source of manuring as it makes the agricultural system sustainable. The leguminous plants make part of this system, mainly because they have better wastes, due to the fact that these plants fix nitrogen from the air and provide part of it to the next crop. Thus, the study of leguminous species is an alternative to reduce the intensive application of chemical fertilizers.

On the other hand leguminous and other species used as cover crop can release and add chemicals to the system. These substances, known as allelochemicals, can cause beneficial or detrimental effect on other species. This phenomenon is known as allelopathy (Rice, 1984) and is important to be observed when a cover crop is inserted, since there is a species-specific effect, which can inhibit both weeds and crop. The investigation of plants with allelopathic activity represents an alternative to the intensive application of pesticides on crops, as it reduces the environmental pollution.

The maize culture gives sustainability to the production systems, mainly through crop rotation (Conselho de Informações sobre Biotecnologia, 2007). As corn is a grain that asks for a high demand of nitrogen, there is a positive answer in its presence. It also has potential to be included in rotations with leguminous crops and can bring a positive impact on their productivity.

Hence, this paper aimed at analyzing leguminous cover crops as dwarf mucuna [Mucuna deeringiana (Bort.) Merr], stylosanthes (Stylosanthes capitata and macrocephala) and dwarf pigeonpea (Cajanus cajan L.), used as cover crops on emergence and development of maize (Zea mays) and weed species of broad and narrow leaves, especially morning-glory (Ipomoea spp.) arrowleaf sida (Sida rhombifolia L.) and beggarsticks (Bidens pilosa). It also aimed at generating and validating information that contributes to the agroecosystem sustainability and allows the incorporation of green manure in production units.

MATERIAL AND METHODS

The experiment was carried out in a farm, western Paraná, in the municipality of Braganey, whose geographic coordinates are 24°49'03" S latitude; 53°07'11" W longitude and 643 m of altitude, during 2007/2008 agricultural year. The soil is classified as Eutroferric Red Latosol, with average annual rainfall of 1,600 mm and average annual temperature of 20 °C. The studied area has been cropped under no-tillage system with soybean/wheat management for about ten years. Soil preparation occurred under minimum tillage (one chiseling + a leveling harrow), since soil was under compaction.

Campo Grande variety was used for stylosanthes seeds, given by Embrapa Beef Cattle, Campo Grande-MS. The seeds of dwarf mucuna and dwarf pigeonpea were acquired at Chopinzinho Seed Company in Chopinzinho city, Paraná. The leguminous were sown with approximately 20 seeds m^{-2} of dwarf mucuna, 50 seeds m^{-2} of dwarf pigeonpea and 70 seeds m^{-2} of stylosanthes. They were sown in an experimental area, in October, 2007, in plots of 4 x 5 m, with one meter between plots. The leguminous seeding was by throwing, incorporated into soil with a hoe, without fertilization.

There were four treatments, three leguminous and one control (fallow). The experimental design was completely randomized with five replications per treatment in the field. The products were sprayed with insecticide metamidafós (Tamaron ®) at a 600 mL ha^{-1} dose, on cover crops whenever necessary, especially to control Diabrotica speciosa, in a total of two pulverizations. In the early flowering stadium, the plants was cut down with a planter machine to cut down the remains, so that straw remained on the plot that corresponds to the treatment.

It was used the hybrid Pioneer 30R32 corn, traditional for cultivation during second harvest. Firstly, the seeds were submitted to germination tests in laboratory, with 92% of germination.

Sowing was with a rattle, on February, 03rd, 2008, on the leguminous remains, 15 days after cutting, in the chosen plots of a no-tillage system. The distance between rows was 80 cm, while six seeds per meter were used for density. Fertilization was 350 kg ha^{-1} urea in the seeding formula 10-20-20 (NPK) and 140 kg ha^{-1} in cover. The control received the same preparation as the other treatments, but its area remained under fallow (development of natural vegetation). The seedling emergence of corn was daily registered, starting at 5th day after sowing and going up to a constant number of seedlings, which occurred 20 days after sowing. According to these data, the emergence speed (ES), the emergence percentage (EM%) and the emergence speed index (ESI) were also determined.

The ES was calculated according to Edmond and Drapala (1958), they consider that the shortest treatment, in relation to seedling emergence, showed the highest emergence rate, therefore, a lower mean of ES:

$$ES = \frac{(N_1 E_1) + (N_2 E_2) + ... + (N_n E_n)}{E_1 + E_2 + ... + E_n}$$

where:
ES = emergence speed;
$E_1, E_2 ... E_N$ = number of normal seedlings, registered on the first count, the second count, (...) until the last one;
$N_1, N_2 ... N_N$ = number of days from sowing until the first, the second, (...) and the last count.

The results were expressed in number of days it takes the seedlings to emerge. The ESI was calculated according to Maguire (1962):

$$ESI = \frac{E_1}{N_1} + \frac{E_2}{N_2} + ... + \frac{E_n}{N_n}$$

where:

ESI = emergence speed index;
$E_1, E_2 ... E_N$ = number of normal seedlings registered on the first count, the second count, (...) until the last one;
$N_1, N_2 ... N_N$ = number of days from sowing until the first, the second, (...) and the last count.

The weed incidence was evaluated during maize development, at 30 and 60 days after sowing and at harvest. The survey of weed incidence consisted of four randomized samples per plot, with a randomized throwing of squares, using a 0.50 x 0.50 m metal frame, so that the internal area was 0.25 m^2.

The weeds found within the frame were counted and separated into broad and narrow leaves, plus the counting and specific identification of B. pilosa species as: morning-glory and arrowleaf sida. The identification was made according to Lorenzi (1994) and Kissmann (1997). Based on the averages of weeds per treatment, the percentage of reduction was calculated in relation to the control. The results were submitted to the analysis of variance and mean comparison was obtained by Scott-Knott test at 5% significance, according to SISVAR software (Ferreira, 2000).

RESULTS AND DISCUSSION

The evaluations of weed incidence during corn development were made for broad and narrow leaves, so that among the broadleaf species, Bidens pilosa, Ipomoea sp and Sida rhombifolia were identified.

The averages of weeds incidence are presented (Table 1) during corn development at 30 and 60 days after sowing only for broadleaf species and Bidens pilosa, since only they differed significantly.

It is observed that, at 30 days after corn sowing, the control showed the greatest number of weeds, for both broadleaf and Bidens pilosa (Table 1). It is also important to emphasize that the treatment with stylosanthes had the lowest number of broadleaf plants and Bidens pilosa, even though the treatment with the least amount of biomass had been left on soil. At 60 days after sowing, for broadleaf, the control has also shown a higher number of weeds when compared to the other treatments, although they have not differed among themselves. For Bidens pilosa, there was no significant difference among treatments, however, the control has shown the highest value of incidence.

In both evaluations, during corn development, the control had the greatest average number of weeds, which meant that the treatments with leguminous as M. deeringiana, C. cajan and S. capitata and macrocephala were efficient in controlling such species. Lorenzi (1984) stated that Mucuna aterrina has a strong and continuous inhibitory effect on Cyperus rotundus and B. pilosa and this action is possibly allelopathic. Weih et al. (2008) highlighted the possibility of using the allelopathic activity as an alternative to the application of chemical control for suppression of weeds in agroecosystems. Favero et al.

(2001) studied the changes in the population of spontaneous plants associated with leguminous cover crops, which are: *Canavalia ensiformis* DC, *Canavalia brasiliensis*, aterrina *Mucuna*, *Dolichos lablab* and *Cajanus cajan*. They concluded that *Canavalia brasiliensis*, followed by *Mucuna aterrina* and *Canavalia ensiformis* DC were the species with the highest biomass yield. *Mucuna aterrina* showed the greatest potential for soil coverage and weeds suppression.

In Table 2, there are significant results for the incidence of broadleaf plants and *Bidens pilosa* during corn harvest.

It is observed that for both broadleaf and *Bidens pilosa*, the control showed the highest number of weed species. These data of weed incidence indicate that the suppressive effect on weed community continues even after corn development. Calegari *et al.* (1992) stated that the important point is that cover crops let free of weeds not only at graze time, but that their wastes remain for longer, in order to prevent such infestation for the next crops. During the analysis of reduction percentage of weed incidence, in relation to the control by leguminous, it was observed that the greatest reduction was caused by *C. cajan*, followed by *M. deeringiana*, *S. capitata* and *macrocephala*, which showed very similar performances.

Table 1. Weeds incidence (plants m^{-2}) and reduction percentage of treatments in relation to control, during corn development in evaluations after 30 and 60 days of sowing. Braganey (PR), 2007/2008.

Evaluation	30 days			60 days		
Treatments	broadleaf	*B. pilosa*	reduction%	Broad-leaf	*B. pilosa*	reduction %
Control	265 a	185 a	0	83 a	11 a	0
M. deeringiana	80 b	40 b	69.8	33 b	4 a	60.2
Cajanus cajan	118 b	75 b	55.5	49 b	8 a	41.0
S.capitata and macrocephala	42 c	11 c	84.2	27 b	5 a	67.5
Coefficient of variance (%)	20.61	24.44	-	34.73	54.28	-
General Mean	126	78	-	48	7	-
F Values	17.77*	22.17*	-	3.15*	1.19*	-

Means followed by the same letter, in column, do not differ among themselves by Scott and Knott test at 5% probability. The presented data are obtained from the original observations, followed by letters obtained from the comparison of means with the transformation in $\sqrt{x+0,5}$.

Table 2. Weed incidence (plants m^{-2}) and reduction percentage of treatments in relation to the control at harvest of corn planted in crop residues of leguminous species. Braganey (PR), 2007/2008.

Treatments	broadleaf	*Bidens pilosa*	% reduction
control	316 a	255 a	0
M. deeringiana	114 b	77 b	63.9
Cajanus cajan	188 b	144 b	40.5
S.capitata e macrocephala	118 b	61 b	62.7
Coef. of Variation (%)	22.18	27.52	-
General Mean	184	134	-
F Values	6.40*	7.49*	-

Means followed by the same letter, in column, do not differ among themselves by Scott and Knott test at 5% probability. These data are results of original observations, followed by the letters obtained from the comparison of means with the transformation in $\sqrt{x+0,5}$.

In all analyses of weed incidence, during maize development, the control showed the highest average values for weeds in broadleaf and *Bidens pilosa* (Tables 1 and 2). This demonstrates that the studied leguminous species have suppressive effect on weed community and such effect may be due to chemical, physical or biological factors (Calegari, 1992; Favero, 2001). There are many studies that have been looked for weed control by cover crops (Fernandes *et al*, 1999, Mateus *et al*. 2004; Trezzi; Vidal, 2004; Tokura, Nobrega, 2005; Balbinot, Bialeski; Backes, 2005; Correia, Durigan; Klink, 2006; Piccolo, 2007), however, there are few records concerning the species studied in this work.

Tokura and Nóbrega (2006) evaluated the allelopathic potential of cover crops as *Triticum* spp, *Avena strigosa*, *Pennisetum glaucum*, *Raphanus sativus* and *Brassica napus* concerning weed development. The authors concluded that, from the studied species, *Brachiaria plantaginea* presented the greatest allelopathic potential and *Chenopodium ambrosioides* the lowest one.

In northern Greece, Dhima *et al*. (2009) carried out a trial regarding the mulch effects of some scented plants that were incorporated and used as green manure for the emergence and growth of weeds and maize. The studied species were: *Foeniculum vulgare, Pimpinela anisum, Ocimum basilicum, Coriander sativum, Coriandrum sativum, Mentha* L., among others. In the field, weed emergence was reduced from 11 to 83% when compared to the control. As in this experiment, corn emergence was not affected by any green manure (Table 3). Uchino *et al*. (2009) also concluded that weeds can be effectively controlled by planting cover crops, including their seed bank.

The values for emergence speed, final emergence percentage and emergence speed index of corn seedlings sowed on crop remains of *M. deeringiana, C. cajan* and *S. capitata macrocephala*, plus the control are shown in Table 3.

According to the analyses in Table 3, there was no statistic difference of treatments concerning: emergence speed (ES), emergence percentage (EM %) and emergence speed index (ESI) of corn seedlings that were under treatments with the studied leguminous species. However, when the ES values were analyzed, it is observed that the control presented the lowest value, i.e, the emergence speed of corn seedlings in the control was larger than in the other treatments. Therefore, even though seed corn of the control had emerged earlier, they showed the lowest emergence percentage (EM %) and consequently the smallest stand (number of plants m^{-2}) at the end of culture (Table 4), but they did not differ statistically among treatments.

According to Ramos *et al*. (2008), straw disposal on soil causes changes on chemical, physical and biological characteristics of soil environment and, depending on the species, the emergence and growth of plants can be affected. Among the changes caused by the waste deposition, Miyazawa *et al*. (2002) observed that for calcitic limestone mobility, applied on soil surface in columns of isolated PVC and with the addition of cover crops wastes, the association of its application with wastes have accelerated the transport of Ca and Mg in soil, among other effects.
According to Carvalho and Nakagawa (2000), the environment changes in which seeds are sown may cause problems or make easy their emergence, since this process depends on water and oxygen availability, in addition to room temperature.

Table 3. Mean values of emergence speed (ES), emergence percentage (EM %) and emergence speed index (ESI) of maize seedlings under treatments with leguminous cover crops and the control. Braganey (PR), 2007/2008.

Treatments	ES (days)	EM (%)	ESI (plants day^{-1})
Control	7.92 a	69 a	3.09 a
M. deeringiana	8.35 a	79 a	3.33 a
Cajanus cajan	8.30 a	72 a	2.99 a
S.capitata and macrocephala	8.34 a	72 a	3.01 a
Coef. of variation (%)	7.33	11.38	19.16
General Mean	8.22	72.95	3.10
F Values	0.59^{ns}	0.31^{ns}	0.34^{ns}

Means followed by the same letter, on the column, do not differ among themselves by Scott and Knott test at 5% probability. The data for emergency % are obtained from the original observations followed by the letters obtained from the comparison of means with the transformation in $\sqrt{x}$.

Table 4. Maize productivity and relative percentage among the treatments with leguminous wastes. Braganey (PR), 2007/2008.

Treatments	Productivity (kg ha^{-1})	Costs (R$)*	Stand (21 days)	relative Production %
Control	4777 a	1600	21 a	100
M. deeringiana	4956 a	1660	24 a	104
C. cajan	4831 a	1610	22 a	101
S.capitata e macrocephala	4418 a	1480	22 a	92
Coef. of variation (%)	13.47	-	22.63	-
General Mean	4746	-	22.36	-
F Values	0.65^{ns}	-	0.28^{ns}	-

Means followed by the same letter in column do not differ among themselves by Scott and Knott test at 5% probability.
* 60 kg bags

Rosa *et al.* (2009), at 60 days after maize sowing, observed that there was a significant difference among treatments with dwarf mucuna and dwarf pigeonpea, which showed the greatest heights in relation to the stylosanthes.

The averages of maize productivity, grown on wastes of leguminous as *M. deeringiana*, *C. cajan* and *S. capitata* and *macrocephala* and the control are shown in Table 4.

The maize productivity was not affected by submission to treatment, so, there was no significant difference. However, there is a numeric lower productivity when corn was grown after stylosanthes. This answer is confirmed in percentage when compared to the control, where *S. capitata* and *macrocephala* was the only treatment in which there was an 8% reduction in productivity in relation to the control (Table 4). Both *C. cajan* and *M. deeringiana* showed an increase in productivity of 1 and 4%, respectively, when compared to the control.

According to the IBGE, 18,745,355 tons of corn were harvested from second harvest in Brazil (IBGE, 2009). In the 2007/2008 second harvest, Paraná received 5,489.330 tons of production in an area of 1,512.078 hectares, which corresponds to almost 3,630 kg ha^{-1}, according to data from the Department of Agriculture and Supply of Paraná (SEAB, 2008). These data show that maize productivity, in this experiment, was above the average in Paraná (Table 4).

The 60 kg bag ranged from R$ 28.00 in December, 2007 to R$ 17.00 in December, 2008 (CONAB, 2009). In June, the price paid to producers by a 60 kg bag was around R$ 20.17 (SEAB, 2009). So, taking into account the production costs (Table 4), it was observed that treatment with *M. deeringiana* showed a gain of R$ 60.00, while *C. cajan* received R$ 10.00 per hectare when compared to the control.

On the other hand, the treatment with *S. capitata* and *macrocephala* decreased in R$ 120.00 per hectare. These differences must be considered by the producer at the moment he chooses a cover crop. In a similar study, Bertin *et al.* (2005) studied cover crops in pre-harvest to maize under no-tillage system. So, when it was planted just after *Crotalaria*, it showed higher production of grains. The *Crotalaria* is a leguminous, as well as the cover crops studied in this work.

In studies of crop rotation and succession with leguminous as *Crotalaria juncea*, the result of corn productivity was 8,362 kilograms ha^{-1}. While for the maize succession, it was 6,806 kilograms ha^{-1} (Penteado, 2007). This corroborates the benefits of using cover crops. Braz *et al.* (2006), in the Brazilian Midwestern region, studied about cover crops as *Brachiaria*; corn + *Brachiaria*; *Cajanus cajan*, *Pennisetum*, *Panicum maximum*, *Sorghum bicolor* and *Stylosanthes,* in wheat culture. They found out that the highest productivities of wheat were when it was grown after the leguminous, as pigeonpea and stylosanthes. The authors also highlighted that the use of leguminous has as advantage the right available nutrients to the next cultures, due to the fast decomposition of wastes.

Suzuki & Alves (2004) analyzed the grain productivity, influenced by soil tillage and cover crops as *atterrimum, Pennisetum americanum, Crotalaria juncea* and *Cajanus cajan* and observed that, when *Pennisetum americanum* was used as cover crop, the no-tillage system gave better response for maize grain yield when compared to the conventional tillage. The cover crops, according to each tillage system, did not show any difference for grain yield of maize.

Carvalho *et al.* (2004) studied soybean grown after green manures under no-tillage and conventional systems, in a soil from Midwest. Among the green manures, there were *Cajanus cajan* and *Mucuna*. The authors concluded that the green manure management during springtime has no effect on soybean productivity in succession, just as it was registered in the present trial, where there was no difference in maize productivity.

In studies that concern about nitrogen supply by leguminous in springtime for maize in succession in the systems of minimum and conventional tillage, Cereta *et al.* (1994) observed that all leguminous have been effective as a nitrogen source for maize, which also provided higher productivity than weeds up to 70%. Among the studied species, *Cajanus cajan* contributed, on average, with 10.7 kg ha^{-1} of nitrogen to corn. Heinrichs *et al.* (2005) found out a higher productivity of corn, 20% higher than the control, when it was grown after *Canavalia ensiformes* L.. But, corn yield was not influenced by the consortium with other green manures, including, dwarf mucuna, dwarf pigeonpea and *Crotalaria juncea* L. sp.

In general, some considerations were possible to be observed in the field about the leguminous plants. The stylosanthes had some difficult in their initial development, mainly due to wheat invasion, which acted out as weed species, providing further invasion by other species. When the cover was established, there were few weeds in the plots, so it is a potential green manure, because it had an efficient weed control, which asks for further researches. The dwarf mucuna and dwarf pigeonpea showed some adaptability in the region of this experiment, so they can be used as cover crops.

The use of green manure during the summer or in spring ensures that crop rotation can be carried out correctly and this guarantees the benefits of using green manure. So, the crop rotation is a fundamental action to guarantee some no-tillage system viability.

Studies regarding the use of crop rotation in Southern region must be stimulated to help the producer on planning at medium and long term, as well as to improve conditions and maintain the soil cover and thus a sustainable no-tillage system, with positive economic consequences for rural producers. It is noteworthy that the most consistent results are obtained through a continuous supply of organic matter to soil at medium and long term, in order to make possible the maintenance or recovery of its fertility.

CONCLUSION

Leguminous plants as green manure did not affect the development of cultivated species and have negative influence on weed community, offering an alternative to the integrated management of species in the green manuring practice and crop rotation under no-tillage system, based on the studied conditions.

ACKNOWLEDGMENTS

To UNIOESTE and CNPq for funding and the scholarships award.

REFERENCES

Balbinot Jr., A.A.; Bialeski, M.; Backes, R.L. 2005. Épocas de manejo de plantas de cobertura do solo de inverno e incidência de plantas daninhas na cultura do milho. Revista Agropecuária Catarinense, 18: 91-94.

Bertin, E. G.; Andrioli, I.; Centurion, J. F. 2005. Plantas de cobertura em pré-safra ao milho em plantio direto. Acta Scientiarum Agronomy, 27: 379-386.

Braz, A. J. A.; Silveira, P. M.; Kliemann, H. J.; Zimmermann, F. J. P. 2006. Adubação nitrogenada em cobertura na cultura do trigo em sistema de plantio direto após diferentes culturas. Ciência e agrotecnologia, 30: 193-198.

Borkert, C.M.; Audêncio, C.A.; Pereira, J. E.; Pereira, R.; Junior, A.O. 2003. Nutrientes minerais na biomassa da parte aérea em culturas de cobertura de solo. Pesquisa Agropecuária Brasileira, 38: 143-153.

Calegari, A.; Mondardo, A.; Bulisani, E.A.; Wildner, L.P.; Costa, M.B.B.C.; Alcantara, P.B.; Miyasaka, S.; Amado, T. J. C. 1992. Adubo verde no sul do Brasil. Rio de Janeiro: ASPTA, pp. 346.

Carvalho, N. M.; Nakagawa, J. 2000. Sementes: ciência, tecnologia e produção. Jaboticabal: FUNEP, pp. 650.

Carvalho, M. A. C.; Athayde, M. L. F.; Soratto, R. P.; Alves, M. C.; Arf, O. 2004. Soja em sucessão a adubos verdes no sistema de plantio direto e convencional em solo de Cerrado. Pesquisa Agropecuária Brasileira, Brasília, 39: 1141-1148.

Cereta, C.A.; Aita, A.; Braida, J. A.;Pavinato,R. L. Salet, R. L. 1994. Fornecimento de nitrogênio por leguminosas na primavera para o milho em sucessão por sistemas de cultivo mínimo e convencional. Revista Brasileira de Ciência do Solo, 18: 215-220.

Conselho De Informações Sobre Biotecnologia. 2007. Milho: A aplicação da biotecnologia na cultura. Boletim Informativo, pp 4.

CONAB. Companhia Nacional de Abastecimento. 2009. Indicadores da agropecuária. Ano XVIII, n. 1. Brasília: Conab, 64 p.

Correia, N. M.; Durigan, J. C.; Klink, U. P. 2006. Influência do tipo e da quantidade de resíduos vegetais na emergência de plantas daninhas. Planta daninha, 24: 245-253.

Dhima, K. V.; Vasilakoglou, I. B.; Gatsis, T. D.; Panou-Philotheou, E.; Eleftherohorinos, I. G. 2009. Effects of aromatic plants incorporated as green manure on weed and maize development. Field Crops Research, 110: 235–241.

Edmond, J. B.; Drapalla, W. J. 1958. The effects of temperature, sana and soil, and acetone on germination of okra seed. Proceedings of the American Society for Horticuticultural Science, Alexandria, 71: 428-443.

Favero C.; Jucksch, I.; Alvarenga, R. C.;.Costa, L. M. 2001. Modificações na população de plantas espontâneas na presença de adubos verdes. Pesquisa Agropecuária Brasileira, 36: 1355-1362.

Fernandes, M. F.; Barreto, A. C.; Emídio Filho, J. 1999. Fitomassa de adubos verdes e controle de plantas daninhas em diferentes densidades populacionais de leguminosas. Pesquisa Agropecuária Brasileira, 34: 1593-1600.

Ferreira, D. F. 2000. Análise estatística por meio do SISVAR (Sistema para Análise de Variância) para Windows versão 4.0. In: Reunião anual da região brasileira da sociedade internacional de biometria, 45, 2000, São Carlos. Anais. São Carlos: UFSCar, pp. 255-258.

Gama-Rodrigues, A. C.; Gama-Rodrigues, E. F.; Brito, E. C. 2007. Decomposição e liberação de nutrientes de resíduos culturais de plantas de cobertura em argissolo vermelho-amarelo na região noroeste fluminense (RJ). Revista Brasileira Ciência Solo, 31: 1421-1428.

Heinrichs, R.; Vitti, G. C.; Moreira, A.; Figueiredo, P. A. M.; Fancelli, A. L.; Corazza, E. J. 2005. Características químicas de solo e rendimento de fitomassa de adubos verdes e de grãos de milho,

decorrente do cultivo consorciado. Revista Brasileira de Ciência Solo, 29: 71-79.

IBGE. Instituto Brasileiro de Geografia e Estatística. 2009. Prognóstico da produção agrícola nacional. Disponível em: < http://www.ibge.gov.br/home/estatistica/indicadores/agropecuaria/lspa/lspa_200812_11.shtm>. Acesso em: 16 jan 2009.

Karam, D.; Mascarenhas, M. H. T; Silva, J. 2008. B. 2008. A ciência das plantas daninhas na sustentabilidade dos sistemas agrícolas: palestras apresentadas no XXVI Congresso Brasileiro da Ciência das Plantas Daninhas e XVIII Congreso de la Asociación Latinoamericano de Malezas. Ouro Preto, MG, 04 a 08 de maio de 2008. Sete Lagoas: SBCPD: Embrapa Milho e Sorgo.

Kissmann, K.G. 1997. Plantas infestantes e nocivas. 2° ed. São Paulo – Basf, pp. 644-649.

Lorenzi, H. 1984. Inibição alelopática de plantas daninhas. In: FUNDAÇÃO CARGILL (Campinas, SP). Adubação verde no Brasil. Campinas: Fundação Cargill, pp. 183-198.

Lorenzi, H. 1994. Manual de identificação e controle de plantas daninhas: plantio direto e convencional. 4 ed. Nova Odessa, Plantarum, pp. 299.

Maguire, J.D. 1962. Speed of germination-aid in selection and evaluation for seedling emergence and vigor. Crop Science, Madison, 2: 176-177.

Mateus, G. P.; Crusciol, C. A. C.; Negrisoli, E. 2004. Palhada do sorgo de guiné gigante no estabelecimento de plantas daninhas em área de plantio direto. Pesquisa Agropecuária Brasileira, 39: 539-542.

Miyazawa M.; Pavan, M. A.; Franchini, J. C. 2002. Evaluationof plant residueson the mobility of surface applied lime. Brazilian archives of biology and technology, 45: 251-256.

Penteado, S. R. 2007. Adubação verde e produção de biomassa: melhoria e recuperação dos solos. Livros Via Orgânica – Campinas, SP, pp. 164.

Piccolo De Lima, G. 2008. Manejo de coberturas vegetais na cultura da soja, 2008, 74f. Dissertação (Mestrado Engenharia Agrícola), Universidade Estadual do Oeste do Paraná.

Ramos, P. N.; Novo, M. C. S. S.; Lago, A. A.; Marin, G. C. 2008. Emergência de plântulas e crescimento inicial de cultivares de amendoim sob resíduos de cana-de-açúcar. Revista Brasileira de Sementes, 30: 190-197.

Rice, E. L. 1984. Allelopathy. 2 ed. New York: Academic press, pp. 422.

Rosa, D. M.; Nóbrega, L. H. P.; Piccolo de Lima, G.; Mauli, M. M.; Tonini, M.; Pacheco, F. P. 2009. Cultura do milho implantada sobre resíduos culturais de leguminosas de verão em sistema plantio direto. In: Di Leo, N.; Montico, S.; Nardón, G. Advances en ingeniería rural 2007-2009. Rosario, pp. 1211.

Seab. Secretaria da Agricultura e do Abastecimento do Paraná. 2008. Área e produção: principais culturas do Paraná. Disponível em :< http://www.seab.pr.gov.br/>. Acesso em: 22 dez 2008.

Seab. Secretaria da Agricultura e do Abastecimento do Paraná. 2009. Preços. Disponível em: <http://www.seab.pr.gov.br/modules/noticias/arti cle.php?storyid=4128#>. Acesso em 09 mar 2009.

Souza, C.M.; Pires, F. B. 2002. Adubação verde e rotação de culturas. Viçosa: UFV, pp. 72

Suzuki, L. E. A. S; Alves, M. C. 2004. Produtividade do milho (Zea mays L.) influenciada pelo preparo do solo e por plantas de cobertura em um Latossolo Vermelho. Acta Scientiarum Agronomy, 26: 61-65.

Tokura, L.K; Nóbrega, L. H. P. 2005. Potencial alelopático de cultivos de cobertura vegetal no desenvolvimento de plântulas de milho. Acta Scientiarum Agronomy, 27: 287-292.

Tokura, L. K.; Nóbrega, L. H. P. 2006. Alelopatia de cultivos de cobertura vegetal sobre plantas Infestantes. Acta Scientiarum Agronomy, 28: 379-384.

Trezzi, M. M.; Vidal, R. A. 2004. Potencial de utilização de cobertura vegetal de sorgo e milheto na supressão de plantas daninhas em condição de campo: II – Efeitos da cobertura morta. Planta Daninha, 22: 1-10.

Uchino, H.; Iwana, K.; Jitsuyama, T.; Yudate, S.; Nakamura, S. 2009. Yield losses of soybean and maize by competition with interseeded cover crops and weeds in organic-based cropping systems. Field Crop Research, 113: 342-351.

Weih A,, U.M.E. Didon A, A. C. Rönnberg-Wästljung B, C. Björkman. M. 2008. Integrated agricultural research and crop breeding: Allelopathic weed control in cereals and long-term productivity in perennial biomass crops. Agricultural Systems, 97: 99–107.

17

Growth Enhancement, Survival and Decrease of Ectoparasitic Infections in Masculinized Nile Tilapia Fry in a Recirculating Aquaculture System

Isabel Jiménez-García[1*], Carlos R. Rojas-García[2**], Carlos N. Castro-José[1], Salim Pavón-Suriano[1], Fabiola Lango-Reynoso[1] and Ma. del Refugio Castañeda-Chávez[1]

[1]*Instituto Tecnológico de Boca del Río. Km 12 Carretera Veracruz-Córdoba, CP. 94290, Boca del Río, Veracruz, Mexico. cncj_2448@hotmail.com, salimset@hotmail.com, fabiolalango@yahoo.com.mx, reyda64@yahoo.com.mx*
[2]*Universidad Católica de Temuco, Casilla 15D, Temuco, Chile. carlosrojas@uct.cl*
***Present address: CCMAR Universidade do Algarve, Campus de Gambelas, 8005-139 Faro – Portugal*
**Corresponding Author: isabel_jimenez@fastmail.fm*

SUMMARY

Under lab conditions, tilapia fry at culture densities of 8 fish/L^{-1} can reach a body weight of 0.5 to 1.0 g after the masculinization phase. In commercial hatcheries, the stocking density is four to six times higher, and consequently the occurrence of ectoparasitic infections also rises. The aim of this study was to examine the growth and survival of masculinized Nile tilapia (*Oreochromis niloticus*) fry in a recirculating aquaculture system (RAS). The fry, which were naturally parasitized by protozoan of the genera *Trichodina*, *Ambiphrya* and *Chilodonella*, weighed 0.013 ± 0.003 g and were reared in replicated tanks (N = 3) during 32 days at density of 18 fish/L^{-1} in the RAS to maintain good water quality, which was achieved especially during the first 22 days of fish rearing. The infection parameters and growth were monitored twice a week. The final fish weight was 1.17 ± 0.6 g and survival 99.5%. The most frequent parasites were *Trichodina* and *Gyrodactylus cichlidarum* (Monogenea). Although nitrogen compounds increased significantly over the last 10 days of fry rearing, final growth and survival were higher than those reported, additionally, the ectoparasitic infections were relatively low.

Key words: *Oreochromis niloticus*; *Gyrodactylus*; *Trichodina*; parasites; tilapia rearing.

INTRODUCTION

Stocking density, food, feeding regimes, water quality and photoperiod are all factors that affect the growth of tilapia fry, yet studies of these are not well documented or are controversial (El-Sayed, 2006). In general, it has been said that masculinized tilapia fry should have a mean weight of 0.4 g after 4 weeks of growth (Rakocy, 2005), or from 0.5 to 1 g (with 20% mortality) after 30 days (Rakocy, 1989).

Despite the potential risk of parasitic infections in pisciculture, information about fluctuations in infection parameters and their effect on the first phases of tilapia fry rearing is scarce, thus the level of risk that cultures are exposed to is unknown. Parasitic loads are assumed to be high during rearing due to the use of management practices that favor them, such as high stocking densities (> 40 fry L^{-1}), which not only increase social stress in fish (El-Sayed, 2006), but also inevitably reduce water quality (Ellis et al., 2002) and increase transmission rates of parasites with direct life cycles (Lafferty and Kuris, 1999), such as some ectoparasites. Within this group, protozoans and monogeneans can affect fish growth (Barker et al., 2001), and they frequently occur in cultures of tilapia fry (El-Sayed, 2006). In addition, such ectoparasites can be vectors of bacterial diseases that affect the health not only of the fish, but also of consumers (Xu et al., 2007), and they might reduce the efficiency of vaccines against Streptococcus iniae, a bacteria that causes heavy losses in tilapia cultures (Martins et al., 2011). The aim of this research was therefore to examine the growth, mortality and ectoparasitic infections of Nile tilapia (Oreochromis niloticus) fry during masculinization.

MATERIALS AND METHODS

The recirculating aquaculture system (RAS) was made of three 18.3 L plastic rectangular tanks, connected to a 60 L up-flow mechanical filter with 11.8 L of gravel (5 to 10 mm φ) as the filter medium. Effluent was treated in a trickling biofilter with 20 L of plastic 3-cm diameter biospheres as the attachment substrate for nitrifying bacteria. The water was recirculated through the RAS with a 1000 L/h^{-1} pump, and the culture tanks were continuously aerated. Between 5 and 10 L of water were replaced daily as a result of losses due to evaporation and siphoning of solid waste.

The experimental fish were fry at seven days post-hatching, with an initial weight of 0.013 ± 0.003 g and total length of 0.87 ± 0.07 cm, cultured in a commercial hatchery at high density in a flow-through system. The original batch of fish had been naturally parasitized by ciliated protozoa of the genera Trichodina, Ambiphrya and Chilodonella, and on day 4 of the study an additional species was recorded, the monogenean Gyrodactylus cichlidarum.

Nine hundred and ninety fry were placed in the three RAS culture tanks at density of 18 fish/L^{-1} and masculinization and the monitoring of parasitic infections began. For the first 28 days, the fish were fed at 20% of their biomass with meal containing 52%

protein and the 17α-methyltestosterone hormone from the manufacturer, and the remaining four days they were fed at 15% of their biomass with tilapia feed containing 32% protein. Food was given from 9:00 a.m. to 5:00 p.m. in nine rations throughout the day. The number of dead fish was recorded daily during feeding times.

A sample of 20 fry from the original batch were examined prior to placing them in the recirculation system to determine the initial parameters of parasitic infection (sampling 1, day 1): prevalence (percentage of fish from a sample positive for a particular parasite species) and mean intensity (mean number of organisms of a particular parasite species per host parasitized with that species in the sample) (Bush et al., 1997).

Parasitological evaluations and biometric measurements were performed twice a week. Eight fry were taken from each tank and euthanized with a transverse cut to the head. An optical microscope (4X and 10X) was then used to observe the skin, the fins and smears prepared with mucus from both sides of the body. The type and number of parasites were recorded. The weight was measured using an analytical balance (0.0001 g readability) and total length was measured with a millimeter ruler with the help of a stereo microscope.

Throughout the study, water temperature and oxygen levels were measured daily using a mercury thermometer and a Hanna HI 83203 photometer, respectively. Ammonia, nitrite and nitrate levels in the water were measured every three days with the same photometer. The pH level was recorded weekly with a Hanna pH211 pH meter.

To compare prevalence, X^2 tests were carried out using the Statistica 7 package (p < 0.05). An analysis of variance was conducted to evaluate the differences between the three tanks with respect to the weight and size of the fish and the number of parasites. As no significant differences were observed for these variables, the data from the tanks were pooled.

RESULTS AND DISCUSSION

Mean values for the physical and chemical parameters of the culture water are shown in Table 1. Oxygen, temperature and pH were within the optimum levels for rearing Nile tilapia (El-Sayed, 2006; Drummond et al., 2009), which stimulates the feeding response and reduces the susceptibility of the fish to disease (Plumb, 1999; Phelps and Popma, 2000). Total ammonia concentration was close to the recommended limit (0.1 mg/L^{-1}) for the first 22 of the 32 days of the culture (0.055 ± 0.058 mg/L^{-1}); however, it increased notably over the final 10 days, and as a result, nitrite and nitrate levels also increased (Table 1). This situation was apparently due to unexpected changes in the water volume or flow of the RAS associated with greater-than-expected water evaporation during the afternoon-night, such that by the next morning, the volume of water was considerably less. Once this situation was corrected by increasing the monitoring and by replenishing the water in the system, the ammonia level started to decrease. With respect to nitrite, it is a compound that can be highly toxic to fish, especially with increased alkalinity and water temperatures above 30 °C (Emerson *et al.*, 1975), conditions that were not present in this study. The tolerance of Nile tilapia to nitrite is also influenced by the size of the fish; fry of 4.4 g have been observed to be more tolerant to nitrite than large fish (90.7 g), and the LC_{50} for 96 h of exposure to nitrites is 81 and 8 mg/L^{-1} for small and large tilapia, respectively (Atwood *et al.*, 2001). Although the increase in ammonia and nitrites was significant over the last 10 days of this study, the levels were relatively low and short-lived, such that the fry never stopped eating and ultimately surpassed the weight and survival recorded in the literature for the masculinization period.

Table 1. Values (mean ± SD) of the physical and chemical parameters of the water during the masculinization of Nile tilapia (*Oreochromis niloticus*) fry (32 days). Values are also provided for two periods: days 1 to 22 and days 23 to 32. Brackets indicate minimum and maximum values observed.

	Days 1 to 32	Days 1 to 22	Days 23 to 32
Oxygen (mg/L)	6.20 ± 0.90	6.50 ± 0.62 (4.8-7.6)	5.53 ± 1.10 (4.0-7.20)
Temperature (°C)	28.7 ± 0.70	28.7 ± 1.0 (26-29.5)	28.8 ± 0.8 (28-30)
pH	7.25 ± 0.50	7.51 ± 0.20 (7.2-7.75)	6.80 ± 0.60 (6.3-7.40)
Ammonia (mg/L^{-1})	0.25 ± 0.33	0.055 ± 0.058 (0-0.15)	0.61 ± 0.34 (0.30-0.99)
Nitrite (mg/L^{-1})	2.54 ± 3.40	0.43 ± 0.41 (0.07-0.90)	5.10 ± 3.91 (1.6-3.90)
Nitrate (mg/L^{-1})	35.06 ± 38.0	9.85 ± 6.73 (0-19.50)	72.90 ± 36.10 (25.3-95.0)

The original batch of fish had three genera of ciliated protozoans from the outset, two of which were only found in the first week of the study, which were *Chilodonella* sp. at sampling 1 (day 1) (prevalence 27%, intensity 3 ± 2 organisms per parasitized fish) and *Ambiphrya* sp. at samplings 1 (day 1) and 2 (day 4) (prevalence 85 and 45%, respectively, intensity 2 ± 1 at both samplings). The most frequent parasites were *Trichodina* sp. and the monogenean *Gyrodactylus cichlidarum*, both of which showed a significant rise (p < 0.05) in prevalence starting on day 25 (Figures 1 and 2).

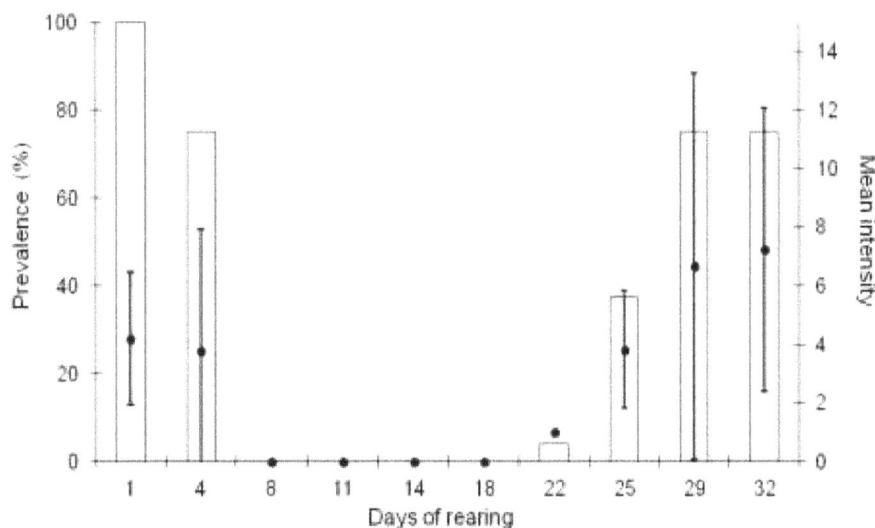

Figure 1. Infection parameters, prevalence (bars) and mean intensity ± SD (points) of *Trichodina* sp. during the masculinization of Nile tilapia (*Oreochromis niloticus*) fry.

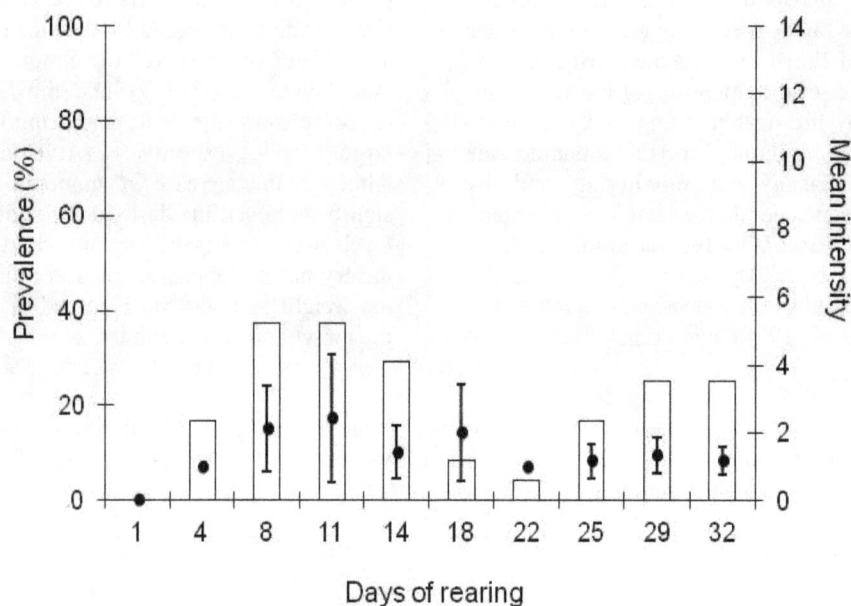

Figure 2. Infection parameters, prevalence (bars) and mean intensity ± SD (points), of *Gyrodactylus cichlidarum* during the masculinization of Nile tilapia (*Oreochromis niloticus*) fry.

The increase in the total number of parasites (*Trichodina* + monogeneans) since day 25 was associated with the increased concentration of ammonia recorded beginning on day 25 of the culture (Figure 3). Infestations of *Trichodina* and monogeneans have been observed to be indicators of deteriorating water quality (e.g., over-population of fish, high ammonia or nitrites, organic pollution or low oxygen) (El-Azez, 1999; Noga, 2000).

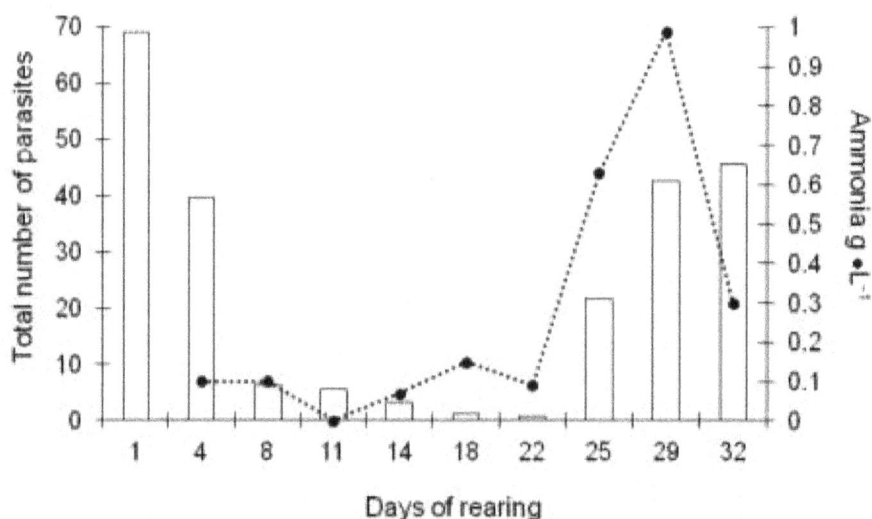

Figure 3. Total number of parasites (protozoans + monogeneans) (bars) and total ammonia concentration (dotted line) in the culture water during the masculinization of Nile tilapia (*Oreochromis niloticus*) fry.

At the end of the study, the fish reached a mean weight of 1.17 ± 0.36 g and a total length of 4.0 ± 0.4 cm (Figure 4), values which are higher than those mentioned by Rakocy (1989) of 0.5 to 1.0 g for a culture density of 8 fry/L[-1] and than those reported by Popma and Lovshin (1995) of 0.15 to 0.80 g for sex-reversed fry over a 30-day period. The stocking density (18 fry/L[-1]) probably played an important role in the growth rates, as at a density of 40 fry/L[-1] the weight, length and survival recorded for sex-reversed fry cultivated at 28 °C were 0.37 to 0.39 g, 2.67 to 2.78 cm, and 64.7 to 70.6%, respectively (Drummond *et al.*, 2009). Lastly, survival (99.5%) was above the 80% reported by Rakocy (1989).

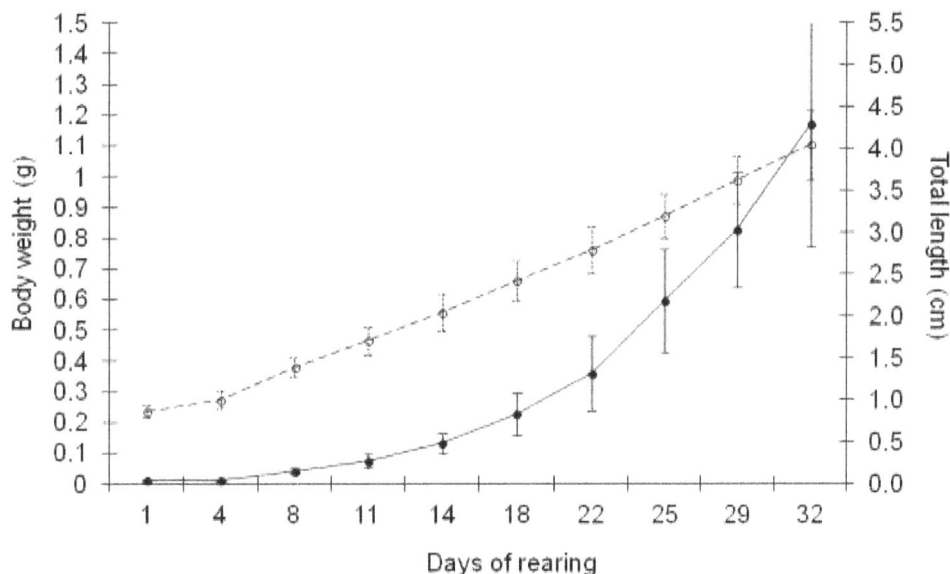

Figure 4. Growth in body weight (dotted line) and total length (solid line) of Nile tilapia (*Oreochromis niloticus*) fry during masculinization. Bars: ± 1 SD.

CONCLUSION

When Nile tilapia fry are cultivated at a density of 18 fish/L^{-1} and with good water quality, they can exceed their expected final weight (1 g), present virtually zero mortality, and naturally control or eliminate their ectoparasites. The rise in parasitic infections associated with a significant increase in nitrogen compounds from days 25 to 32 of the culture reinforces the importance of maintaining these parameters under control.

ACKNOWLEDGEMENTS

The authors are grateful to the Dirección General de Educación Superior Tecnológica of the Secretaría de Educación Pública (DGEST-SEP), México, for the funding granted to IJG to carry out this study (PROMEP/103.5/09/4061), and for the scholarships provided to CNCJ and SPS (Programa de Integración de Estudiantes de Licenciatura a Proyectos de Investigación).

REFERENCES

Atwood, H.L., Fontenot, Q.C., Tomasso, J.R., Isely, J.J. 2001. Toxicity of nitrite to Nile tilapia effect of fish size and environmental chloride. North American Journal of Aquaculture. 63:49-51.

Barker, D.E., Cone, D.K., Burt, M.D.B. 2001. *Trichodina murmanica* (Ciliophora) and *Gyrodactylus pleuronecti* (Monogenea) parasitizing hatchery-reared winter flounder, *Pseudopleuronectes americanus* (Walbaum): effects on host growth and assessment of parasite interactions. Journal of Fish Diseases. 25:81-89.

Bush, A.O., Lafferty, K.A., Lotz, J.M., Shostak, A.W. 1997. Parasitology meets ecology on its own terms: Margolis et al. revisited. Journal of Parasitology. 83:575-593.

Drummond, C.D., Solis, M.L.D., Vicenti, B. 2009. Growth and survival of tilapia *Oreochromis niloticus* (Linnaeus, 1758) submitted to different temperatures during the process of sexual reversion. Ciencia e Agrotecnologia, Lavras. 33:895-902.

Ellis, T., North, B., Scott, A.P., Bromage, N.R., Porter, M., Gadd, D. 2002. The relationships between stocking density and welfare in farmed rainbow trout. Journal of Fish Biology. 61:493-531.

El-Azez, H.H.M.A. 1999. Trichodiniasis in farmed freshwater tilapia in Eastern Saudi Arabia. J. KAU: Marine Science. 10:157-168.

El-Sayed, A.F. 2006. Tilapia Culture. CAB International, UK.

Emerson, K., Russo, R.C., Lund, R.E., Thurston, R.V. 1975. Aqueous ammonia equilibrium calculations: effect of pH and temperature. Journal of the Fisheries Research Board of Canada. 32:2379-2383.

Lafferty K., Kuris, A. 1999. How environmental stress affects the impacts of parasites. Limnology and Oceanography. 44:925-931.

Martins, M.L., Shoemaker, C.A., Xu, D., Klesius, P.H. 2011. Effect of parasitism on vaccine efficacy against *Streptococcus iniae* in Nile tilapia. Aquaculture. 314:18-23.

Noga, E. J. 2000. Fish Disease-Diagnosis and Treatment. Iowa State University Press, USA.

Phelps, R.P., Popma, T.J. 2000. Sex reversal of tilapia. In: Costa-Pierce, B.A. and Rakocy, J.E. (eds.). Tilapia Aquaculture in the Americas, Vol. 2. The World Aquaculture Society, Baton Rouge, Louisiana, USA. pp. 34-59.

Plumb, J.A. 1999. Health maintenance and principal microbial diseases of cultured fishes. Iowa State Univ. Press, USA.

Popma, T.J., Lovshin, L.L. 1995. Worldwide Prospects for Commercial Production of Tilapia. International Center for Aquaculture and Aquatic Environments, Department of Fisheries and Allied Aquacultures. Auburn University, Alabama, USA.

Rakocy, J.E. 1989. Tank Culture of Tilapia. Southern Regional Aquaculture Center, Publication 282, Texas Agricultural Extension Service, Texas A&M University, USA.

Rakocy, J.E. 2005. Cultured Aquatic Species Information Programme. Oreochromis niloticus. In: FAO Fisheries and Aquaculture Department, Rome. www.fao.org/fishery/culturedspecies/Oreochromis_niloticus/en (Consulted: 26 February 2012).

Xu, D.H., Shoemaker, C.A., Klesius, P.H. 2007. Evaluation of the link between gyrodactylosis and streptococcosis of Nile tilapia *Oreochromis niloticus* (L.). Journal of Fish Diseases. 30:233-238.

18

Primary Production Variables of *Brachiaria* Grass Cultivars in Kenya Drylands

**Susan A. Nguku[1*], Donald N. Njarui[2], Nashon K.R. Musimba[1],
Dorothy Amwata[1] and Eric M. Kaindi[1].**

[1]South Eastern Kenya University, P.O. Box 170-90200 Kitui, Kenya

*[2]Kenya Agriculture and Livestock Research Organisation, Katumani, P.O. Box 340,
Machakos, Kenya,
* Corresponding author*

SUMMARY

The study was conducted to evaluate the primary production variables of *Brachiaria* grass cultivars in semi arid regions of Eastern Kenya. *Brachiaria* cultivars *B. decumbens* cv. Basilisk, *Brachiaria* hybrid Mulato II, four *Brachiaria brizantha* cultivars Marandu, Xaraes, Piata, MG4 and *Brachiaria humidicola* cv Llanero were assessed with reference to their field establishment and growth rates. *Chloris gayana* cv. KATR3 and *P. pupureum* cv. Kakamega I were included as controls. Plant numbers, heights, spread, plant tiller number and plant cover were monitored for the initial 16 weeks following seedling emergence. A standardization cut was done on all the plots at week 16 and dry matter yields determined. All growth parameters measured varied significantly ($p<0.05$) among the cultivars. *Chloris gayana* cv. KATR3 recorded the highest plant numbers (48plants/m^2). Llanero recorded highest plant spread (146.9 cm) and tiller number (31tillers/plant). Napier had the highest plant height (103.8cm), cover (94.9%) and average dry matter yields (5430kg /ha). The results demonstrate that *Brachiaria* grasses are capable of establishing themselves in the semi arid regions of Eastern Kenya. It is recommended that the experiment be conducted for a longer period of time to determine their growth capacity during dry spells and pests and diseases that can hinder establishment and production.

Keywords: *Brachiaria*; grass; Livestock production; Climate change; Forages; Establishment.

INTRODUCTION

Livestock form an important avenue for rural development and provide the bulk of meat consumed in Kenya. One of the most important constraints on livestock production is nutrition, especially during the dry season when forage quality and quantity is low (Orodho, 2007). Grazing systems are most affected by climate change because of their dependence on climatic conditions, their natural resource base and their limited adaptation opportunities (Aydinalp and Cressor, 2008). Climatic impacts are expected to be more severe in arid and semi arid grazing systems at low latitudes where higher temperatures and lower rainfall are expected to reduce yields on rangelands and increase land degradation (Thornton, 2010). Current weather projections for East Africa indicate that temperature will increase between 1.3°C and 2.1°C depending on climate models by 2050 (Waithaka et al., 2013). The rise in temperature will increase evaporation and cause loss of soil moisture hence increasing the plant moisture requirements (Waithaka et al., 2013). Impacts on livestock production systems include productivity losses due to temperature increase, change in water availability and alteration in fodder quality and quantity. Prolonged droughts often leave many livestock keepers poorer and unsure of reliable livestock feed source. A drought experienced in 2008/9 in the country affected approximately 10 million people; a third of the country's population with massive losses in livestock occurring in the Northern Frontier (Gullet et al., 2011).

Grass is one of the most important sources of nutrients for domesticated ruminants during a large part of the year (Taweel et al., 2005) and is more easily accessible, better in taste and quicker in digestion than shrubs and trees (Quraishi, 1999). Animals do not compete with humans for grasses as food and they are therefore a cheap and economical feed source. According to Herrera (2004), pasture turn out to be an appropriate source of food for ruminants, mainly in countries of tropical climate such as Cuba due to the high number of species that can be used and the possibility of cultivating them throughout the year. The developmental morphology of plants defines their architectural organization, influences their palatability and accessibility to herbivores, and affects their ability to grow following defoliation (Briske, 1986). Tillers increase the chance of survival and the available forage resource of grasses and tiller numbers are an indicator of resource use efficiency by different grass species (Laidlaw, 2005).The weight of a plant's tillers will determine its productivity (Nelson and Zarrough, 1981). Plant cover on the other hand is important because it chokes weeds and also serves to protect the soil from

erosion agents. Pasture species which grow fast and tall are more efficient in use of resources and therefore, are more competitive. Such species eventually shade out the other species if planted in mixed stands thereby, suppressing their growth (Opiyo, 2007; Mganga, 2009; Ogillo, 2010).

There is a great potential to improve pastures through breeding by increasing diversity through germplasm introductions to the Kenyan environment to boost the forage resource base for livestock (Orodho, 2007). Forage grasses commonly found growing in the semi arid regions of Kenya include Buffel grass (*Cenchrus ciliaries*), Rhodes grass (*Chloris gayana*), *Panicum maximum*, Masai love grass (*Eragrostis superba*), Horse tail (*Chloris roxbhurgiana*) and *Enteropogon machrostachys* (Orodho, 2007). These grasses' nutritional and yield status decline with changing climatic conditions in the year making them incapable of meeting the needs of livestock (Gitunu et al., 2003). Napier grass which is the most commonly grown fodder grass by dairy farmers has been encountering disease and pest attacks that are rendering it vulnerable (Orodho, 2006). The *Brachiaria* grass spp. is a perennial grass native to East and Central Africa and has been introduced into humid tropical regions of Latin America, Southeast Asia, and northern Australia where it has revolutionized grassland farming and animal production. Whereas their potential in their native land remains largely unexploited, in their adopted homes in South America and Asia, there have been several research and development efforts to improve the productivity, nutritive values and other agronomic characteristics of these grasses (Ndikumana and de Leeuw, 1996).

The study sought to examine the primary production variables of seven *Brachiaria* grass cultivars in Semi arid Kenya. These included four *Brachiaria brizantha* cultivars namely; MG4, Xaraes, Piata and Marandu; *Brachiaria* hybrid cv. Mulato II; *Brachiaria humidicola* cv. Llanero and *Brachiaria decumbens* cv. Basilisk. *Chloris gayana* cv KATR3 and Napier grass (Kakamega I) were grown alongside these cultivars to serve as controls.

MATERIALS AND METHODS

Site

The study was conducted at the KARI (Kenya Agriculture Research Institute) now KALRO (Kenya Agriculture and Livestock Research Organisation (KALRO) Katumani- Machakos, Kenya (10 58′S, 370 28′E). Elevation is 1600m above sea level and the mean temperature is19.60C (Njarui et al., 2003). The soils are Luvisols, low in nitrogen and phosphorus with a PH of 6.5 (Aore and Gitahi, 1991). The mean annual rainfall is 717 mm, with a bimodal pattern, the long rains (LR) occurring from March-May and the short rains (SR) from October-December with two dry seasons (June-September; January-February). During the time of the experiment, the total rainfall recorded was 571 mm. This figure includes rainfall data collected in February and March, 2014. Average temperature ranged between 15.30C -26.20C.

Experimental design and treatment

The experiment was run within the period of October, 2013 at the onset of the short rainy season to March 2014 at the onset of the long rainy season. Site selection and laying of plots was done in October whereas sowing of seed was done on 11th November, 2013. Data collection begun at 4 weeks post seedling emergence and ended at 16 weeks post seedling emergence. *Brachiaria brizantha* cv. Marandu, *B. brizantha* cv. Xaraes, *B. brizantha* cv. Piata, *B. brizantha* cv. MG4, *B. decumbens* cv. Basilisk, *B. humidicola* cv. Llanero, *Brachiaria* hybrid cv. Mulato II, *C.gayana* cv. KATR3 and Napier grass were tested for field establishment, growth and thereafter for dry matter yields. The experimental design was a randomized complete block design with 4 replications. Individual plot sizes were 5m x 4m with a 1m path between plots and 1m path between the blocks. The seeds were drilled in furrows of about 2cm deep on a well prepared seedbed at an inter row spacing of 0.5m, giving 10 rows in each plot. Triple super phosphate (TSP, 46 % P) fertilizer was applied to the soil prior to sowing of the seed at the rate of 50.8kg P/ha in the planting rows. Sowing was done manually by placing the seeds in the furrows and covering them with a thin layer of soil. The grass seed was sown at rates of 5kg/ha. For Napier grass 3 splits were planted at intervals of 1m apart in holes dug 15cm deep after adding TSP at the rates of 50.8kg P/ha. The trials were kept weed free throughout the experiment by hand weeding and slashing inter row spaces to reduce weed competition within the replications. Standardization cuts were carried out in the sixteenth week at the onset of the long rains and dry matter yields established.

Data collection

Plant attributes (Plant height, plant counts, tiller numbers, plot cover and plant spread) were recorded at week 4, 8, 12 and 16 after seedling germination. A standard cut was done at week sixteen at the onset of the long rainy season (March-May) and dry matter yields established.

Plant counts: Number of plants was determined by counting plants in a 1m x 1m fixed quadrat placed over two rows.

Plot cover: Percentage plot cover was established by using a quadrat of 1m x1m subdivided into 25 squares of 0.2m x 0.2m as described by (Njarui and Wandera, 2004). The percent canopy cover of Napier grass was determined using the dot method as described by (Sarrantonio, 1991).

Plant spread: For spread, the plant diameter was measured from one edge to the other of each of 4 randomly selected and tagged plants. This was done using a metre ruler.

Plant height: Plant height was measured on the primary shoot from the soil surface to the base of the top-most leaf using a metre rule as described by (Rayburn and Lozier, 2007).This was done on the same four plants tagged.

Number of tillers: The number of tillers for the same 4 tagged plants was counted and recorded. Total tiller number per tuft on each measurement occasion was defined as the sum of total tiller number at previous measurement and number of tillers formed after previous measurement.

Standardization: The onset of the long rainy season (March-May, 2014) marked the end of the establishment and primary production period. This was at week16 post seedling emergence. After recording plant counts, plant cover, plant height, tiller numbers and plant spread, the grasses were cut to a stubble height of 5cm in a randomly selected area of 4m² within the plots as described by (Tarawali *et al.,* 1995). A fresh weight of all the harvested material was recorded after which sub samples of these were weighed and recorded. The sub samples were dried in an oven for 72 hours at temperatures of 65⁰C after which the dried sample weights were recorded. The oven-dry weights were used to calculate dry matter (DM) yield per plot which was then extrapolated to kg/ha. These oven dried samples included the leaves and stems harvested at 5cm stubble height.

Statistical analyses

Data on agronomic parameters and dry matter yields of forage samples were subjected to ANOVA based on the model designed for a randomized complete block design (RCBD) according to (Gomez and Gomez, 1984).To compare significant differences in response variables, ANOVA analysis was done using SAS package (SAS, 2001). Duncan's Multiple Range Test was carried out for subsequent comparison of means as described by (Steel and Torrie, 1986).

RESULTS

Climatic data

Figure 1 below shows the rainfall data for the site recorded during the period of January 2012–June 2014 presenting 5 rainy seasons: long rains (March–May), short rains (October–December), short dry season (January–February) and long dry season (June–September). Rainfall for the long rains in 2014 is given for 3 months, March– May. During the short rainy season of 2013, maximum rainfall was experienced in the month of December. During the long rainy season of 2014, maximum rains were experienced in March which was also above the Short Term Average (STA). The months of April and May recorded lower rainfall which was also below the short term average. The temperatures in almost all months were similar to the short term average (Fig.2).

Plant population

Changes in plant numbers over time are shown in table 1. Plant population means for all the cultivars were significantly ($p<0.05$) different. At week 16, *Chloris gayana* (Kat R3) recorded highest plant numbers at 48.5plants/m^2. MG4 (27.3 plants/m^2), Mulato II (23.8 plants/m^2), Marandu (20.8plants/m^2) and Basilisk (24 plants/m^2) recorded similar plant population. Plant population for Xaraes (12 plants/m^2), Piata (8.3 plants/m^2) and Llanero (16 plants/m^2) were lowest but similar ($p>0.05$).

Plant spread

Table 2 shows the mean values for spread for the cultivars. Mean plant spread for the cultivars were significantly different ($p<0.05$) and increased progressively from week 4 (4.1 cm) to week 16 (65.8 cm). At week 16 Plant spread for Llanero was highest at 146.9cm and lowest for Mulato II at 40.7cm. Napier (72.2cm) recorded second highest mean plant spread which was similar to mean plant spread for MG4 (58.6cm), Basilisk (60.4cm) and *C.gayana* KATR3 (60cm). Initially at week 4, Marandu showed highest mean spread at 10cm but by week 12 it was among the lowest in spread. Mulato II maintained the lowest mean plant spread throughout this period (2.3 - 40.7cm).

Plant cover

Plot cover generally increased for all the cultivars as shown in Table 3. Cover means were significantly different for the cultivars ($p<0.05$) during this period. All the cultivars except Piata and Xaraes recorded high and similar plot cover at week 16. Only Piata had less than 50% plot cover at week 16.

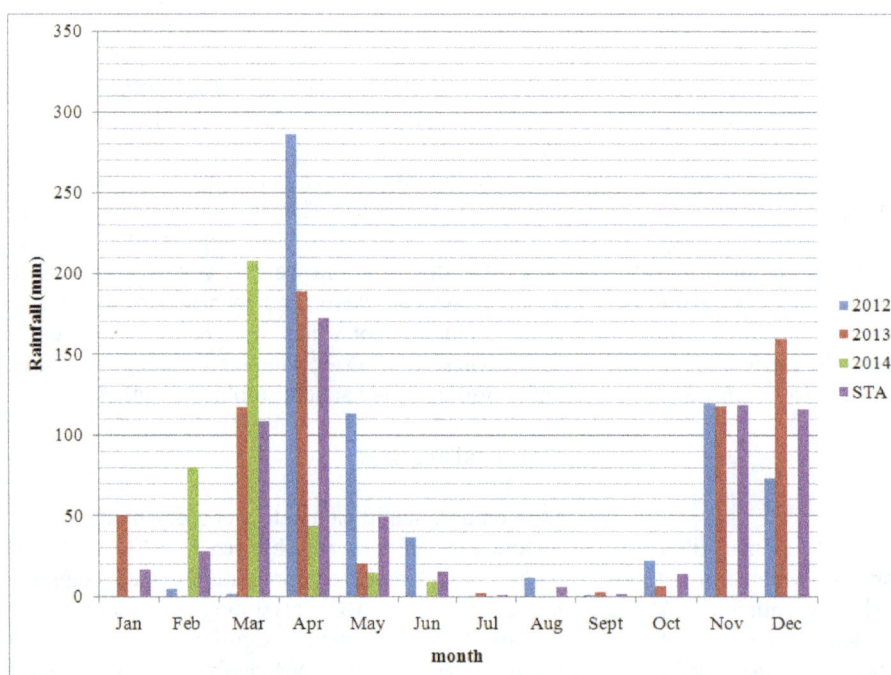

Figure1: Monthly rainfall recorded data at experimental site from January, 2012 to June, 2014

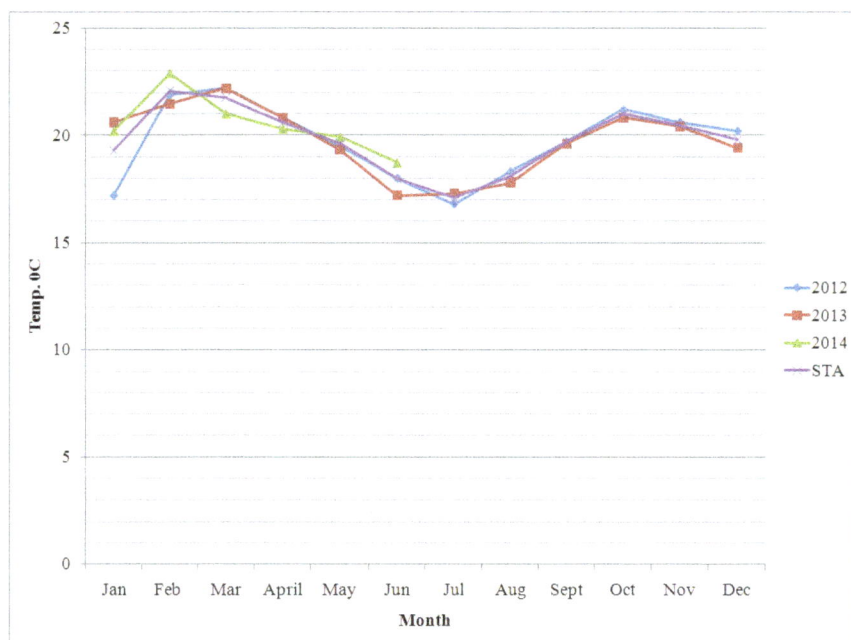

Figure 2: Mean monthly temperature at experimental site from January 2012 to June 2014

Table 1: Mean plant population (plants/ m2) of the grass cultivars during field establishment and growth

Cultivar	Week4	Week 8	Week 12	Week 16
Llanero	14[bac]	17[ced]	17[cbd]	16[cbd]
MG4	22.3[a]	27.3[b]	27.3[b]	27.3[b]
Marandu	16.5[ba]	20.8[cbd]	21.8[cb]	20.8[cb]
Piata	7.3[c]	8.3[fe]	8.3[ed]	8.3[ed]
Xaraes	10.5[bc]	11.8[fed]	12[ced]	12[ced]
Mulato II	19.5[ba]	23.8[cb]	23.8[b]	23.8[b]
Basilisk	18[ba]	19.5[cbd]	24.0[b]	24.0[b]
Kat R3	22.5[a]	48.5[a]	48.5[a]	48.5[a]
Napier	-	4[f]	4[e]	4[e]
Mean	16.3	20.1	20.7	20.5
SE	±1.0	±1.0	±1.2	±1.2

Column means with similar superscripts are not significant ($p<0.05$)

Table 2: Mean plant spread (cm) of the grass cultivars during field establishment and growth

Cultivar	Week4	Week 8	Week 12	Week 16
Llanero	3.6[b]	49.4[a]	107.7[a]	146.9[a]
MG4	4.8[b]	16.3[b]	38.1[cb]	58.6[cb]
Marandu	10.0[a]	12.1[b]	29.2[cb]	49.7[cd]
Piata	3.0[b]	10.0[b]	31.2[cb]	56.4[c]
Xaraes	3.6[b]	12.1[b]	29.4[cb]	47.4[cd]
Mulato II	2.3[b]	9.8[b]	24.5[c]	40.7[d]
Basilisk	3.2[b]	10.1[b]	36.6[cb]	60.4[cb]
Kat R3	2.3[b]	13.5[b]	34.7[cb]	60.0[cb]
Napier	-	16.3[b]	47.1[b]	72.2[b]
Mean	4.1	16.6	42.1	65.8
SEM	±0.6	±0.8	±2.1	±1.5

Column means with the same superscript are not significantly different ($p<0.05$).

Table 3: Mean Plant cover (%) of the grass cultivars during field establishment and growth

Cultivar	Week4	Week 8	Week 12	Week 16
Llanero	8.0^{bc}	25.0^{bac}	71.5^a	81.0^{ba}
MG4	13.0^a	32.0^{ba}	51.3^{ba}	74.0^{ba}
Marandu	10.0^{ba}	19.0^{bac}	47.0^{bc}	74.0^{ba}
Piata	5.0^{bc}	16.0^c	34.0^{bc}	49.0^c
Xaraes	8.0^{bc}	25.0^{bac}	39.0^{bc}	62.0^{bc}
Mulato II	8.0^{bc}	18.0^{bc}	45.0^{bc}	70.0^{ba}
Basilisk	13.0^a	33.0^a	31.0^{bc}	71.0^{ba}
Kat R3	9.0^b	13.0^c	44.0^{bc}	69.0^{ba}
Napier	-	15.8^c	24.5^c	84.9^a
Mean	9.3	21.9	43.0	70.5
SEM	±0.4	±1.5	±2.6	±2.0

Column means with the same superscript are not significantly different (p<0.05)

Plant tiller number

Mean tiller number increased progressively for all cultivars and there were significant differences among the cultivars (p<0.05) as shown in Table 4. The *Brachiaria* spp recorded generally higher tiller numbers than both *C.gayana* and Napier throughout the growth period. Mean tiller numbers were highest for Marandu, MG4 and Basilisk at week 4. At week 16, Llanero (30.5tillers/plant) recorded highest tiller numbers but Marandu (16.8 tillers/plant) was among the lowest in tiller recruitment. MG4 (24.5tillers/plant), Piata (25.5tillers/plant), Xaraes (25.5tillers/plant), Mulato II (23.8tillers/plant) and Basilisk (20.5tillers/plant) also recorded high and similar tiller numbers with Llanero at week 16.

Plant height

Mean plant heights for the cultivars generally increased and were significantly different (p<0.05) throughout the growth period as shown in Table 5 below. At week 16, Napier recorded the highest mean plant heights (103.8cm) and Llanero lowest at 6cm. Among the *Brachiaria* cultivars MG4 (63.4cm) recorded higher plant heights and although second after Napier (103.8cm), it's height was not significantly different (p>0.05) from *C.gayana* cv. Kat R3 (52.8cm).

Dry matter yields

Figure 3 shows the dry matter yields of the cultivars at week 16. Dry matter yields at week 16 represented primary production There were significant differences (p<0.05) between the cultivars for dry matter yields. Napier (5430KgDM/ha) recorded highest dry matter yields followed by MG4 (4583.4 Kg DM/Ha) and Mulato II (4050.2 Kg DM/Ha). The lowest yields were recorded for Llanero at 2282 Kg DM/Ha though this value was similar to that of *C.gayana* KATR3 (2741 Kg DM/Ha), Marandu (2596 Kg DM/Ha) and Xaraes (2335 Kg DM/Ha).

Table 4: Mean plant tiller number of the grass cultivars during field establishment and growth

Cultivar	Week4	Week 8	Week 12	Week 16
Llanero	3.5^b	9.5^{ba}	16.5^{ba}	30.5^a
MG4	4.8^a	12.3^a	16.8^a	24.5^{ba}
Marandu	4.8^a	8.0^{bc}	11.8^{bc}	16.8^{bc}
Piata	2.0^c	5.5^c	12.8^{bac}	25.5^{ba}
Xaraes	3.3^b	9.3^{ba}	14.3^{ba}	25.5^{ba}
Mulato II	3.2^b	8.5^{bc}	14.8^{ba}	23.8^{ba}
Basilisk	4.0^{ba}	7.8^{bc}	12.3^{bac}	20.5^{bac}
Kat R3	3.4^b	6.8^{bc}	11.8^{bc}	17.8^{bc}
Napier	-	8.0^{bc}	8.3^c	10.6^c
Mean	3.6	8.4	13.2	21.7
SEM	±0.1	±0.4	±0.5	±1.2

Column means with the same superscript are not significantly different (p<0.05).

Table 5: Mean Height (cm) of the grass cultivars during field establishment and growth

Cultivar	Week4	Week 8	Week 12	Week 16
Llanero	2.3^c	3.3^b	3.9^d	6.0^f
MG4	5.0^a	4.5^b	37.3^b	63.4^b
Marandu	4^{ba}	3.4^b	10.2^{dc}	20.4^{de}
Piata	2.9^{bc}	3.3^b	14.6^c	29.8^{dc}
Xaraes	3.9^{ba}	4.3^b	12.6^{dc}	24.9^{dce}
Mulato II	2.1^c	3.0^b	7.9^{dc}	14.3^{fe}
Basilisk	4.6^a	3.7^b	12.8^c	34.9^c
Kat R3	2^c	2.3^b	7.3^{dc}	52.8^b
Napier	-	44.4^a	67.3^a	103.8^a
Mean	3.3	8.0	19.3	38.9
SEM	±0.1	±0.4	±0.9	±1.4

Column means with the same superscript are not significantly different (p<0.05).

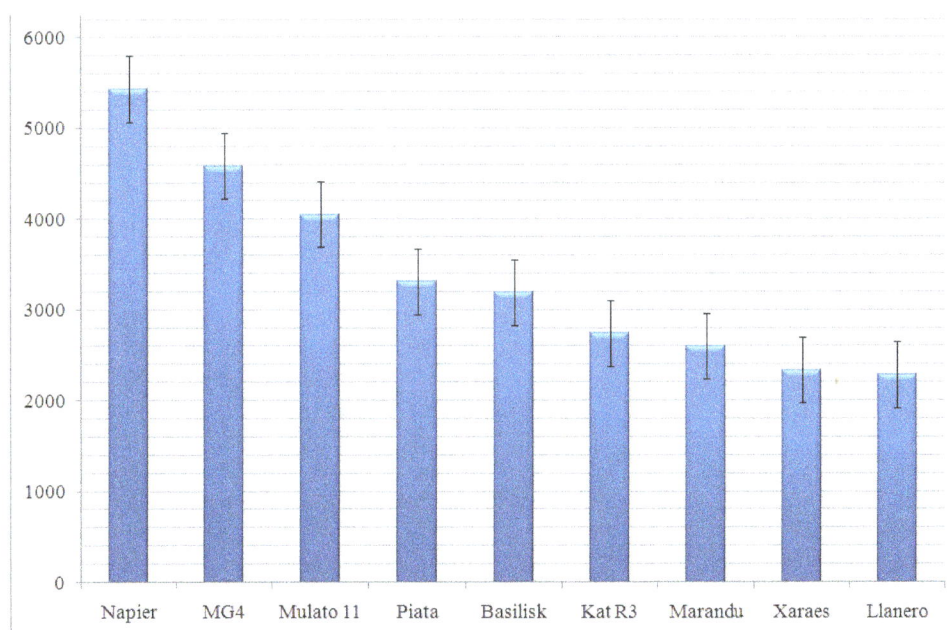

Figure 3: Dry matter yields in Kg/ha at primary production

DISCUSSION

The study demonstrated considerable variation in establishment of the grass cultivars. The differences in plant population can be attributed to species differences in seed germination rates, seedling establishment and survival. Plant numbers were highest for *C. gayana* KATR3. According to (Cook *et al.,* 2005), *Chloris gayana* is likely to have the greatest number of seed (7 250 000 to 9 500 000 per kg of seed). This explains the higher plant counts for *C.gayana* as compared to the other cultivars. Seed proportion of *C.gayana* has been shown to have a significant effect on agronomic traits (Yisehak, 2008). Higher plant populations for *Brachiaria brizantha* cultivars MG4 and Marandu, can be

explained by their higher germination percentages and accompanying seedling vigour (Nguku, 2015).

Vegetative growth (height, spread and tiller number) generally increased for all the cultivars and can be attributed to the morphological and physiological differences among the cultivars. The rapid spread of the cultivars indicates that they can play an important role in quick soil stabilization for erosion control and can be utilized in the stabilization of terrace banks in semi-arid areas. Plant spread can be attributed to individual growth habits of the cultivars. *Brachiaria humidicola* is a strongly stoloniferous and rhizomatous perennial grass, forming a dense ground cover and *B. decumbens* has good ground cover, aggressive growth and decumbent habit as reported by ook *et al* (2005). Napier being a fodder crop and gigantic in nature would naturally out do the other grasses when it comes to spread. Cook *et al* (2005) further report that the ability of *C .gayana* to spread is because it produces stolons which creep over the ground, developing roots at the nodes and that Marandu has some allelopathic effect which even reduces seedling recruitment of its own seed. This can explain the initial high spread of Marandu and the decreased vigour in this attribute relative to other cultivars by week 16.

MG4 and Napier were taller than the other plants and also produced high dry matter yields at week 16. Studies by Tessema *et al* (2003) concur that increasing foliage height in Napier grass increased biomass yield. *Chloris gayana* KATR3 on the other hand has a higher proportion of stem relative to leaf by week sixteen which could be the reason for lower dry matter yields. The vertical growth habit of Napier, MG4 and *C. gayana* cv. KATR3 explain why they are tallest by week 16 relative to other cultivars. (Opiyo, 2007; Mganga, 2009 and Ogilo, 2010) report that, pasture species which grow fast and tall are more efficient in use of resources and therefore, are more competitive. Such species eventually shade out the other species if planted in mixed stands thereby, suppressing their growth. *Brachiaria humidicola* cv. Llanero's decumbent habit explains why it is the shortest at week 16.

All cultivars but Piata attained over 50% plant cover by week 16. The high plant cover for both *B. brizantha* and *B. decumbens* cultivars could be attributed to their growth habits. *Brachiaria brizantha* is more tufted in terms of growth habit and *Brachiaria decumbens* more decumbent and this makes them form a dense plant cover (Cook *et al.,* 2005). Mulato II on the other hand produces vigorous cylindrical stems, some with a semi-prostrate habit capable of forming roots at the nodes when they come into contact with the soil (Vendramini *et al.,* 2011) making them attain high plot cover (70%). Studies in Honduras indicate that Mulato II establishes rapidly, and was able to achieve 85% ground cover at 2 months (Cook *et al.,* 2005). Cook *et al* (2005) further report that Llanero has a strongly stoloniferous growth habit and this causes it to have good ground cover. Napier on the other hand is a tall, tufted, rhizomatous perennial, very coarse and robust, in dense clumps. Its giant nature naturally makes it occupy a larger area relative to the other grasses hence the higher plot cover (Bogdan, 1977). *Chloris gayana* is a tufted perennial that also has a stoloniferous growth habit making it have a high plot cover. It produces stolons which creep over the ground, rooting at the nodes, and also produces

abundant seed to give rise to new plants (Cook *et al.,* 2005).

Llanero, MG4, Mulato II, Piata, Basilisk and Xaraes had higher tillering ability than the rest of the cultivars. This is an indication of the ability of the cultivars to recover faster after defoliation. Hiernaux *et al* (1994) found that plant tillering early in the life of the stand compensated for low plant density that resulted from drought or intense grazing. Marandu was not able to maintain a high tiller recruitment which could be attributed to allelopathic effect exhibited by the cultivar (Cook *et al.,* 2005). Cook *et al* (2005) further reports that tillering in Llanero can be attributed to its growth habit. According to Halim *et al* (2013) taller varieties of Napier tend to have fewer tillers but produce higher DM yields compared with shorter varieties which recruit higher tiller numbers and have higher nutritive values.

There were differences in dry matter yields among the grasses which can be attributed to their genotypic and phenotypic differences. Despite low plant numbers in some cultivars all the grasses persisted during the duration of study. On the basis of dry matter yield the best is Napier grass whereas the most promising among the *Brachiaria* cultivars is MG4. Napier out yielded the other grasses confirming studies by Humphreys (1994) and Skerman and Riveros (1990) of its potential for high dry matter yields relative to other tropical grasses. Herbage yield of Napier grass may however be affected by the harvesting day after planting. Generally, as grass ages, herbage yield is increased due to the rapid increase in the tissues of the plant (Minson, 1990). Mulato II was second to MG4 among the *Brachiaria* grass cultivar in dry matter yields. Dry matter yields for Mulato II can be largely attributed to its large size leaves (15-2" long) and thick stems (1-1.5" width) (Guiot and Meléndez, 2003). Mean primary dry matter yields for *C. gayana* though low were found to be similar to *Brachiaria* cultivars Piata and Basilisk which ranked third and fourth respectively in primary production. Although Marandu had high establishment rates this did not parallel its dry matter yields at primary production as reported by studies by Rao *et al* (1998) of rapid establishment accompanied by high DM production.

A longer period of study is recommended for evaluation of the *Brachiaria* cultivars for age at senescence and subsequent productivity.

CONCLUSION

The grass cultivars depended solely on the short rains with no irrigation water added. All the species established and persisted for the duration of the study.

Although *C. gayana* KATR3 is superior in terms of plant population and Napier grass superior in plant cover and height, the *Brachiaria* species perform better in terms of spread, plant cover, plant tiller number and even height and The *Brachiaria* species performed better in plant attributes like spread (Llanero), plant cover (MG4, Xaraes, Llanero) and tiller recruitment (MG4 and Xaraes). Napier and MG4 are both tall varieties and also produced comparatively higher dry matter yields compared to the other cultivars demonstrating their potential to utilize available plant resources. The tillering and spreading ability of Llanero is an added advantage for soil cover and protection as well as being an animal feed. This is also true for all the *Brachiaria* cultivars in this study.

Acknowledgement

This study is a collaborative undertaking between KARI and Bioscience for eastern and Central Africa/ International Livestock Research Institute (BecA/ILRI) and was funded by Swedish International Development Agency (Sida). We also want to appreciate the cooperation from various KALRO-Katumani and KALRO- Muguga staff for their support and team work during project execution.

REFERENCES

Aore, W.W. and Gitahi, M.M. 1991. Site characterization of ACIAR Project Experimental Sites (Machakos and Kitui districts).A Provisional Report. Kenya Soil Survey. Miscellaneous Paper No. M37.

Aydinalp, C. and Cressor, M.S. 2008. The Effects of Global Climate Change on Agriculture. American–Eurasian Journal of Agriculture and Environmental Science, 3, 672-676.

Bogdan, A. 1977. Tropical Pasture and Fodder Plants. London: Longman Group Limited. Pp233-244.

Briske, D. D. 1986. Plant response to defoliation:morphological consideration and allocation priorities. In P. L. Eds. Joss, Rangelands: A Resource Under Siege (pp. 425-427). United Kingdom: Cambridge University Press.

Cook, B. G.; Pengelly, B. C.; Brown, S. D.; Donnelly, J. L.; Eagles, D. A.; Franco, M. A.; Hanson, J.; Mullen, B. F.; Partridge, I. J.; Peters, M. and Schultze-Kraft, R. 2005. Tropical Forages. Brisbane Australia. Brisbane Australia: Csiro, Dpi&F(Qld), CIAT And ILRI.

Gitunu, A.M., Mnene W.N., Muthiani, E.N., Mwacharo, J.M., Ireri R., Ogillo, B.P and Karimi S.K. 2003. Increasing the productivity of livestock and natural resources in semi-arid areas of Kenya. A case study from the southern Kenya rangelands. Proceedings for end of programme conference, agriculture/livestock research support programme, phase II held from 11-12 November 2003 at KARI headquarters. Nairobi, Kenya: Kenya Agriculture Research Institute.

Gomez, K. A. And Gomez A. A. 1984. Statistical Procedures For Agricultural Research. 2nd Edition. New York, USA: John Wiley and Sons.

Guiot, J.D. and Meléndez, F. 2003. Pasto Mulato: Excelente alternativa paraproducción De Carney Leche En Zonastropicales. Instituto para el desarrollo de Sistemas de Producción del trópico húmedo de Tabasco. Villahermosa. Mexico

Gullet, A., Kisia, J. and Mungou, T. 2011. Dought appeal 2011. Alleviating human suffering. Nairobi, Kenya: Kenya Red Cross Society .

Halim, R. A., Shampazuraini, S. and Idris, A.B. 2013. Yield and Nutritive Quality of Nine Napier Grass Varieties in Malaysia. Malaysian Journal of Animal Science 16, 27-44.

Herrera, R. S. 2004. Photosynthesis: Tropical Grasses, Contribution to Physiology, Establishment, Biomass Yield, Biomass Production, Seed Production and Recycling of Nutrients. Ica, La Habana : Editions:Edica p.27.

Hiernaux, P., De Leeuw, P.N. and Diarra. L. 1994. Modeling Tillering of Annual Grasses as a Function of Plant Density: Application to Sahelian Rangelands Productivity and Dynamics. Agricultural Systems. 46, 121-139.

Humphreys, L. 1994. Tropical Forages: Their Role in Sustainable Agriculture. Harlow, UK: Longman, Pp. 414.

Inc., S. 2001. Sas/Stat. Users Guide Version 8.2. North Carolina: Sas Institute Inc.Cary.P 3884.

Laidlaw, A. 2005. The Relationship Between Tiller Appearance in Spring and Contribution of iDry-Matter Yield In Perennial Ryegrass (*Lolium perennel*.) Cultivars Differing in Heading Date. Grass And Forage Science, 60 , 200-209.

Mganga, K. 2009. Impact of Grass Reseeding Technology on Rehabilitation of Degraded Rangelands: A case study of Kibwezi district, Kenya. Nairobi, Kenya: MSc Thesis, University of Nairobi.

Minson, D. 1990. Forage In Ruminant Nutrition. San Diego, California: Academic Press.Pp 482.

Ndikumana J and de Leeuw P.N. 1996. Regional Expertice with *Brachiaria*: Sub-Saharan Africa. In J. M. Miles, *Brachiaria*: Biology, Agronomy, and Improvement (pp. 247-258). Cali, Colombia: CIAT and EMBRAPA.

Nelson, C.J. and Zarrough, K.M. 1981. Tiller Density And Tiller Weight as Yield Determinants of Vegetative Swards. In Wrigth,C. Pasture Production (pp. 25-29). Hurley: British Grassland Society.

Nguku, S. 2015. An Evaluation of *Bracharia* Grass Cultivars Productivity In Semi Arid Kenya. Kitui, kenya: Unpublished MSc. Thesis, .

Njarui D.M.G., Mureithi, J.G. ,Wandera F.P. and Muinga, R.W. 2003. Evaluation of Four Forage Legumes as Supplementary Feed for Kenya Dual-Purpose Goat in the Semi-Arid Region of Eastern Kenya. Tropical and Subtropical Agroecosystems 3, 65-71.

Njarui, D.M.G. and Wandera, P.J. 2004. Effect of Cutting Frequency on Productivity of Five Selected Herbaceous Legumes and Five Grasses in Semi-Arid Tropical Kenya. Tropical Grasslands, 38, 158-166.

Ogillo, B. P. 2010. Evaluating Performance of Range Grasses under Different Micro Catchments and Financial Returns from Reseeding in Southern Kenya. Nairobi, Kenya: Unpublished Msc Thesis, University of Nairobi.

Opiyo, F. 2007. Land treatment effects on morphometric characteristics of three grass species and economic returns from reseeding in Kitui District, Kenya. Nairobi, Kenya: University of Nairobi, Kenya: Unpublished MSc Thesis.

Orodho, A. 2007. Forage Resource Profiles- Kenya. Country Pasture. Nairobi, Kenya: Food and Agriculture Organisation.

Orodho, A. 2006. The Role and Importance of Napier Grass in the Smallholder Dairy Industry in Kenya. Kitale, kenya: Food and Agriculture Organisation.

Quraishi, M. A. 1999. Range Management in Pakistan. Faisalabad: University Of Agriculture.

Rao, I.M., Miles, J.W. and Granobles, J.C. 1998. Differences in Tolerance to Infertile Acid Soil Stress among Germplasm Accessions and Genetic Recombinants of the Tropical Forage Grass Genus, *Brachiaria*. Field Crops Research, 59, 43-52.

Rayburn,E.B. and Lozier J.D. 2007. Alternative Methods of Estimating Forage Height And Sward Capacitance. In E. A. Rayburn, Pastures can be Cross Calibrated. St. Paul Minnesota. USA: Plant Management Network.

Sarrantonio, M. 1991. Methodologies For Screening Soil Improving Legumes. In M. Sarrantonio, Methodologies For Screening Soil Improving Legumes (p. 310). Kutztown, Pa 19530 Usa: Rodale Institute Research Center.

Skerman, P.J. and Riveros, F. 1990. Tropical Grasses. Fao Plant Production and Protection Series No. 23. Rome, Italy: Food and Agriculture Organisation.

Steel, R. D. G. and Torrie, J. H. 1986. Principles and Procedures of Statistics. A Biometrical Approach. 5th Ed.Ll. New-York : Ed. Mcgraw-Hill International Book Company .

Tarawali, S. A., Tarawali,G., Larbi A. and Hanson, J. 1995. Initial Screening Methods. In G. T. Tarawali, Methods For the Evaluation of Forage Legumes,Grasses and Fodder Trees for use as Livestock Feeds (pp. 10-11). Nairobi, Kenya: International Livestock Research Institute.

Taweel, H.Z., Tas, B.M., Smit, H.J., Elgersma, A., Dijkstra, J. and Tamminga, S. 2005. Improving The Quality Of Perennial Ryegrass (*Lolium Perenne* L.) For Dairy Cows By Selecting For Fast Clearing and/or Degradable Neutral Detergent Fiber. Livestock Production Science, 96, 239-248.

Tessema Z., Bears R.M.T. and Yami, A. 2003. Effect of Plant Height at Cutting and Fertilizer on Growth of Napier Grass (*Pennisetum Purpureum*). Tropical Science, 42, 57-61.

Thornton, P. K. 2010. Livestock Production: Recent Trends, Future Prospects. Philosophical Transactions of the Royal Society B, 365, 2853–2867.

Vendramini J., Sellers, B., Sollenberger, L.E., and Silveira, M. 2011. Mulato Ii (Brachiaria Sp.) Ss Agr 303. Florida, USA: Uf/Ifas Extension.

Waithaka M., Nelson G.C., Thomas T.S. and Kyotalimye, M. 2013. Overview. In N. G. Waithaka M., East African Agriculture and Climate Change.A Comprehensive Analysis.Ifpri Issue Brief,76 (pp. 8-30). Washington DC: International Food Policy Research Institute.

Yisehak, K. 2008. Effect of Seed Proportions Of Rhodes Grass (*Chloris Gayana*) And White Sweet Clover (*Melilotus Alba*) at Sowing onn Agronomic Characteristics and Nutritional Quality. Livestock Research For Rural Development 20, 2. http://www.lrrd.org/lrrd20/2/yise20028.htm

19

Response to Simple Superphosphate and Top-Phos Fertilizer on Wheat in an Oxisoil

P. V. D. Molin[1], L. Rampim[1*], F. Fávero[1], M. do Carmo Lana[1],
M. V. M. Sarto[1], J. S. Rosset[1], D. Mattei[1], P. S. Diel[1]
and R. N. D. Molin[1]

[1] *State University of West Paraná, Unioeste, CCA/PPGA, Pernambuco Street No. 1777, P.O. Box 9, Zip Code 85960-000, city of Marechal Cândido Rondon, Paraná state, Brazil. E-mail: paulo_vi7or@hotmail.com; rampimleandro@yahoo.com.br*
*Corresponding author

SUMMARY

The limitation of natural resources coupled with growing demand for fertilizers to sustain increased crop productivity ally to meet world demand for food intensifies the search for greater efficiency in fertilizer use. The objective of this study was to evaluate the response of phosphorus fertilization on wheat plants, noting its influence on the agronomic and nutritional characteristics of wheat and chemical attributes soil. The experiment was conducted in a greenhouse with two sources of single superphosphate (P - 18% P_2O_5 and top-phos - 22/28% P_2O_5) and P_2O_5 five doses (0, 50, 100, 150 and 200 kg ha^{-1}), in 2 dm^3 pots. Shoot dry matter give maximum point buildup dose of 151.25 kg ha^{-1} P_2O_5 and root dry matter point is the maximum dose of 165 kg ha^{-1} P_2O_5 independent of fertilizer used. Application of P_2O_5 levels with superphosphate fertilizer and top-phos increase levels of pH and available soil P and P content, K and S in the leaf tissue. Recommend both, simple superphosphate with 18% P_2O_5 as top-phos, considered with 28% P_2O_5 for phosphate fertilizer in wheat crop, by selecting the fertilizer that provide best value for money, in this case the superphosphate.

Key words: *Triticum aestivum*; Available phosphorus; Adsorption of phosphorus; Fertilizer; Organomineral.

INTRODUCTION

Wheat (*Triticum aestivum* L.) is a cereal widely used in food and feed. Due to its ease of adaptation is much cultivated in subtropical regions as tropical, with the most cultivated cereal in production volume (Caierão, 2009), with approximately 651 million tons of grain (Usda, 2012).

To keep the food supply, the increase in productivity of existing agricultural areas is one of the (Pinto-Zevallos and Zarbin, 2013) solutions. Thus, it becomes important to conduct studies which will help minimize the effect of the factors causing yield loss and depreciation of the quality of the wheat, as the occurrence of pests and diseases (Gallo *et al.*, 2002; Kimati *et al.*, 1997), problems with soil fertility (Raij, 2011) and mineral nutrition (Fernandes, 2006; Malavolta, 2006).

Among the nutritional factors stand positive responses to the use of nitrogen (N) (Freitas *et al.*, 1994, 1995) and phosphorus (P) (Oliveira *et al.*, 1984; Camargo and Merriman, 1987; Clark, 1990; Souza, 1996). In the case of P, is cited as responsible for achieving high productivity of wheat (Gargantini *et al.*, 1958) and is essential in the early stages of plant growth (Malavolta, 2006), the adequacy of the pH with liming. (Bataglia *et al.*, 1985, Souza, 1996). According to Lopes and William (2000), 50% of P is assimilated by plants when at pH 6.0 and 100% at pH 6.5 and 7.0. In other words, soil pH can influence the efficiency of use of applied P, and consequently affect the productivity and profits of the farmer.

Among the essential macronutrients for plants, P is one of the least required by plants, but it is the most used in fertilization by low levels available in the soil and also for being the one that most often limits crop production (Faquin, 2005). The low availability to the plants is due to high content of aluminum and iron oxides in the soil, which has a high phosphorus adsorption capacity (Kliemann and Santos, 2005), mainly evidenced in acidic soils. Different answers for the use of soil P and fertilizer among cultivars have been reported, so that Coimbra et al. (2014) identified different behavior between corn hybrids with tolerance to low soil P for some hybrids, while others are more responsive to phosphorus fertilization, an increase in productivity.

The site of application of fertilizer in the soil can also influence fertilization, since the greater the dilution of P in the soil, the greater its interaction with soil particles, which is observed when performing broadcast fertilization, reduced condition with the application localized fertilizer at sowing (Resende *et al.*, 2007). In the case of the P, localized application is most suitable, not only due to its interaction with the oxides of iron and aluminum in the soil, but its way of absorption, diffusion (Lopes and Guilherme, 2000), related to their low mobility in soil. To Vilar and Vilar (2013) the plants have several ways to alleviate the problems arising from P deficiency, such as changes to obtain, absorb and retain P to be used, as well as the symbiotic mycorrhizal associations increasing the area of land use and releasing phosphatases acting on solubilization of P.

Single superphosphate is obtained by reaction between finely ground rock phosphate and sulfuric acid. P low solubility of phosphate present in the natural undergoes changes in its chemical structure, giving product of the highest water solubility and highest agronomic efficiency. This immediately makes available P in soil in adequate amounts from the start of the crop cycle, providing the conditions for correct formation of the root system of the plants (Novais *et al.*, 2007). Sandy soils have higher P availability after phosphate fertilizers compared to soils with increasing levels of clay (Machado *et al.*, 2011).

Top-phos is present in a complex formulation that protects the P element fixing with aluminum, iron and calcium, making this nutrient more available and usable for the plants, with the complex provides greater root development. Thus, the top-phos can also benefit from the absorption of water and other nutrients, particularly P, due to the larger contact area with the roots (Timac Agro, 2012).

Thus, the aim of this study was to evaluate the response of phosphorus fertilization on wheat plants using two phosphate fertilizer superphosphate and top-phos in Oxisoil, noting its influence on the agronomic and nutritional characteristics of wheat and chemical attributes soil.

MATERIAL AND METHODS

The experiment was conducted in a greenhouse at the Horticulture and Protected Cultivation Mário César Lopes Station, belonging to the State University of West Paraná – Unioeste, campus Marechal Cândido Rondon - PR, with geographical coordinates 54° 22' W and 24° 46' S, and an average altitude of 420 meters.

The experimental design was completely randomized, with four replications in a factorial scheme 2 x 5, with two sources of single superphosphate (P - 18% P_2O_5, 18% Ca and 10% of S and top-phos - 22/28% P_2O_5, 17% Ca and 5% S) and five P rates (0, 50, 100, 150 and 200 kg ha^{-1} P_2O_5). For the top-phos was considered the ratio of 28% P_2O_5 which includes the total concentration of P fertilizers, since 6% is considered slow release (TIMAC AGRO 2012).

The experiment was conducted in pots and deployed for up to 5 dm^3 of soil. Fertilizer application was done before planting by mixing the ground fertilizer with the soil contained in pots. The soil was collected in the municipality in Jesuítas - PR, ranked Oxisoil clayey (Embrapa, 2013) with the following chemical properties: pH in $CaCl_2$ = 4.92; COT = 5.43 g dm^{-3}, P = 2.40 mg dm^{-3}, K^+ = 0.70 $cmol_c$ dm^{-3}, Ca^{2+} = 5.84 $cmol_c$ dm^{-3}, Mg^{2+} = 2.22 $cmol_c$ dm^{-3}, H + Al = 2.51 $cmol_c$ dm^{-3}, Al^{3+} = 0 $cmol_c$ dm^{-3}, SB = 8.76 $cmol_c$ dm^{-3}, CEC = 11.27 $cmol_c$ dm^{-3} and V = 77.73%.

Sowing was held on June 14, 2011, using supervised seeds of wheat cultivar Mirante. Six seeds per pot of wheat, subsequently performing thinning, leaving four plants per pot were sown. The control of soil moisture was through irrigation of the plants keeping the soil at field capacity. There was no need for fungicide application in shoots. To control thrips (*Thrips* sp.), The application of the insecticide thiamethoxam + lambda cyhalothrin at a dose of 40 mL ha^{-1} performed on August 5, 2011.

Phenological stage of elongation of wheat plants, four plants per pot were cut at ground level, with a separate shoot of the root system. The roots were washed to remove adhering soil, then the mass of shoots and roots were placed in an incubator at 60°C with forced air until constant mass, so that the shoot dry matter was determined (SDM, grams) and root dry matter (RDM, grams).

Subsequently, the SDM was ground and subjected to the nitric-perchloric digestion and determination of the levels of P, K, Ca, Mg and S (g kg^{-1}), according to the method described by EMBRAPA (2009). Content of P in shoots depending on the RDM, obtained by multiplication between the SDM and the P levels obtained in shoots was calculated. With the collection of plants, three samples were remove of soil from each pot for later determination of pH_{CaCl2} and P exchangeable (mg dm^{-3}) (EMBRAPA, 2009).

Data by treatments were subjected to analysis of variance for variables, and regression for P doses when significant, selecting higher R^2 to linear and quadratic effect. When required fertilizer sources was significant, the comparison of means was performed by F-test, that conclusion for just two levels by factor doses (top-phos and single superphosphate). Statist analysis was used SAEG 8.0 software (SAEG, 1999).

RESULTS AND DISCUSSION

Growth characteristics

For the growth characteristics in Table 1, the mean effect was observed only for doses tested independent of superphosphate fertilizer or top-phos, no difference between the fertilizers for both shoot dry matter (SDM) and root dry matter (RDM).

The SDM showed quadratic growth with maximum point to the dose of 151.25 kg ha^{-1} P_2O_5 independent of the use of SFS or top-phos, since there was no interaction between fertilizer. Accumulation was 1.073 g shoot dry matter of wheat plants in the elongation phase (Figure 1a). Thus the application of P through different products did not affect the response of plants to accumulate dry matter, corroborating data presented in a study by Resende *et al.* (2007), which tested the application of different P sources, different solubilities and different methods of application on corn plants, a similar increase in production was observed.

For RDM, seen in Figure 1b, was found similar to the SDM behavior with quadratic growth, the point of maximum dry matter accumulation was obtained with the dose of 165 kg ha^{-1} P_2O_5, independent of fertilizer used. Zanini *et al.* (2009) to use the form of triple superphosphate as P for forage species also found quadratic response to fertilization with the point of maximum accumulation of RDM with the dose of 170 mg kg^{-1} soil P to *Brachiaria brizantha*.

Both SDM increased as the RDM with the application of superphosphate and top-phos observed in the elongation stage of wheat plants are related to the immediate immediate effects of these phosphate fertilizers, since they exhibit high solubility (Kliemann and Santos, 2005), showing the use of both phosphorus deficient soils on the ground. On the other hand, when reaching limit accumulation of SDM and RDM probably exceeded the critical level of P for plants, primarily occurring luxury consumption by P (Figure 2a) without passing in increased dry matter in wheat, indicating the limit doses of fertilizer to perform a correction. This was, from the optimal doses of P, the phosphorus can adversely affect the sustainability and cost-effectiveness of fertilizer recommendation.

The point of maximum accumulation of SDM showed levels five times higher compared to the control without applying P. Cunegato *et al.* (2011) studied the application of P levels on SDM of wheat, observed that the treatments without P addition had reduced growth in Planosolo compared to plant growth with higher doses of up to 512 mg kg^{-1} of P soil, starting plateau accumulation of SDM close to 100 g kg^{-1} of P soil.

Table 1. F values, coefficient of variation (CV%), production of shoot dry matter (SDM) and root dry weight (RDM) in the elongation of wheat crop, and pH CaCl$_2$, phosphorus exchangeable (P) and P content in soil in terms of levels of phosphorus (P$_2$O$_5$) with fertilizer super simple (SFS) and top-phos. Marechal Cândido Rondon – PR, 2011

Fertilizer doses	SDM	RDM	pH CaCl$_2$	P exchangeable	P content
Top-phos	--------------- g ----------------		------------	--- mg dm^{-3} ---	---- mg ----
--- kg P$_2$O$_5$ ---					
0	0.179	0.194	4.71	4.74	0.0004
50	0.507	0.415	5.72	6.03	0.0015
100	1.036	0.680	5.28	9.30	0.0038
150	1.054	0.573	5.45	17.76	0.0055
200	1.062	0.599	5.44	19.79	0.0071
Simple superphosphate					
--- kg P$_2$O$_5$ ---					
0	0.253	0.245	5.23	4.27	0.0005
50	0.891	0.693	5.18	6.17	0.0028
100	1.110	0.686	5.57	9.86	0.0044
150	1.040	0.609	5.64	14.90	0.0056
200	0.957	0.607	6.11	15.41	0.0057
Fertilizer					
Top-phos	0.7676	0.4922	5.32	11.52	0.004
Simple superphosphate	0.8502	0.5680	5.55	10.12	0.004
Medium values					
	0.81	0.53	5.43	10.82	0.004
F values					
Fertilizer	1.07 ns	2.56 ns	3.63 ns	0.19 ns	0.23 ns
Doses	16.76 **	11.55 **	5.08 **	23.50 **	60.49 **
Linear	47.32 **	23.02 **	4.18 *	45.14 **	237.69 **
Quadratic	18.75 **	19.16 **	4.31 *	0.17 ns	3.33 ns
Fertilizer x Doses	1.07 ns	1.17 ns	3.02 *	0.44 ns	2.54 ns
C.V. (%)	31.11	28.23	6.79	22.18	23.67

* e **: significant at 5% and 1%, respectively, by F test; ns: not significant at the 5% level by F test.

Leaf tissue of nutrients

For the levels of foliar nutrients in Table 2, significant effects for the independent tested doses of superphosphate fertilizer or top-phos was observed with no difference between fertilizer for both the concentration of the nutrients P, K and S as to the contents of P in shoots of wheat plants. However, for Ca and Mg nutrients, a significant effect for fertilizers.

P concentration in leaf tissue, observed in Figure 2a, was influenced positively with the addition of P$_2$O$_5$ levels with fertilizer. Can evaluate that until the dose

of 200 kg ha^{-1} P$_2$O$_5$ gave the highest content of foliar P, proving the plant uptake of P added. However, has not reached the point of maximum accumulation in leaf tissue, other words, would be able to use higher dose to evaluate the content of P. Frandoloso et al. (2010) in an experiment with corn to test two types of phosphate fertilizers, found an increase in foliar P concentration up to a dose of 247 kg ha^{-1} of P$_2$O$_5$ for triple superphosphate and until the dose of 194 kg ha^{-1} P$_2$O$_5$ for phosphate rock.

Rosolem and Nakagawa (2005) evaluating nutrient content in leaves and grains of oat due fertilization with phosphorus and potassium, also found increased levels of nutrients flag leaf with phosphate. This increase may be a reflection of the increase in plant development provided by the addition of P in soil with low content of element (Nakagawa *et al.*, 2003), as in this work with 2.4 mg dm^{-3} in the soil and may result in greater root formation (Malavolta *et al.*, 2006), thus favoring the absorption and accumulation of nutrients.

For the P content in shoots (Figure 2a), linear growth with the doses for both fertilizer was observed,

demonstrating that the phosphorus, regardless of its source, made available soil P (Figure 4b), being absorbed by plants wheat. Thus, even while a reduction of SDM (Figure 1a) remained linear increase of P content in the tested doses, because the increase in the availability of soil P (Figure 4b) has kept the P uptake by wheat plants identified by the P content in the leaf tissue (Figure 2a), without reaching maximum level with available soil P provided with doses up to 200 kg ha^{-1} de P_2O_5.

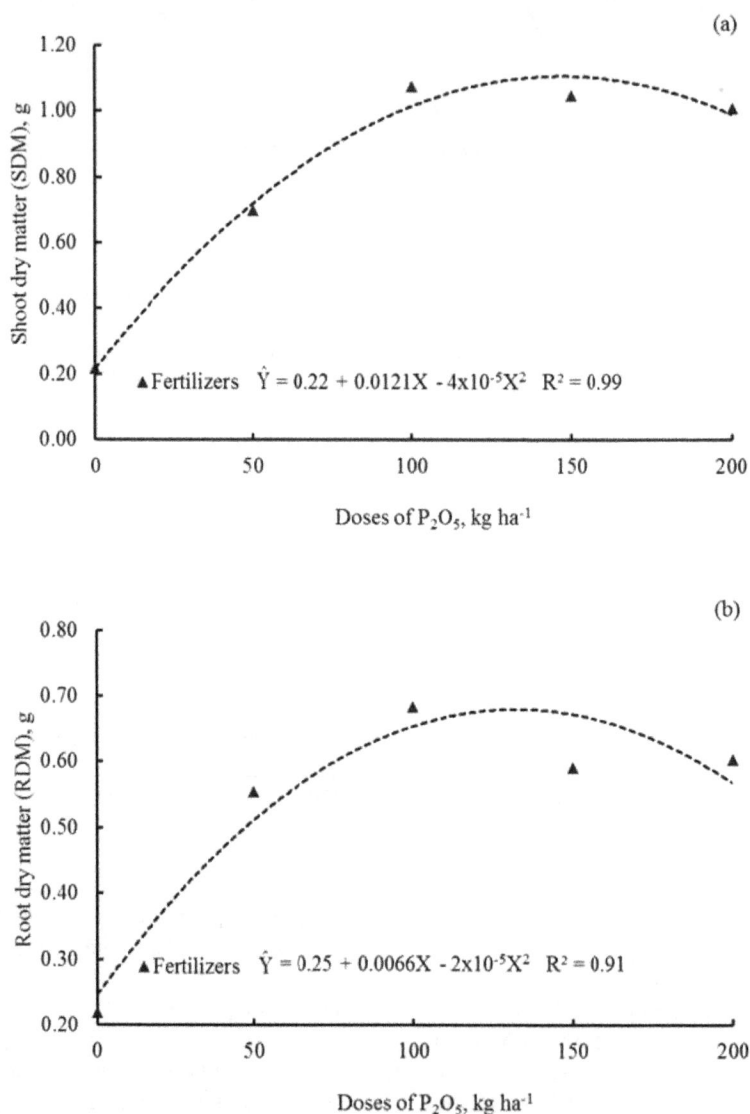

(a)

(b)

Figure 1. Production of shoot dry matter (SDM) (a) and root dry matter (RDM) (b) the elongation of wheat crop in function of P rates applied to average superphosphate fertilizer and top-phos. Marechal Cândido Rondon – PR, 2011.

Table 2. F values, coefficient of variation (CV%) and phosphorus (P), potassium (K), sulfur (S), calcium (Ca) and magnesium (Mg), and P content in shoots in wheat during elongation of wheat crop in terms of levels of phosphorus (P_2O_5) with superphosphate fertilizer (SFS) and top-phos. Marechal Cândido Rondon – PR, 2011

Fertilizer doses	P	K	S	Ca	Mg
Top-phos	---------------------------------- g kg^{-1} ----------------------------------				
--- kg P_2O_5 ---					
0	1.08	18.67	2.56	47.77	17.32
50	1.33	22.15	2.91	34.21	12.61
100	1.58	26.09	1.96	33.64	12.53
150	2.12	28.69	2.47	29.32	10.69
200	2.79	31.20	4.82	31.64	13.25
Simple superphosphate					
--- kg P_2O_5 ---					
0	0.88	17.62	2.01	30.42	11.76
50	1.35	23.64	2.32	32.23	10.39
100	1.64	25.16	2.53	26.04	10.85
150	2.27	28.94	2.89	32.14	12.12
200	2.50	31.25	3.23	28.66	9.85
Fertilizer					
Top-phos	1.78	25.36	2.94	35,32 a	13,29 a
Simple superphosphate	1.73	25.32	2.60	29,90 b	11,00 b
Medium values					
	1.75	25.34	2.77	32.61	12.14
F values					
Fertilizer	0.84 ns	0.01 ns	1.28 ns	5.28 **	8.96 **
Doses	11.91 **	11.12 **	4.46 **	2.15 ns	2.49 ns
Linear	92.50 **	43.80 **	10.60 **	5.98 *	-
Quadratic	0.90 ns	0.54 ns	4.83 *	2.27 ns	-
Fertilizer x Doses	0.39 ns	0.11 ns	1.62 ns	2.09 ns	2.24 ns
C.V. (%)	44.79	17.11	35.12	22.86	19.93

* e **: significant at 5% and 1%, respectively, by F test; ^{ns}: not significant at the 5% level by F test. Means followed by the same capital letter in the line do not differ by the F test.

In Figure 3 it can be seen that the dose from 56,8 Kg ha^{-1} de P_2O_5 was no increase in foliar S content, probably due to both fertilizers having sulfur in the formulation (Figure 3b), with a concentration of 16% the super simple, and 5% in the top-phos. In fact, according Novais *et al.* (2007), superphosphate can be recommended to provide sulfur fertilization in the row for annual crops in soils with chemical deficiency in this attribute. The presence of S in both fertilizer and increase of S in the leaf tissue, evidence the possibility to use them to provide part or all of the need for S to wheat crop. On the other hand, according to Raij (2011) and Rampim et al. (2013), the need for high doses of S can be provided through the gypsum being recommended dose of this product to clay soils with high P adsorption capacity (Vilar *et al.*, 2010).

Concentration of K in the sheet was affected at doses of P_2O_5, increase being detected through the use of phosphate fertilizers doses (Figure 3a). This may be related to the establishment of more appropriate nutritional conditions, resulting in an increase in SDM and RDM (Figure 1a and Figure 1b), and leaf P content (Figure 2a). Since, P is predominantly absorbed and translocated in plants under anionic form ($H_2PO_4^-$), the absorption of positive companion ion to the charge balance in the plant making required and potassium (K^+) (Faquin, 2005). Moreover, it may have been an increased uptake of K in the soil before the increased metabolic activity in the higher P levels, reflecting mainly the increase of SDM and RDM. According Faquin (2005), K acts on cell expansion and P in metabolic processes, increased metabolism to occur has consequently increased cellular growth, increased photosynthetic activity and increased formation of ATP, raising the application K.

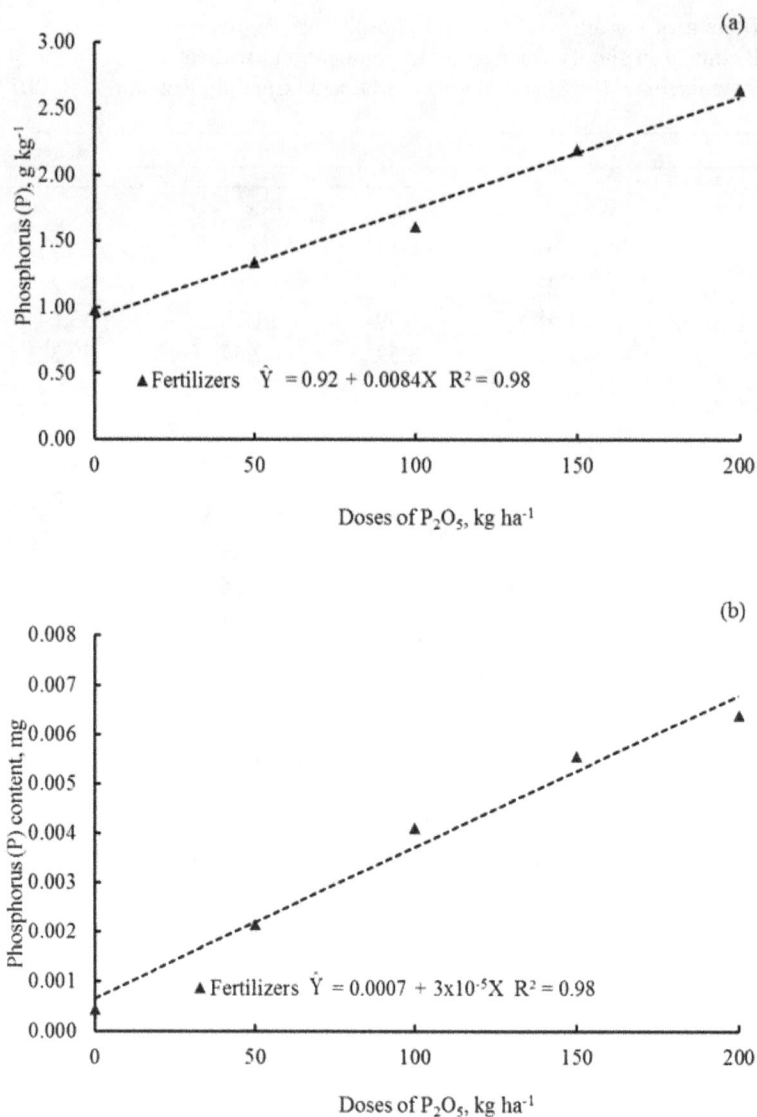

Figure 2. P (a) leaf tissue foliar and P content (b) in shoot elongation in wheat crop in terms of rates of phosphorus fertilizer applied to the average of superphosphate and top-phos. Marechal Cândido Rondon – PR, 2011.

For the calcium and magnesium leaf, there was significant difference between the tested products, and in both cases, the top-phos fertilizer increased concentration of Ca and Mg higher than superphosphate (Table 2). The uptake of Ca and Mg fertilization provided by the top-phos may be related to complexing present in the fertilizer. When the linear absorption of P with doses of P_2O_5 occurs may be due to increased Ca absorption, which can originate both fertilizer as soil and Mg in the soil that the complexing may have increased plant uptake because keeping the soil moisture at field capacity, and at work, enhances the absorption of P (Costa *et al.*, 2006.; Costa *et al.*, 2009) found in Figure 2a.

Thus, moisture can facilitate the mobility of complexes of Ca and Mg in the soil and can be absorbed into the transport stream (Novais *et al.*, 2007).

Chemical properties in the soil

For soil chemical properties, it was found a significant effect for the interaction between fertilizer and the doses for variable soil pH, while for the P exchangeable in the soil was a significant effect only for independent fertilizer doses tested, not being observed difference between them (Table 1).

In Figure 4a we observed a linear increase in soil pH with doses of phosphate fertilizers for both the SPS and for the top-phos; so that the SFS showed higher increase in pH per unit of added fertilizer in the top-phos. Thus, with the increase of phosphorus fertilization on soil acidity reduction was observed

and may be related to the presence of calcium in the formulation, which can raise the base saturation and consequently minimize the effect of the acidity due to the dilution effect, aimed at reducing the hydrogen ion activity in the soil (Novais et al., 2007; Raij, 2011).

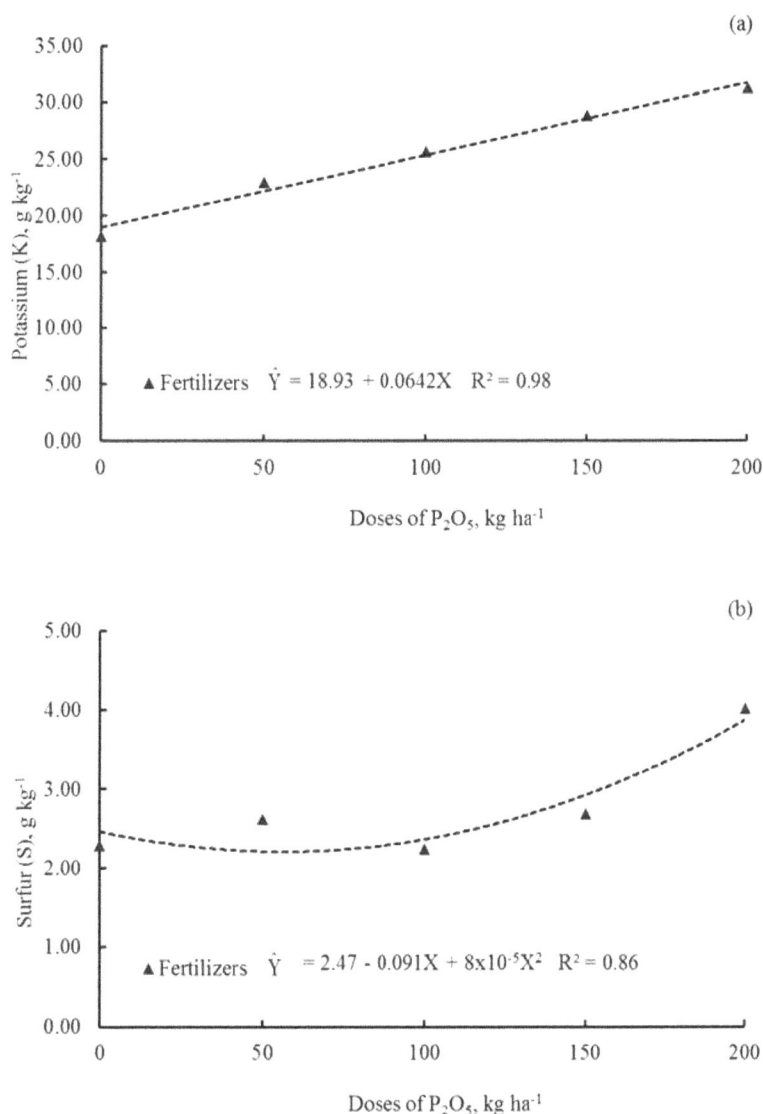

Figure 3. K (a) and S (b) leaf tissue foliar in shoot elongation in wheat crop in terms of rates of phosphorus fertilizer applied to the average of superphosphate and top-phos. Marechal Cândido Rondon – PR, 2011.

With increasing doses of P_2O_5, showed an increase in the amount of available P in soil (Figure 4b), other words, at the dose of 200 kg ha^{-1} was higher P_2O_5 concentration of P in soil, but not reaching the point of maximum increment in soil, regardless of fertilizer applied, since P exchangeable in soil was 2.40 mg dm^{-3}. Dallamea et al. (2005) studying the source of P,

triple superphosphate, found that the treatments had very low initial concentrations of P, had a lower response in the construction of P levels in the soil. This is due to the soil, possibly because they possess a greater number of sites for immobilization of P, which were not fulfilled, so part of P added to the soil binds to iron and aluminum oxides, remaining adsorbed and not expressing the same increase so pronounced in the levels of P (Vilar *et al.*, 2010). In the same study, in treatments that had high initial concentrations of P, were more responsive in available soil P with increasing doses of P_2O_5. This fact, as this paper demonstrates the direct relationship between P added to the soil by fertilization, providing direct increase in soil P and contribute to enhance the foliar P, requiring doses higher than those used to achieve the optimal level in the soil.

Fontoura *et al.* (2010) also found that the phosphorus in wheat, oats, corn, barley and soybeans with different sources (single superphosphate, triple superphosphate and rock phosphate) increased P exchangeable in the soil that had initially 6.9 mg dm^{-3} P in the middle of the layers 0-0.10 and 0.10-0.20 m depth in Oxisol alumínico. Moreover detected maximum productivity with the dose of 135 kg ha^{-1} P_2O_5 using triple superphosphate and increase linear in soil with rock phosphate to test the dose of 160 kg ha^{-1} of P_2O_5.

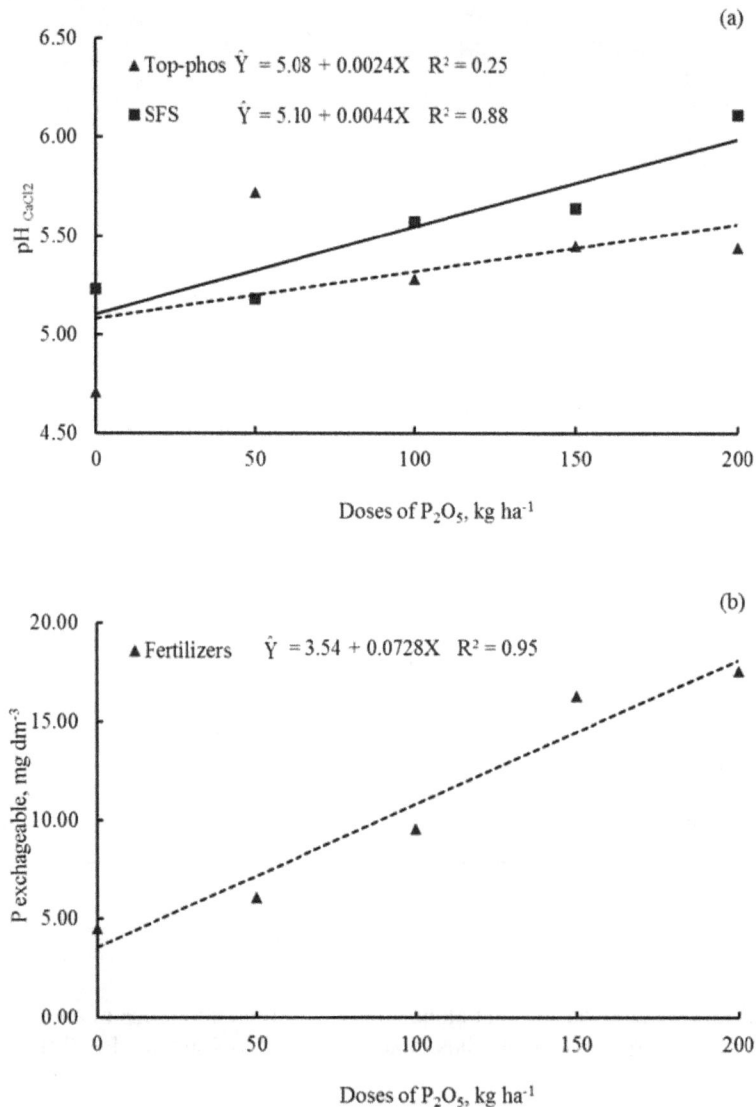

(a)

Top-phos $\hat{Y} = 5.08 + 0.0024X$ $R^2 = 0.25$

SFS $\hat{Y} = 5.10 + 0.0044X$ $R^2 = 0.88$

(b)

Fertilizers $\hat{Y} = 3.54 + 0.0728X$ $R^2 = 0.95$

Figure 4. pH in $CaCl_2$ and phosphorus (P) exchangeable in the soil in elongation of wheat crop in terms of rates of phosphorus applied to single superphosphate fertilizer and for top-phos. Marechal Cândido Rondon – PR, 2011.

To achieve high levels of available P in the soil at the higher doses of fertilizers, considered above those required by the wheat crop (Embrapa, 2009), was observed decreasing effect on the accumulation of yield dry matter from the dose of 151.25 kg ha^{-1} P$_2$O$_5$ independent of the use of SFS or top-phos. Which may indicate the dose that reached P concentrations in soil that may be feasible to perform only maintenance fertilization. Pavinato and Cereta (2004) showed a low response in grain yield to fertilization with P and K in soil with high levels of these nutrients, showing that under these conditions, the application to launch or online, the winter crop, would be used to fertilize in assembly need to culture in succession summer, optimization and concentration occurring in just one application during cultivation.

At high levels of soil chemical attributes, Altmann (2012) suggests the use of fertilizer system, increasing operational efficiency in systems with succession or rotation with wheat culture, directing the necessary fertilizer for crops used for the application only in the winter season, performing the supply deficient nutrients to increase productivity and enhance the production of waste (Fonseca et al., 2011), and thus increase the cycling of nutrients and consequently improve systems such as tillage, making it more sustainable.

In soil with low P concentrations, as seen in the soil of this study, according to Freitas et al. (1995), use of tolerant genotypes to acidity, more responsive to the application of lime and phosphorus, can make possible the wheat crop in soils low in P, with smaller investments. Thus, the increase in the concentration of P in the soil (Figure 4b) and consequently leaf P (Figure 2a), may give rise in productivity, as evidenced by the response observed in dry matter, both in shoots and in roots (Figures 1). Above all, the fertilization correction can be performed either with single superphosphate as with top-phos, and obtained similar level of available soil P (Figure 4b).

In soil with low P concentrations, as seen in the soil of this study, according to Freitas et al. (1995), use of tolerant genotypes to acidity, more responsive to the application of lime and phosphorus, can enable the wheat crop in soils low in P, with smaller investments. Thus, the increase in the P exchangeable in the soil (Figure 4b) and consequently leaf P (Figure 2a), may give rise in productivity, as evidenced by the response observed in dry matter, both in shoots and in roots (Figures 1). Above all, the fertilization correction can be performed either with single superphosphate as with top-phos, and obtained similar level of available soil P (Figure 4b).

Viability of fertilizer use

Considering the results of this work can make a comparison of the economic viability of these two fertilizers. The top-phos cost was R$ 1,600.00 a ton while superphosphate cost R$ 824.00. Considering that the top-phos has 28% P$_2$O$_5$ in their formulation and superphosphate has 18% P$_2$O$_5$ and also that the dose of greater accumulation of SDM (Figure 1) was 151.25 kg ha^{-1} P$_2$O$_5$: for top-phos would require 539 kg ha^{-1}, totaling R$ 858.00 per hectare for single superphosphate would require 838 kg ha^{-1} to R$ 701.00 per hectare. Thus, it was recommended to use fertilizer more cost-effective, which may vary according to the cost of each fertilizer in each region of cultivation of wheat.

CONCLUSION

Shoot dry matter gave maximum point build up dose of 151.25 kg ha^{-1} P$_2$O$_5$ and root dry matter point was the maximum dose of 165 kg ha^{-1} P$_2$O$_5$ independent of fertilizer.
Application of P$_2$O$_5$ levels with superphosphate fertilizer and top-phos increased levels of pH and available soil P and P content, K and S in the leaf tissue.
Recommend both, simple superphosphate with 18% P$_2$O$_5$ as top-phos, considered with 28% P$_2$O$_5$ for phosphate fertilizer in wheat crop, by selecting the fertilizer that provide best value for money, in this case the superphosphate.

Acknowledgements
For Coordination of Improvement of Higher Education Personnel (CAPES/PNPD), National Council for Scientific and Technological Development (CNPq) and Araucaria Foundation for Scientific and Technological Development of Parana (Araucaria Foundation).

REFERENCES

Altmann, N. 2012. Adubação de sistemas integrados de produção em plantio direto: resultados práticos no Cerrado. IPNI – International Plant Nutrition Institute. Informações Agronômicas, 140:1-20.

Bataglia, O.C., Camargo, C.E.O., Oliveira, O.F., Nagay, V., Ramos, V.J. 1985. Resposta à calagem de três cultivares de trigo com tolerância diferencial ao alumínio. Revista Brasileira de Ciência do Solo, 9(2):139-147.

Caierão, E. Sistemas de Produção: Embrapa Trigo. 2009. Versão Eletrônica. Set 2009. Disponível em: <http://sistemasdeproducao.cnptia.embrapa.

br/FontesHTML/Trigo/ Cultivo de Trigo/introducao.htm>. Acesso em: 20 mar 2013.

Camargo, C.E.O., Felício, J.C. 1987. Trigo, triticale e centeio: avaliação da eficiência ao fósforo e tolerância à toxicidade ao alumínio. Bragantia, 46 (2):203-215.

Clark, R.B. 1990. Physiology of cereals for mineral nutrient uptake, use and efficiency. In: Baligar, V.C., Duncan, R.R. Crops as enhancers of nutrient use. New York, Academic Press, 1990. p. 131-209.

Coimbra, R.R., Fritsche-Neto, R., Coimbra, D.B., Naoe, L.K., Cardoso, E.A., Raoni, D., Miranda, G.V. 2014. Relação entre tolerância do milho a baixo teor de fosforo no solo e responsividade a adubação fosfatada. Bioscience Journal, 30(2):332-339.

Costa, J.P.V., Barros, N.F., Albuquerque, A.W., Moura Filho, G., Antos, J.R. 2006. Fluxo difusivo de fósforo em função de doses e da umidade do solo. Revista Brasileira de Engenharia Agrícola e Ambiental, 10(4):828–835.

Costa, J.P.V., Bastos, A.L., Reis, L.S., Martins, G.O., Santos, A.F. 2009. Difusão de fósforo em solos de Alagoas influenciada por fontes do elemento e pela umidade. Revista Caatinga, 22(3):229-235.

Cunegatto, G., Poletto, S., Souza, E., Vahl, L. 2011. Concentração mínima de fosforo em plantas de trigo. XX Congresso de Iniciação Científica da UFPEL.

Dallamea, R.B.C., Amado, T.J.C., Eltz, F.L.F., Cubilla, M., Wendling, A., Fülber, R., Graminho, D.H. 2005. Níveis de fósforo no solo submetido a diferentes doses de adubação fosfatada sob sistema plantio direto na região sudoeste do Paraguai. XXX Congresso Brasileiro de Ciência do Solo, Recife.

Embrapa - Empresa Brasileira De Pesquisa Agropecuária. 2009. Manual de análises químicas de solos, plantas e fertilizantes. 2.ed. Brasília, 628p.

Embrapa - Empresa Brasileira De Pesquisa Agropecuária. 2013. Sistema brasileiro de classificação de solos. 3.ed. Brasília, Embrapa, 353p.

Faquin, V. 2005. Nutrição mineral de plantas. Lavras:UFLA/FAEPE, 186p.

Ferdandes, M.S. 2006. Nutrição mineral de plantas. Sociedade Brasileira de Ciências do Solo, Viçosa, 432p.

Fonseca, A.F.; Caires, E.F., Barth, G. 2011. Fertilidade do Solo e Nutrição de Plantas no Sistema Plantio Direto. 1a. edição. Ponta Grossa: Associação dos Engenheiros Agrônomos dos Campos Gerais e Universidade Estadual de Ponta Grossa, 327p.

Fontoura, S.F., Vieira, R.C.B., Bayer, C., Ernani, P.R., Moraes, R.P. 2010. Eficiência técnica de fertilizantes fosfatados em Latossolo sob plantio direto. Revista Brasileira de Ciência do Solo, 34(6):1907-1914.

Frandoloso, J.F., Lana, M.C., Fontaniva, S., Czycza, C.C. 2010. Eficiência de adubos fosfatados associados ao enxofre elementar na cultura do milho. Revista Ceres, 57(5):686-694.

Freitas, J.G., Camargo, C.E.O., Ferreira Filho, A.W.P., Castro, J.C. 1995. Eficiência e resposta de genótipos de trigo ao nitrogênio. Revista Brasileira de Ciência do Solo, 19:229-234.

Freitas, J.G., Camargo, C.E.O., Ferreira Filho, A.W.P., Pettinelli Júnior, A. 1994. Produtividade e resposta de genótipos de trigo ao nitrogênio. Bragantia, 53(2):281-290.

Gallo, D., Nakano, O., Silveira Neto, S., Carvalho, R.P.L., Batista, G.C., Berti Filho, E., Parra, J.R.P., Zucchi, R.A., Alves, S.B., Vendramin, J.D., Marchini, L.C., Lopes, J.R.S., Omoto, C. 2002. Entomologia agrícola. Piracicaba: FEALQ, 920p.

Gargantini, H., Conagin, A., Purchio, M.J. 1958. Ensaio de Adubação N - P - K em Cultura de Trigo. Bragantia, Boletim técnico do estado de São Paulo, 17(2).

Kimati, H., Amorim, L., Bergamin Filho, A., Camargo, L.E.A., Rezende, J.A.M. 1997. Manual de fitopatologia: Doenças das Plantas Cultivadas. ed. São Paulo: Editora Agronômica Ceres Ltda. v. 2, 705p.

Lopes, A.S., Guilherme, L.R.G. 2000. Uso Eficiente de Fertilizantes e Corretivos Agrícolas: Aspectos Agronômicos. Boletim Técnico 4-Associação Nacional para Difusão de Adubos, São Paulo.

Machado, V.J., Souza, C.H.E., Andrade, B.B., Lana, R.M.Q., Korndorfer, G. 2011. Curvas de disponibilidade de fósforo em solos com diferentes texturas após aplicação de doses crescentes de fosfato monoamônico. Bioscience Journal, 27(1):70-76.

Malavolta, E. 2006. Manual de nutrição mineral de plantas. São Paulo: Editora Agronômica Ceres, 638p.

Nakagawa, J., Cavariani, C., Bicudo, S.J. 2003. Adubação nitrogenada, fosfatada e potássica em aveia-preta. Cultura Agronômica, 12(1):125-141.

Nakagawa. J., Rosolem C.A. 2005. Teores de nutrientes na folha e nos grãos de aveia-preta em função da adubação com fósforo e potássio. Bragantia, 64(3):441-445.

Novais, R.F., Alvarez V., V.H., Barros, N.F., Fontes, R.L.F., Cantarutti, R.B., Neves, J.C.L. 2007. Fertilidade do solo. Sociedade Brasileira de Ciências do Solo, Viçosa, 1017p.

Oliveira, O.F., Camargo, C.E.O., Ramos, V.J. 1984. Efeito do fósforo sobre os componentes de produção, altura das plantas e rendimentos de grãos em trigo. Bragantia, 43(1):31-44.

Pavinato, P.S., Ceretta, C.A. 2004. Fósforo e Potássio na sucessão trigo/milho: épocas e formas de aplicação. Ciência Rural, 34(6):1779-1784.

Pinto-Zevallos, D.M., Zarbin, P.H.G.A. 2013. Química na agricultura: perspectivas para o desenvolvimento de tecnologias sustentáveis. Química nova, 36(10):1509-1513.

Raij, B. 2011. Fertilidade do solo e manejo de nutrientes. Piracicaba: Internacional Plant Nutrition Institute, 420p.

Rampim, L., Lana, M.C., Frandoloso, J.F. 2013. Fósforo e enxofre disponível, alumínio trocável e fósforo remanescente em latossolo vermelho submetido ao gesso cultivado com trigo e soja. Semina: Ciências Agrárias, 34(4):1623-1638.

Resende, A.V., Neto, A.E.F. 2007. Aspectos relacionados ao manejo da adubação fosfatada em solos do cerrado. 1.ed. Planaltina, Embrapa Cerrados, 32p.

Saeg. 1999. Sistema para análises estatísticas. Versão 8.0. Viçosa, MG, Universidade Federal de Viçosa.

Santos, E.A., Kliemann, H.J. 2005. Disponibilidade do fósforo de fosfatos naturais em solos de cerrado e sua Avaliação por extratores químicos. Pesquisa Agropecuária Tropical, 35:139-146.

Souza, P.G.A. 1996. Resposta diferencial à calagem e ao fósforo de três cultivares de trigo (Triticum aestivum L.) com diferentes graus de tolerância ao alumínio. Jaboticabal, Tese (Doutorado) - Faculdade de Ciências Agrárias e Veterinárias/UNESP, 137p.

Timacagro. 2012. TOP-PHOS expedition. Disponível em: < http://www.timacagro.com.br/top-phos-expedition/> Acesso em 04 set. 2012.

Usda – United States Department of Agriculture. Produção mundial de trigo. Novembro 2012. Disponível em: http://www.abitrigo.com.br/pdf/PROD-TRIGO.pdf. Acervo 2012.

Vilar, C.C., Vilar, F.C.M. 2013. Comportamento do fósforo em solo e planta. Campo Digital: Revista de Ciências Exatas e da Terra e Ciências Agrárias, 8(2):37-44.

Vilar, C.C., Costa, A.C.S., Hoepers, A., Souza Júnior, I.G. 2010. Capacidade máxima de adsorção de fósforo relacionada a formas de ferro e alumínio em solos subtropicais. Revista Brasileira de Ciências do Solo, 34(4):1059-1068.

Zanini, F.H., Schultz, T.A., Castagnara, D.D., Oliveira, P.S.R., Neres, M.A. 2009. Adubação fosfatada sobre a produção de matéria seca de forrageiras tropicais. Synergismuss cyentifica UTFPR, 4(1).

20

Assessment of Chemical Composition and *In Vitro* Degradation Profile of some Guinea Savannah Browse Plants of Nigeria

S.A. Okunade[*], O.A. Isah, R.Y. Aderinboye and O.A. Olafadehan

Animal Production Technology Department, Federal College of Wildlife Management, P.M.B 268, New Bussa, Nigeria.
Department of Animal Nutrition, Federal University of Agriculture, Abeokuta, Nigeria
Department of Animal Science, University of Abuja, Abuja, Nigeria
Email: saokunade2013@gmail.com
[*]Corresponding author

SUMMARY

The study was conducted to estimate the nutritive value of six indigenous browse fodders *(Etanda africana, Piliostigma thonningii, Detarium microcarpum, Daniellia oliveri, Pterocarpus erinaceus, and Afzelia africana)* by the evaluation of chemical composition, anti-nutritional factors and *in vitro* gas characteristics. All samples (g/100g DM) had high CP (12.6–24.7), moderate fibre concentrations (NDF, 34.7–54.6; ADF, 19.7–35.2 and lignin, 7.36–12). There were significant differences ($P < 0.05$) in NDF, ash, ether extract, hemicellulose, cellulose and mineral concentrations among the browse fodders. Except for condensed tannins which were similar among the browse fodders, other anti-nutritional factors were different ($P < 0.05$). The relative feed values of the selected legume browses ranged from 114.43 in *E. africana* to 202.94 in *A. africana*. Gas volume (ml/200mg DM), methane (ml/200mg DM), methane/total gas volume (v:v), metabolisable energy (MJ/kg DM), organic matter digestibility (%), short chain fatty acids (µmol) and *in vitro* dry matter degradability (%) ranged from 19- 34, 8.66-11.33, 0.29-0.46, 4.53-6.48, 35.73-49.06, 0.15-0.43 and 46-67 respectively. Results show that the browse species have good nutrient profile, low and safe levels of anti-nutritional factors and relatively high degradability which qualify them as suitable feed supplements to low quality basal diets for ruminants.

Key words: browse fodder; chemical composition; relative feed value; anti-nutritional factors; in vitro gas degradability profile.

INTRODUCTION

The relevance of evaluating the nutritional value of indigenous shrubs, trees and browse plants is evident (Topps, 1992; Nherera *et al.*, 1999; Cerillo and Juarez, 2004) as their foliage can make important contributions to the protein and energy consumption of ruminant animals. This is particularly important during the critical dry season when forage availability and quality are severely limited and affected.

The nutritive value of a ruminant feed is determined by the concentration of its chemical components, as well as the rate and extent of digestion. *In vivo* determination of nutritive value most accurately reflects feed nutritive value. However, *in vivo* techniques are strenuous; it is difficult to assess large number of samples. Also, determining the digestibility of feeds *in vivo* is laborious, expensive; requiring large quantities of feed, and is largely unsuitable for single feedstuff thereby making it unsuitable for routine feed evaluation (Getachew *et al.*, 2004). There are a number of *in vitro* techniques available to evaluate the nutritive value of feeds at relatively low cost.

The leaves of the evergreen tree and shrub are used as emergency food by ruminants in the Guinea savanna region of northern Nigeria where it is customary to feed non-conventional feedstuffs such as browse species (Olafadehan, 2011). However, there is little information on their nutritive values. Chemical composition, in combination with *in vitro* digestibility and ME content can be considered useful indicators for preliminary evaluation of the potential nutritive value of previously uninvestigated shrub and tree leaves (Ammar *et al.*, 2005).

The rational use of fodder and browse requires accurate information about the nutritive value of these alternative feed resources. Along with the information on the nutrient content, the presence of anti-nutritional compounds is of special interest for this sort of feedstuff, as high concentrations of tannins are found in many browse trees (Dube *et al.*, 2001). These compounds can impair the digestive utilization of the feed ingested by the animal and even induce toxicity when consumed above the threshold level. There is therefore the need for continuous screening of non-conventional browse plants to identify those with good potentials as livestock fodder which can serve as alternatives to those species that have already been evaluated.

Current chemical analytical techniques do not reflect the biological effects of tannin therefore the use of *in vitro* techniques has been proposed to supplement the chemical analysis (Nsahlai *et al.*, 1994). The gas production technique has proved to be efficient in determining the nutritive value of feeds containing anti-nutritive factors (Siaw *et al.*, 1993). The objective of this study was to investigate the nutritive value of foliage from selected tropical browse species based on chemical analysis, quantification of anti-nutritional factors and in vitro gas production technique.

MATERIALS AND METHODS

Experimental site

This study was carried out at the Federal College of Wildlife Management, New Bussa, Niger State. It is located between latitudes 7^0 $80'$ and 10^0 $00'$N and longitudes 4^0 $30'$ and 4^0 $33'$ E and has mean annual temperature of about 34^0C with relative humidity of about 60%.

Collection of forages and processing

Fresh leaves from the branches of six selected browse plant species (*Etanda africana, Piliostigma thonningii, Daniellia oliveri, Detarium microcarpum, Pterocarpus erinaceus and Afzelia africana*) were harvested from several stands in the rangeland of the Federal College of Wildlife Management New Bussa, Niger state, Nigeria and its environment between April and May 2013. Samples from fresh foliage of the selected browse plants were oven-dried to constant weight at 60°C for 72 hours to determined dry matter. Dried sample were milled to pass through 1mm screen for subsequent laboratory analysis after stored in air tight container.

Chemical analysis

Samples of green forages were analyzed according to the standard methods of AOAC (2002) for dry matter (DM), crude protein (CP), ether extract (EE) and ash. Neutral detergent fibre (NDF), acid detergent fiber (NDF) and acid detergent lignin (ADL) were determined as described by Van Soest *et al.* (1991). Hemicellulose and cellulose were estimated as the difference between NDF and ADF and ADF and ADL, respectively. Non-fiber carbohydrates were estimated as $100-CP-NDF-EE-ash$ (Sniffen *et al.*, 1992). Condensed tannins were determined according to the procedures of Makkar (2005), saponins by Babayemi *et al.* (2004), phytate as phytic acid using the method of Maga (1982) and oxalate was determined by the method of (AOAC, 2002). Calcium, magnesium, potassium and phosphorus were analysed using atomic absorption spectrophotometer.

In vitro gas production study

The *in vitro* gas production was determined according to Menke and Steingass (1988). Three Red Sokoto goats fed a mixed diet of *Pannicum maximum* (60% DM) and concentrates (40% DM) were used. The concentrate feed consisted of 40% corn, 10% wheat offal, 10% palm kernel cake, 20% groundnut cake, 5% soybean meal, 10% dried brewers' grain, 1% common salt, 3.75% oyster shell and 0.25% fish meal. Feeds were offered in two equal meals at 07:00 and 18:00 h, respectively, to the goats. The animals had free access to water and mineral licks. Rumen fluid was collected from the goats with the use of suction tube prior to morning feeding. The collected rumen liquor was strained through four layers of cheese cloth and kept at 39°C. All laboratory handling of rumen fluid was carried out under a continuous flow of carbon IV oxide. Samples (200 mg) of the oven dry and milled leaves were accurately weighed into 100 ml glass syringes fitted with plungers. *In vitro* incubation of the samples was conducted in triplicates. Syringes were filled with 30 ml of medium consisting of 10 ml of rumen fluid and 20 ml of buffer solution (g/liter of 1.985 $(Na_2)HPO_4$ + 1.302 KH_2PO_4 + 0.105 $MgCl_2.6H_2O$ + 1.407 NH_2HCO_3 + 5.418 $NaHCO_3$ + 0.390 Cystene HCl + 0.100 NaOH) and three blank samples containing 30 ml of medium (inoculums and buffer) only were incubated at the same time. The syringes were placed in a rotor inside the incubator (39°C) with about one rotation per min. The gas production was recorded at 3, 6, 9, 12, 18, 24, 36 and 48 h. At post incubation period, 4 ml of (10M) Sodium hydroxide (NaOH) was dispensed into the each incubated sample. Sodium hydroxide was added to absorb carbon-dioxide that was produced during the process of fermentation and the remaining volume of gas was recorded as methane according to the report of Fievez *et al.* (2005). The average of the volume of gas produced from the blanks was deducted from the volume of gas produced from sample.

In vitro dry matter degradability (IVDMD)

After 24h digestion, the samples were transferred into test tubes and centrifuge for 1hour in order to obtain the residues which were then filtered using Whatman No 4 filter paper by gravity and the residues placed in crucible for drying at 65°C for 24h. The dry residues were weighed and digestibility calculated using the equation as follows:

IVDMD (%) = Initial DM – DM residue-Blank)* 100
Initial DM Input

Statistical analysis and calculations

Data were subjected to one way of ANOVA in completely randomized design using version 9.1 of SAS software (SAS Institute, 2003). Significance difference between individual means was separated by Duncan's procedure of the same software. Mean differences were considered significant at $P < 0.05$.

Relative feed value (RFV) of the legume tree leaves was calculated from the estimates of dry matter digestibility (DMD) and dry matter intake (DMI) according to Rohweder *et al.* (1978). Following are the equations used:

DMD (%) = 88.9 – (0.779 – ADF %)
DMI (as % of body weight) = 120/NDF %
RFV = (DMD% × DMI %)/1.29

Forage quality was determined using the standard assigned by Hay Market Task Force of American Forage and Grassland Council (Table 1). Metabolisable energy (ME, MJ/Kg DM) and organic matter digestibility (OMD%) of the incubated samples were estimated at 48 hr post gas collection with equation according to Menke and Steingass (1988), while short chain fatty acid (SCFA, µmo) at 24 h post gas collection was computed using linear equation by Makkar (2005).

ME =2.20 + 0.136GV + 0.057CP + 0.0029CF

OMD = 14.88 + 0.889 GV + 0.45 CP + 0.651 ash
(Menke and Steingass, 1988)

SCFA(µmol)=0.0222GV (at 24 hr) – 0.00425
(Makkar, 2005)

where: Total gas volume (GV) is expressed as (ml/0.2 g DM) CP and ash as g/kg DM. CP and CF are crude protein and crude fibre of the incubated samples respectively.

RESULTS

Chemical composition and fibre fraction

There were differences ($P < 0.05$) in the chemical composition of the browse fodders except for the DM and OM (Table 2). Crude protein (CP) contents were highest (P < 0.05) in *A. africana* (24.64 g/100 g DM) and least in *D. microcarpum* (12.64 g/100 g DM). Ash and ether extract contents varied ($P < 0.05$) from 4.40 in *D. microcarpum* to 8.99 g/100g DM in *P. thonningii* and 37.45 in *D. oliveri* to 63.8 g/100 g DM in *A. africana*, respectively. Non-fibre carbohydrate values were highest ($P < 0.05$) in *D. microcarpum* (31.22 g/100 g DM). There were variations ($P < 0.05$) in the NDF, ADF, cellulose and hemicellulose contents of the browse species. Acid detergent lignin was highest ($P < 0.05$) in *P. erinaceus* (12.00 g/100 g DM).

Macro mineral concentration

Ca concentration was different ($P < 0.05$) among the browse fodders, the value ranged between 9.0 g/kg DM in *D. oliveri* to 5.9 g/Kg in *E. africana*. Highest ($P < 0.05$) P concentration was observed in *P. erinaceus* (6.7 g/kg DM), while *D. microcarpum* (2.8 g/Kg DM) had the least (P < 0.05) concentration. K concentrations were higher ($P < 0.05$) in *D. oliveri* (9.6 g/kg DM) while Mg concentration was highest in and *P. erinaceus*, (5.8 g/kg DM)) relative to other browse fodders (Table 3).

Anti-nutritional factors concentration

Plant phytochemical components (saponins, phytate and oxalate) in the browse species forages were different ($P < 0.05$) except for the tannin concentration levels (Table 4).

Relative feed value (RFV)

Estimated DMD and DMI of the browse fodders ranged (P < 0.05) from 61.50% in *D. microcarpum* to 73.53% in *A. africana* and 2.23% in *P. thonningii* to 3.55% in *A. africana*, respectively (Table 5). *A. africana* had the best RFV value (202.94%), while *E. africana* had the least RFV (114.43%). The grade of the legume browses ranged from 4 to prime in the following order: *D. microcarpum* (4), *E. africana* (3), *P. thonningii* (2), *D. oliveri* (2), *P. erinaceus* (1) and *A. africana* (prime), respectively, according to the standard assigned by Hay Market Task Force of American Forage and Grassland Council.

Table 1. Legume, grass and legume-grass mixture quality standards

Quality standard[a]	CP	ADF (DM %)	NDF (DM %)	RFV[b]
Prime	>19	<31	<40	>151
1	17- 19	31-40	40-46	151-125
2	14 - 16	36-40	47-53	124-104
3	11 - 15	41-42	54-60	102.87
4	8 - 10	43-45	61-65	86.75
5	<8	>45	>	<

[a]Standard assigned by Hay Market Task Force of American Forage and Grassland Council;
[b]Relative Feed Value (RFV) – Reference hay of 100 RFV contains 41% ADF and NDF and 53% NDF

Table 2. Chemical composition and fibre fraction (g/100 g DM) of the selected browse foliages

Browse species	DM	OM	CP	Ash	EE	NFC	NDF	ADF	ADL	HC	CEL
E. africana	91.66	86.39	20.33[b]	5.27[bc]	4.37[b]	15.47[e]	54.56[a]	33.76[a]	7.73[b]	20.79[b]	26.03[a]
P. thonningii	91.83	82.84	14.23[c]	8.99[a]	3.89[b]	18.03[d]	53.86[a]	27.11[b]	8.98[b]	26.75[a]	18.15[b]
D. microcarpum	89.52	85.11	12.64[cd]	4.40[c]	5.18[ab]	31.22[a]	46.53[ab]	23.65[b]	7.36[b]	22.87[b]	16.31[b]
D. oliveri	91.83	84.04	14.00[c]	7.79[ab]	3.75[b]	27.08[b]	49.80[ab]	26.27[b]	9.36[b]	23.62[b]	16.91[b]
P. erinaceus	92.97	85.39	18.42[b]	7.63[ab]	4.73[b]	20.00[d]	45.79[ab]	35.15[a]	12.00[a]	15.38[c]	23.15[bc]
A. africana	90.79	83.57	24.69[a]	7.22[ab]	6.38[a]	23.78[c]	34.36[b]	19.73[c]	8.07[b]	14.63[c]	11.66[c]
Mean	91.43	84.56	17.49	6.83	4.72	22.59	48.28	27.61	8.91	20.67	18.49
SEM	1.24	1.25	0.69	0.69	0.73	1.39	1.75	1.71	0.98	1.93	1.60

[abc] means within the same column with different superscripts are significantly different ($P < 0.05$).
DM = dry matter, OM, organic matter; CP = crude protein, EE = ether extract, NFC, Non-fiber carbohydrates; NDF, neutral detergent fibre, ADF = acid detergent fibre, ADL = acid detergent lignin, HC = hemicellulose, CEL = cellulose,

Table 3: Macro mineral composition (g/100 g DM) of the selected browse foliages

Browse species	Calcium	Phosphorus	Ca:P	Magnesium	Potassium
Etanda africana	5.9[c]	4.4[b]	1.3[d]	3.8[c]	8.3[b]
Piliostigma thonningii	8.2[ab]	3.7[b]	2.2[b]	3.9[c]	8.1[b]
Detarium microcarpum	7.5[b]	2.8[c]	2.7[a]	3.2[cd]	7.2[b]
Daniellia oliveri	9.0[a]	4.4[b]	2.0[b]	2.5[d]	9.6[a]
Pterocarpus erinaceus	6.0[c]	6.7[a]	0.9[d]	5.8[a]	8.3[b]
Afzelia africana	7.7[b]	4.1[b]	1.9[bc]	5.0[b]	8.3[b]
Mean	7.4	4.5	1.6	4.0	8.3
SEM	1.5	1.2	0.5	0.01	0.15
Recommended requirement range[b]					
Minimum	1.9	1.2	1:1	0.1	5
Maximum	8.2	4.8	2:1	2.5	10

Means within the same column with different superscripts are significantly different ($P < 0.05$).
[b] Recommended range of mineral elements (for all classes of ruminants) as suggested by the National Research Council and summarized by McDowell (1997).

Table 4: Anti-nutritional concentration (g/kg DM) of the selected browse foliages

Browse species	Tannin	Saponin	Phytate	Oxalate
Etanda africana	0.60	2.08[a]	3.79[ab]	0.43[ab]
Piliostigma thonningii	0.34	0.57[b]	4.81[ab]	1.26[ab]
Detarium microcarpum	0.81	1.06[ab]	7.82[a]	1.58[ab]
Daniellia oliveri	0.50	0.38[b]	6.42[ab]	1.48[ab]
Pterocarpus erinaceus	0.40	2.02[a]	5.38[ab]	0.26[b]
Afzelia africana	0.90	0.59[b]	0.92[b]	1.79[a]
Mean	0.59	1.12	3.89	1.13
SEM	0.41	0.54	1.01	0.49

Means within the same column with different superscripts are significantly different ($P < 0.05$).

Table 5: Dry matter digestibility, dry matter intake and relative feed value of selected browses

Browses	DMD (%)	DMI (% BW)	RFV	Quality standard
Etanda africana	61.52[c]	2.35[b]	114.43[c]	3
Piliostigma thonningii	67.78[b]	2.23[b]	117.14[c]	2
Detarium microcarpum	61.50[c]	2.44[b]	116.76[c]	4
Daniellia oliveri	68.43[b]	2.48[b]	132.69[bc]	2
Pterocarpus erinaceus	70.47[ab]	2.64[ab]	145.60[b]	1
Afzelia africana	73.53[a]	3.55[a]	202.94[a]	Prime
Mean	67.21	2.62	143.26	-
SEM	1.73	0.31	19.99	-

Means within the same column with different letters are significantly different (p < 0.05).
DMD = dry matter digestibility (%), DMI = dry matter intake (% of body weight), RFV = relative feed value

In vitro gas production

In vitro degradation profile of the different foliages is presented in Table 6. The final net gas volumes at 48 h after incubation and CH_4 production were significantly different ($P < 0.05$) among the browse fodders. Ratio of methane/total gas volume (v: v)

produced which indicates the methanogenic property of the browses was least ($P < 0.05$) in *A. africana*. The estimated ME (MJ/kg DM), OMD% and SCFA (µmol) were higher ($P < 0.05$) in *A. africana* compared to other browse fodders. Figure 1 shows the *in vitro* gas production pattern of the browses.

Table 6: *In vitro* gas production (ml/200 mg DM), Metabolizable energy (ME MJ/kg DM), Organic matter (OMD %), Short chain fatty acids (SCFA µmol), Methane (CH_4 ml/200g DM) production of the browse species

Browse species	NGV (48 h)	CH_4	CH_4/total gas (v:v)	ME	OMD	SCFA	IVDMD
Etanda Africana	19.00[c]	8.66[b]	0.46[a]	5.43[b]	40.49[b]	0.29[c]	46[b]
Piliostigma thonningii	26.00[b]	9.33[ab]	0.36[b]	5.28[b]	41.36[b]	0.32[b]	63[a]
Detarium microcarpum	21.00[c]	9.33[ab]	0.44[ab]	4.81[c]	32.11[d]	0.15[d]	59[a]
Daniellia oliveri	20.70[c]	8.70[b]	0.42[ab]	4.53[c]	35.73[c]	0.19[d]	63[a]
Pterocarpus erinaceus	34.00[a]	11.33[a]	0.38[b]	5.28[b]	45.32[b]	0.32[b]	66[a]
Afzelia Africana	33.50[a]	11.00[a]	0.29[c]	6.48[a]	49.06[a]	0.43[a]	67[a]
SEM	1.06	0.44	0.03	0.15	1.07	0.02	2.0

Means in the same column with different superscripts are significantly different ($P < 0.05$)

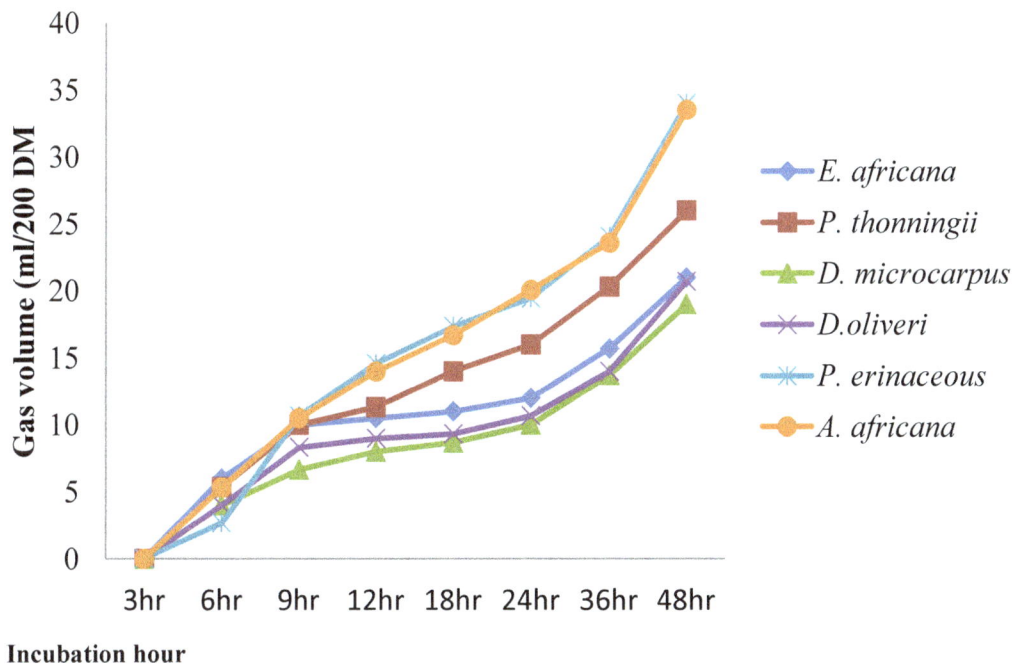

Figure 1: *In vitro* gas production characteristics

DISCUSSION

The range of the CP content of the browse fodders is in agreement with previous reports (Osuga *et al.*, 2006; Bouazza *et al.*, 2012). The lowest value of CP (12.64 g/100 g DM) for *D. microcarpum* is well above the range of 7.0–8.0 g/100 g DM suggested as critical limit below which intake of forages by ruminants and rumen microbial activity would be adversely affected (Van Soest, 1994). The CP content of the browse fodders is an indication of their nutritional quality since CP content is a very important index of nutritional quality of a feed. This justifies their use as supplements to poor quality natural pastures and crop residues. The NFC contents of the browse species should be adequate to stimulate NH_3–N utilization in the rumen (Tylutki *et al.*, 2008). The optimal concentration of NFC is important in ruminant diets to avoid acidosis and other metabolic problems. Diets with excess NFC can cause ruminal upsets and health problems (Nocek, 1997). The fibre fraction contents of the plant species were generally moderate and within the limits established by NRC (1978) for ruminant animals for ensuring proper digestion and rumination. The mean NDF and ADF values of 48.28 and 27.61 g/100 g DM were low to moderate when compared with low quality roughages, which ruminants can effectively degrade (Arigbede and Tarawali, 1997). The low to moderate fibre contents of the browse fodders suggests their high nutritive value since fibre plays a significant role in voluntary intake and digestibility. The range of cellulose concentration shows that the fodders have the potentials to support intestinal movement, proper rumen function and promote dietary efficiency. Humphreys (1991) opined that the higher the hemicellulose fraction, the higher is the feed value.

Most of the foliages contained high Mg relative to requirements (Table 3). Therefore, deficiencies of these elements seem unlikely in ruminants maintained solely on these foliages. Ca concentration compares favourably with other Nigerian browse species (Yusuf et al., 2013) and within the recommended range (McDowell, 1997) except for D. oliveri which was higher. Phosphorus in the plant foliages was comparable to values reported by Topps (1992) for several browse plants and was within the required range for ruminants (McDowell, 1997) except for P. erinaceus. The Ca:P ratio compared to those recommended for ruminants (McDowell, 1997, Ahn et al., 1989), which may not create problem with vitamin D metabolism (ARC, 1984). Ca and P are very important for growing ruminants and too much P compared to Ca can lead to urinary calculi. The content of K was within the required range for ruminants (McDowell, 1997). All the foliages contained high Mg relative to requirements. It appears that the deficiencies of the studied macrominerals seem unlikely in ruminants maintained solely on these foliages. Most of the minerals except P are present in more than adequate levels when compared with NRC recommendation for ruminant animals (NRC, 1978).

The levels of CTs recorded in this study are much below the range of 60 to 100 g/kg DM, considered to depress feed intake and growth (Mbomi et al., 2011). Therefore, the browse species contained CTs at levels beneficial to ruminants because CTs at low level produce mild or low protein binding effect (Olafadehan, 2013). Similarly, CT-containing forage minimizes methane emission by ruminants (methane mitigation), in addition to other benefits, when not included at a high proportion of the diet (Bodas et al., 2012; Cieslak et al., 2012). Saponin levels in all the samples were lower than the tolerable level of 15-20 g/kg DM reported for goats (Onwuka, 1983), which suggests the levels reported herein are not likely to affect nutritional potentials of the browses to ruminants. Feedstuffs containing saponin have been shown to act as defaunating agents (Teferedegne, 2000) and capable of reducing methane production. Both the phytate and oxalate concentrations are lower than the values reported by Fadiyimu et al. (2011). The low level of phytate in the forages indicates the potential of the browse fodders to make minerals available to the ruminants since phytates bind

minerals like Ca, Mg Fe and Zn, interfere with their metabolism and cause muscular weakness and paralysis. Oxalates have been implicated for decreased availability of Ca during digestion (Norton, 1994). However, the low content of oxalate in the experimental fodder may not interfere with Ca utilization.

The mean estimated DMD and DMI compared favourably with mean values of 65.07% and 3.47%, respectively, reported by Hanlin et al. (2011) for several tropical legumes in China. However, the overall RFV is lower than mean value (181.00), while the quality standard followed the same trend as reported by the same authors. All the selected six legume browses except D. microcarpum may be regarded as having high relative nutritive value for ruminants in Nigeria based on this ranking.

Although, gases produced during rumen fermentation process are waste products and of no nutritive value to the ruminant, gas production tests are used routinely in feed research because gas volumes are related to both the extent and rate of substrate degradation (Blummel et al., 1997). The cumulative volume of the gas production by the browse fodders increased with increasing hour of the in vitro incubation (Fig. 1). The significant differences among the browse foliages for their in vitro gas production and fermentation characteristics are in agreement with previous studies on tropical browses in Kenya (Abdulrazak et al., 2000; Osuga et al., 2006) and Nigeria (Isah et al., 2012). In the current study, all the browse foliages generally had moderate gas production potential. The in vitro gas production pattern of the forages shown in Fig. 1 indicates that more degradation of DM was still possible beyond 48 h. Many factors such as the nature and level of fibre, the presence of secondary metabolites and potency of the rumen liquor for incubation have been reported (Babayemi et al., 2004) to determine the amount of gas to be produced during fermentation. In the current study, it appears the secondary metabolites more than fibre content influence in vitro gas production and hence degradability. Higher gas production and the extent of in vitro fermentation of P. erinaceus and A. africana suggests that these substrates are of higher nutritional value than the other browse species, in agreement with Isah et al. (2012). The low gas production of E. africana could be attributed to its high amount of saponin (Table 6). Saponin is known to deter the activities of bacteria in the rumen (Babayemi et al., 2004). Although, high methane implies an energy loss to the animal, forage with a higher degradability will lead to more intensive fermentation in the rumen (Rinne et al., 1997) and thereby increase in CH_4 production. This possibly explained why *P. erinaceus* and *A. africana* had the higher CH_4 production compared with other browse fodders. However, when ranking plants according to their methanogenic property, the proportion of methane-to-total gas ratios is more relevant than absolute methane formation: a low value for this proportion indicates a low methanogenic potential of the digestible part of the feed, i.e. fewer methane production per unit net gas volume production (Moss *et al.*, 2000). In the light of this, *A. africana*, with the least (0.29 ml/200 mg DM) CH_4 production value (on the basis of methane-to-total gas volume ratio), was the most promising fodder regarding a low methanogenic potential, while *E. africana* with the highest CH_4 production on the basis of methane-to-total gas volume ratio may contribute most to the green house effect if fed solely to ruminant animals. A mutual relationship exists between total gas production and ME, OMD and SCFA. The estimation of the ME values is imperative for purposes of ration formulation and to set economic value of feeds for other purposes. The values of the ME agree with that of Isah *et al.* (2012), but higher than (3.28 - 3.83 MJ/kg) value recorded by Getachew *et al.* (2002). The highest ME of *A. africana* coupled with its highest CP suggests that such fodder may enhance microbial protein synthesis as it may promote better synchronization of fermentable energy and degradable N in the rumen (Olafadehan, 2014). The browse foliages generally had relatively high IVOMD, but comparable to the values reported for some tropical browse species (Anele *et al.*, 2008). This relatively high IVOMD is an indication of the high nutritive value of the browse foliages when used in ruminant feeding. Lower fibre fractions in *A. africana* relative to other browse fodders may have resulted in the higher values for IVOMD and SCFA (Van Soest, 1994). The IVDMD values obtained in the present study except for *E. africana* were found to be slightly higher than IVDMD values reported by Mokonnen *et al.* (2009) for multipurpose fodder tree and shrub species in central highlands of Ethiopia. Low level of tannins and moderate levels of NDF and ADF may be responsible for generally high IVDMD in fodders (Njidda, 2014) since, tannins in the NDF and ADF fractions are tightly bound to the cell wall and cell protein and seem to be involved in decreasing digestibility (Reed *et al.*, 1990). In reality, when IVDMD falls below 550 g/kg there is physical limitation on the rate of eating and the rate of digestion and passage through the gastrointestinal tract is restricted while live weight loss becomes inevitable (SCA, 1990). Based on this, the generally high IVDMD demonstrates the high nutritive potential of the browses when used in livestock feeding.

CONCLUSION

Results show that all the selected browses are rich in crude protein as well as macro minerals, have moderate fibre level and low concentrations of anti-nutritional factors. The generally high *in vitro* degradability and estimated dry matter digestibility suggest their nutritive potential as alternative low cost sources of good protein supplements to poor quality roughages for ruminant feeding especially during the dry season. The browse species are thus promising fodders that can be used for sustainable ruminant production in the tropics. Further research should, however, be conducted to establish their fodder value in *in vivo* trials.

REFERENCES

Abdulrazak, S.A., Fujihara T., Ondiek, J. K., Orskov, E. R. 2000. Nutritive evaluation of some acacia tree leaves from Kenya. Animal Feed Science Technology. 85: 89–98.

Ammar, H., Lopez, S. and Gonzalez, J. S. 2005. Assessment of the digestibility of some Mediterranean shrubs by *in vitro* techniques. Small Ruminant Research. 119: 323-331.

Anele, U.Y., Arigbede, O.M., Olanite, J.A., Adekunle, I.O., Jolaosho, A.O., Onifade, O.S., Oni, A.O. 2008. Early growth and seasonal chemical composition of three indigenous multipurpose tree species (MPT); Moringa oleifera, Millettia griffoniana and Pterocarpus santalinoides in Abeokuta, Nigeria. Agroforestry System. 73: 89-98.

ARC. 1984. The Nutrient requirements of ruminant livestock, supplement No. 1. Report of the Protein group of the Agricultural Research Council Working Party on Nutrient Requirements of Ruminants. CAB, Farnham Royal, UK.

AOAC. 2002. Official methods of analysis 16th edition. Association of Official Analytical Chemists. Arlington, Virginia, U.S.A.

Babayemi, O.J., Demeyer, D. and Fievez, V. 2004. In vitro fermentation of tropical browse seeds in relation to their secondary metabolites. Journal of Animal Science Technology. 13 suppl. 1:31-34.

Blümmel, M., Makkar, H.P.S and Becker, K. 1997. *In vitro* gas production: A Technique Revisited. Journal of Animal Physiology and Animal Nutrition. 77: 24-34.

Bodas, R., Prieto, N., García-González, R., Andrés, S., Giráldez, F.J. and López, S. 2012. Manipulation of rumen fermentation and methane production with plant secondary metabolites. Animal Feed Science Technology. 176: 78–93

Bouazza, L., Bodas, R., Boufennara, S., Bousseboua, H. and López, S. 2011. Nutritive evaluation of foliage from fodder trees and shrubs characteristic of Algerian arid and semi-arid areas. Journal of Animal and Feed Sciences. 21: 521–536.

Cerillo, M.A. and Juarez, R.A.S. 2004. *In vitro* gas production parameters in cacti and tree species commonly consumed by grazing goats in a semiarid region of North Mexico. Livestock Research for Rural Development. 16 (4). www.cipav.org.co/lrrd/lrrd16/4/cerr16021.htm

Cieslak, A., Zmora, P., Pers-Kamczyc, E., Szumacher-Strabel, M., 2012. Effects of tannins source (Vaccinium vitis idaea L.) on rumen microbial fermentation *in vivo*. Animal Feed Science Technology. 176: 102–106.

Dube, J.S., Reed J.D., Ndlovu L.R., 2001. Proanthocyanidins and other phenolics in Acacia leaves of Southern Africa. Animal Feed Science Technology. 91: 59-67.

Fadiyimu, A.A., Fadjemisin, A.N. and Alokan, J.A. 2011. Chemical composition of selected browse plants and their acceptability by WAD sheep. Livestock Research for Rural Development. 23: 12-17

Fievez, V., Babayemi, O.J. and Demeyer, D. (2005). Estimation of direct and indirect gas production in syringes: A tool to estimate short chain fatty acid production requiring minimal laboratory facilities. Animal Feed Science and Technology. 123-124: 197-210.

Getachew, G., Makkar, H.P.S. and Becker, K. 2002. Tropical browses: Contents of phenolic compounds, in vitro gas production and stoichiometric relationship between short chain fatty acid and in vitro gas production. Journal of Agricultural Science. 139: 341-352.

Getachew, G., De Peters, E.J. and Robinson, P.H. 2004. In vitro gas production provides effective method for assessing ruminant feeds. California Agriculture 58, 1-12.

Hanlin, Z, Mao L, Xuejuan Z, Tieshan, X and Guanyu, H. 2011. Nutritive value of several tropical Legume Shrubs in Hainan Province of China. Journal of Animal and Veterinary Advances. 10: 1640-1648.

Humphreys, L.R. 1991. Tropical forages: their role in sustainable agriculture. Longman Scientific and Technical, Essex, UK, pp. 414.

Isah, O.A., Omorogiuwa, L.E. and Okunade, S.A. 2013. Chemical evaluation of some browse plants eaten by local breeds of goats in Edo State, Nigieria. Pacific Journal of Science and Technology. 14 (1): 406-412.

Makkar, H.P.S.2005. *In vitro* Gas methods for evaluation of feeds containing phytochemicals. Animal Feed Science Technology, (123-124):291-302

Mbomi, S.E., Ogungbesan, A.M., Babayemi, O.J. and Nchinda, V.P. 2011. Chemical Composition, Acceptability of Three Tephrosia Species And Use Of *Tephrosia purperea* as Supplement for Grazing Animals in the Western Highlands of Cameroon. Journal of Environmental Issues and Agriculture in Developing Countries. 3 (3):132.

McDowell, L.R. 1997. Minerals for Grazing Ruminants in Tropical Regions. Bulletin of the Institute for Food and Agricultural Sciences, University of Florida, Gainesville, USA. 81pp.

Menke, K.H., Steingass, H., 1988. Estimation of the energetic feed value obtained from chemical analysis and gas production using rumen fluid. Animal Research Development. 28, 7–55.

Mokonnen, K., G. Glatzel and M. Sieghardt, 2009. Assessment of fodder values of 3 indigenous and 1 exotic woody plant species in the highlands of central Ethiopia. Mountain Research Development. 29: 135-142.

Moss, A.R., Jouany, J.P., Newbold, J. 2000. Methane production by ruminants: its contribution to global warming. Animal Zootechnology. 49, 231–253.

Nherera, F. V, Ndlovu, N. R, and Dzowela. B H. 1999. Relationships between *in vitro* gas production characteristics, chemical composition and *in vivo* quality measurements in goats fed tree fodder supplements. Small Ruminant Research. 31, 117-126.

Njidda, A. A. (2014). Determining dry matter degradability of some semi-arid browse species of north- eastern Nigeria using the *in vitro* technique. Nigerian Journal of Basic and Applied Science 18(2): 160-167

Nocek, J.E. 1997. Bovine acidosis: Implications on lameness. Journal of Dairy Science. 80:1005-1028.

Norton, B. W. 1994. Tree legumes as dietary supplements for ruminants. In: Gutteridge R C, Shelton H.M. (eds). Forage Tree Legumes in Tropical Agriculture. CAB International, UK. pp. 177-191

Nsahlai, I. V., Siaw, D.E.K.A. and Osuji, P.O. (1994). The relationship between gas production and chemical composition of 23 browses of genus Sesbania. Journal of Science and Food Agriculture. 65: 13-20.

Olafadehan, O.A. 2011. Changes in Haematological and Biochemical Diagnostic Parameters of Red Sokoto Goats fed Tannin-rich

Pterocarpus erinaceus forage diets. Veterinarski arhiv 81 (4), 471-483.

Olafadehan, O.A. 2013. Feeding value of Pterocarpus erinaceus for growing goats. Animal Feed Science Technology, 185, 1–8.

Olafadehan, O.A., Okunade, S.A. and Njidda, A. A. 2014. Evaluation of bovine rumen content as a feed ingredient for lambs. Tropical Animal Health and Production. DOI 10.1007/s11250-014-0590-9 (In press).

Onwuka, C.F. 1983. Nutritional evaluation of some Nigerian browse plants in the humid tropics. Ph.D. Thesis. University of Ibadan: Ibadan, Nigeria.

Osuga, I.M., Abdulrazak, S.A., Ichinohe, T. and Fujihara, T. 2006. Rumen degradation and in vitro gas production parameters in some browse forages, grasses and maize stover from Kenya. Journal of Food, Agriculture and Environment. 4(2): 60-64.

Rohweder, D.A., Barnes, R.F and Jorgensen, N. 1978. Proposed hay grading standard based on laboratory analyses for evaluating quality. Journal of animal science. 47: 747-759.

Rinne, M., Jaakkola S. and Huhtanen, P. 1997. Grass maturity effects on cattle fed silage-based diets. 1. Organic matter digestion, rumen fermentation and nitrogen utilization. Animal Feed Science Technology. 67, 1–17.

SAS. 2003. Statistical Analytical Systems, Users Guide Version 6), SAS Institute Inc., Cary, North Carolina, USA.

Sca, 1990. Feeding standards for australian livestock. Csiro publications, East Melbourne, Australia.

Siaw, D.E.K.A., Osuji, P.O. and Nsahlai, I.V. (1993).

Evaluation of multipurpose tree germplasm: the use of gas production and rumen degradation characteristics. Journal of Agricultural Science. (Cambridge) 120: 319-330.

Sniffen, C.J., O. Connor J.D., Van Soest P.J., Fox D. G. and Russell J.B. 1992. A Net Carbohydrate and Protein System for Evaluating Cattle Diets. II. Carbohydrate and protein availability. Journal of Animal Science. 70:3562-3577.

Teferedegne, B. 2000. New perspectives on the use of tropical plants to improve ruminant nutrition. Proceeding of Nutrition Society. 59, 209 – 214.

Topps, J.H. 1992. Potential, composition and use of legume shrubs and trees as fodder for livestock in the tropics review. Journal of Agricultural Science. 118, 1-18.

Tylutki, T.P., Fox, D.G., Durbal, V.M., Tedeschi, L.O. and Russell et al., J.B. 2008. Cornell net carbohydrate and protein system: A Model for Precision Feeding of Dairy Cattle. Animal Feed Science Technology. 143: 174-174.

Van Soest, P.J. 1994. Nutritional ecology of ruminants. 2nd edition. Cornell University Press.

Yusuf, K.O., Isah, O.A., Arigbede, O.M., Oni, A.O. and Onwuka, C.F.I. 2013. Chemical composition, Secondary Metabolites, In Vitro Gas Production Characteristics and Acceptability Study of Some Forages for Ruminant Feeding in South-Western Nigeria. Nigerian Journal of Animal Production. 40 (1): 179-190.

21

Nutrient Consumption and Digestibility of Sugar Cane Diets Supplemented with Soybean Meal or Urea

A. H. N. Rangel[1], J. M. S. Campos[2], S. C. Valadares Filho[3], G. S. Difante[1], D. M. Lima Júnior[4], L. P. Novaes[1], M. G. Costa[1]

[1]*Unidade Acadêmica Especializada em Ciências Agrárias, Universidade Federal do Rio Grande do Norte – UFRN, 59000-000, Macaíba, RN, Brazil. Corresponding autor: Adriano Henrique do Nascimento Rangel. ; e-mail:adrianohrangel@yahoo.com.br .*

[2] *Unidade Acadêmica de Garanhuns, Universidade Federal Rural de Pernambuco – UFRPE, Garanhuns, PE, Brazil.*

[3] *Universidade Federal de Viçosa - UFV, Viçosa, MG, Brazil.*

[4] *Campus Arapiraca, Universidade Federal de Alagoas - UFAL, Arapiraca, AL, Brazil*
Corresponding author

SUMMARY

The present work was developed aiming to verify the effect on intake and digestibility of diets based on sugar cane, whose CP levels were supplemented with concentrate based on soybean meal or different levels of urea, fed to dairy cows. Twelve Holstein dairy cows were used, arranged in three 4 x 4 Latin squares distributed according to the lactation period. After analysing the variance, we proceeded to compare the non-orthogonal contrasts. There was no difference (P>0.05) for the consumption of dry matter/material (DM), organic matter (OM), crude protein (CP), ether extract (EE), total carbohydrates (CHO), non-fibrous carbohydrates (NFC), neutral detergent fiber (NDF), and DM (g/kg $^{0.75}$) between diets. Between different levels of urea, a linearly decreasing effect for the consumption of NDF kg/day and NDF (% body weight) was verified. There was no difference (P>0.05) for apparent digestibility coefficients of DM, OM, CP, EE, NDF, and NFC when comparing different diets. There was linear increase between treatments of urea for CHO digestibility and total nutrient digestive. The soybean meal in the concentrate is not necessary in diets based on sugarcane supplemented with urea. The inclusion of urea is not necessary in diets based on sugar cane supplemented with a base concentrate of soybean meal for dairy cows producing 20 kg/day.

Key words: Dairy cows; Protein nutrition; *Saccharum officinarum*; Tropical forage.

INTRODUCTION

Among the options for supplementary forage, sugar cane has a consolidated position. In simulations comparing the sources of roughage for the herd, sugar cane is often suggested as an alternative that satisfies the most interesting conditions. Few plants have received as much special attention as sugarcane, which has been studied extensively with large investments in research targeted at culture and nutrition of animals with a view to formulating diets with it (Schmidt and Nussio, 2005; Siqueira et al., 2012). According to Landell et al. (2002), approximately 500 000 hectares of sugar cane are destined for animal feed, primarily for dairy herds (Freitas et al., 2011; Carvalho et al., 2011).

Using a dynamic and mechanistic model of digestion able to predict the absorption of nutrients in cattle fed diets based on sugar cane, Pereira and Collao-Saenz (2004) simulated the response of 200 and 300 kg heifers to the inclusion of urea in their diets with dietary levels ranging from 0 to 1 kg per 100 kg of sugar cane. The authors concluded that nothing is gained in the flow of nutrients absorbed nor available to the animal's organism when supplementation exceeds 50 grams / day, equivalent to 300 g of urea per 100 kg of raw sugar cane, i.e. 1% dry matter, considering there was a content of 30% dry matter.

As sugar cane varies with the variety, crop year, and stage of maturity, among others, Preston (1977) recommended a simple method to estimate the urea level to be added to the sugar cane with the formula: urea in sugar cane (g urea / kg of raw sugar cane) = 0.6 Brix (94.8 to 1.12 Brix) / (100 - Brix). The level of 1% corresponds to 17 ° Brix. Considering the increase in sugar yield of new varieties of sugar cane used for sugar industries (Smith et al., 2012; Nassif and Martin, 2013) which are available for use by cattle, perhaps today the need for adding urea would be no less, but greater than 1%, that is 1.15 and 1.25%. If this is verifiable, it would be an economically beneficial tool for producers.

Thus, an assessment is needed in order to test levels of urea in diets based on sugar cane for dairy cows of higher production potential, the results of which are still generally insufficient in Brazil. One of the current challenges being to increase participation levels of sugar cane for cows producing 20 to 25 liters. For a full understanding of the advantages and disadvantages of using a supplement, it's necessary to get information that goes beyond the production and composition of milk, consumption, and digestibility of nutrients.

The present work was developed aiming to verify the effect on intake and digestibility of diets based on sugar cane, whose CP levels were supplemented with concentrate based on soybean meal or different levels of urea, fed to dairy cows.

MATERIALS AND METHODS

The experiment was conducted at the Unidade de Ensino, Pesquisa e Extensão de Bovinocultura Leiteira (UEPE-GL), Departmento de Zootecnia (DZO), the Universidade Federal de Viçosa (UFV).

The Viçosa city is located in the Zona da Mata, State of Minas Gerais, 649 m altitude, geographically defined by the coordinates of 20°45'20'' south latitude and 42°52'40''west longitude. The climate is Cwa, according to the classification proposed by Köppen, with two defined seasons: dry from April to September, and wet from October to March. The average annual rainfall is 1341.2 mm. The mean maximum and minimum temperatures are 26.1 and 14.0°C respectively.

Twelve Holstein cows were used, purebred and crossbred, distributed in three 4 X 4 Latin squares design, according to the lactation period. The animals were subjected to four treatments in which raw volume sugar cane was used (Saccharum officinarum, L., RB range 73-9735), whose protein content was adjusted to a concentrate based on soybean meal and three other diets containing 0.4, 0.8, and 1.2% of urea with ammonium sulfate mixture (9:1), based on in natura forage. Concentrated sodium bicarbonate and magnesium oxide (2:1) was added to all diets. Diets were formulated to be isonitrogenous, according to the NRC (2001) for dairy cows with 600 kg of body weight (BW), producing 20 kg/day of milk with 3.5% fat content in milk (Table 1 and Table 2). The amount of concentrate was 1 kg for every 2 kg of milk produced, which corresponded to the total diet forage volume:concentrate ratio of 45:55, at the beginning of the experiment. The adjustment in the concentrate supply was made in the fifth and tenth day of each adjustment period. The Table 1 presents the proportions of the ingredients used in the concentrated mixture. The chemical composition of sugar cane and the concentrate used is shown in Table 2.

Table 1. Proportion of ingredients of the concentrated feed, expressed as a percentage of dry matter

Ingredients, g/100 g	Soybean Meal	Level of urea (%) in raw sugar cane		
		0.4	0.8	1.2
Corn meal	37.58	37.71	45.48	53.05
Soybean meal	31.45	0.00	0.00	0.00
38% Cottonseed meal	0.00	31.28	23.24	15.40
Wheat bran	27.27	27.27	27.27	27.27
Mg of sodium bicarbonate	1.09	1.09	1.09	1.09
Mineral mix	2.60	2.65	2.92	3.19
Total	100.00	100.00	100.00	100.00

[1] 67% sodium bicarbonate and 33% magnesium oxide; [2] Dicalcium phosphate (22.99, 9.16, 11.91, 14.12 %), limestone (42.81, 55.04, 50.03, 45.95 %), common salt (32.83, 34.24, 31.21, 28.73 %), sulphur flowers (0.95, 1.10, 0.78, 0.52%), zinc sulphate (0.3424, 0.3421, 0, 3389; 0.3361 %), copper sulfate (0.0515, 0.0981, 0.0999, 0.1012 %), potassium iodate (0.0037, 0.0038, 0.0035, 0.0033 %).

Table 2. Levels of average analytical fractions obtained for sugar cane and experimental concentrates.

Items	Concentrates		Level of urea (%) in raw sugar cane		
	Sugar cane	Soybean meal	0.4	0.8	1.2
Dry matter (%)[1]	29.29	87.95	87.83	88,75	88,34
Organic matter[1]	95.53	95.24	94.63	95.23	95.81
Crude Protein[1]	2.47	23.34	21.58	18.79	16.07
INNDF[2*]	43.72	15.31	18.03	18.95	21.28
INADF[2*]	25.12	6.31	6.02	5.06	4.92
Ether extract[1]	0.70	2.75	2.75	2.80	2.84
Total Carbohydrates[1]	90.36	70.15	70.30	73.63	76.89
NDF[1*]	45.69	21.35	29.39	27.28	25.22
Non-fiberous Carbohydrates[1]	44.67	48.80	40.91	46.35	51.67
ADF[1*]	24.85	9.06	10.47	9.35	8.25
Lignin[1]	7.10	1.93	2.50	2.25	2.00

* Insoluble Nitrogen in Neutral Detergent (INNDF); Insoluble Nitrogen in Acid Detergent (NIADF); Insoluble Fiber in Neutral Detergent (NDF) and Insoluble Fiber in Acid Detergent (ADF).
1 Values in percentage of MS.
2 Values as a percentage of total nitrogen.

The experiment consisted of four periods, each lasting 17 days each, with the first ten days of diet adaptation and the other for assessment of consumption, nutrient digestibility, milk production and composition, and the variation of weight.

The total digestible nutrients (TDN) were calculated according to Weiss (1999), by the equation: TDN (%) = DRP + DNDF + DNFC + 2.25 DEE, where DRP = digestible crude protein; DNDF = digestible neutral detergent fiber; DNFC = digestible non-fiber carbohydrates; and DEE = digestible ether extract.

The animals were housed in individual Tie Stalls, where they were fed *ad libitum* feed twice a day daily, at 8 and 17 h o'clock. The quantities of food provided and the treatment for consumption were weighed daily. Daily monitoring of consumption was made in order to keep the remaining food on the order of 10% of the total offered based on natural materials. At feeding time during the experimental period, samples of food and leftovers, which were placed in plastic bags and frozen for subsequent analysis were taken.

The preparation of composite samples of supplied feed and daily leftovers and analysis of dry matter (DM), organic matter (OM), mineral matter (MM), total nitrogen compounds (NT), nitrogen insoluble in neutral detergent (NIND) nitrogen insoluble in acid detergent (NIAD), ether extract (EE), neutral detergent fiber (NDF), acid detergent fiber (ADF) and lignin (LIG) followed the specifications described by Silva and Queiroz (2002).

The total carbohydrates (TC) were calculated according to Sniffen *et al.* (1992), wherein CHO = 100 - (% crude protein +% fat +% ash) with NFC being obtained by the formula NFC = 100 - [(% CP - % CP urea + % urea) +% EE +% MM].

The total amount of excreted fecal DM used to evaluate the apparent digestibility of foods was estimated through the indigestible acid detergent fiber (iADF), obtained after ruminal feed incubation, leftovers, and feces were put in Ankom bags (filter bag 57) for a period of 144 hours, following adaptation of the technique described by Cochran *et al.* (1986). Feces were collected in the 13th and 16th days of each experimental period, always before morning and afternoon milking, and placed in plastic bags which were stored in a freezer at -15°C and at end of the collection period a composite sample was made per animal based on the dry weight in air.

After variance analysis, we proceeded to compare the sum of squares for treatments in non-orthogonal contrasts related to concentrate based on soybean meal against urea levels, and the effects of linear and quadratic order relating to varying levels of urea through the Scheffé test. For all statistical procedures, 0.05 was adopted as the critical level of probability type I error.

RESULTS

Dry matter consumption did not differ (P>0.05) among diets. Diet with concentrate based on soybean meal showed an average consumption of 18.05 kg DM/day, while diets with urea showed an average 19.19 kg DM/day (Table 3).

The lack of differences in consumption of OM, EE, CHO, NFC, and TDN (kg/day) may be explained by similar DM consumption and approximately similar composition of the components of the experimental diets. However, the consumption of CP and NDF differed (P<0.05) between diets and urea levels.

Consumption of NDF, in kg/day and BW% was lower (P<0.05) for treatment with concentrate based on soybean meal compared to treatments with levels 0.4 and 0.8% urea. Between diets in which urea was used in its composition, there was a decreasing linear effect (P<0.05) as it raised the level of urea.

Table 3. Means and coefficients of variation (CV) and contrasts obtained for the daily intake of dry matter (DM), organic matter (OM), crude protein (CP), ether extract (EE), total carbohydrates (CHO), fiber in neutral detergent (NDF), non-fiber carbohydrates (NFC), total digestible nutrients (TDN) obtained for diets with sugar cane supplemented with concentrate based on soybean meal (SBM) or three levels of urea

Items	Diets with sugar cane				CV (%)	Contrasts[a]				
	FS	Levels of urea (%)				SBM *vs* 0.4%	SBM *vs* 0.8%	SBM *vs* 1.2%	L	Q
		0.4	0.8	1.2						
Consumption (kg/day)										
DM	18.05	19.55	19.41	18.60	7.53	ns	ns	ns	ns	ns
OM	17.30	18.51	18.34	17.52	7.48	ns	ns	ns	ns	ns
CP	2.23	2.76	2.72	2.66	11.28	**	**	**	ns	ns
EE	0.34	0.37	0.37	0.35	9.47	ns	ns	ns	ns	ns
CHO	14.72	15.37	15.25	14.50	6.88	ns	ns	ns	ns	ns
NDF	5.83	7.01	6.79	6.30	7.02	**	**	ns	**	ns
NFC	8.89	8.36	8.45	8.19	7.10	ns	ns	ns	ns	ns
TDN	10.71	10.81	11.25	11.13	21.04	ns	ns	ns	ns	ns
Consumption (% BW)										
DM	2.95	2.98	2.97	2.95	4.93	ns	ns	ns	ns	ns
NDF	0.98	1.17	1.11	1.02	7.90	**	**	ns	**	ns
Consumption (g/kg$^{0.75}$)										
DM	145.67	147.82	147.71	146.97	4.68	ns	ns	ns	ns	ns

[a] FS vs. U, L and Q = contrasts of the comparison between soybean meal and different levels of urea and linear and quadratic effects associated with the level of urea, respectively.
Ns: not significant.
** P <0.05.

In Table 4, the estimated values of protein and energy requirements are presented for lactating cows with average body weight of 600 kg and average daily production of 20 kg at 3.5% fat and weight gain of 0.30 kg/day, according to the NRC (2001).

The digestibility of nutrients did not differ between diets with concentrate based on soybean meal compared to diets containing different levels of urea, for DM, OM, CP, EE, NDF, and NFC (Table 5).

The digestibility of total carbohydrates (DCHO) and total digestible nutrients observed (TDN_{obs}) were higher (P<0.05) for the diet containing soybean meal in relation to the diet with 0.4% urea. For diets in which urea was used in its composition, there was increased linearity (P<0.05) with increasing levels of urea to DCHO and TDN_{obs} (Table 5). The TDN_{obs} was higher (P<0.05) when comparing the concentrate based on soybean meal to the level of 0.4% urea.

With the increase of levels of urea content in the diet, there was a significant effect on TDNobs (P<0.05).

DISCUSSION

Diet with concentrate based on soybean meal showed an average consumption of 18.05 kg DM/day, while diets with urea showed an average 19.19 kg DM/day. The literature points out when the percentage of concentrate in sugar cane based diets reached 60% dry base, there was DM consumption similar to that found in this study (Costa *et al.*, 2005; Oliveira *et al.*, 2007). Also the values recommended by the NRC (2001) for DM, 18.30 kg/day and 3.05% of body weight for dairy cow weighing 600 kg body weight, producing 20 kg of milk corrected to 3.5% fat and gaining approximately 0.300 kg/day, are similar to the average values found in this study, of 18.90 kg/day and 2.96% of body weight.

Table 4. Values observed and requirements of crude protein (CP) and total digestible nutrients (TDN), according to the NRC (2001) for lactating cows with 600 kg of body weight, producing 20 kg/day on average with 3.5% fat, with weight gain of 0.30 kg/day, expressed as kg/day.

Items	Requirements	Sugar cane Diets			
		Soybean	Level of urea (%)		
			0.4%	0.8%	1.2%
CP (kg/day)	2.62	2.23	2.76	2.72	2.66
Difference		-0.39	+ 0.14	+0.10	+ 0.04
TDN (kg/day)	10.55	10.72	10.82	11.25	11.14
Difference		+0.17	+0.27	+0.70	+0.59

Table 5. Means, coefficients of variation (CV) and contrasts obtained for the coefficient of digestibility of dry matter (DDM), organic matter (DOM), crude protein (DCP), ether extract (DEE), total carbohydrates (DCHO) fiber in neutral detergent (DNDF) and non-fiber carbohydrates (DNFC), obtained for the diets supplemented with concentrate based on soybean meal (SBM) or three levels of urea.

Items	Sugar cane diets				CV (%)	Contrasts[a]				
	FS	Levels of urea (%)				SBM *vs* 0.4%	SBM *vs* 0.8%	SBM *vs* 1.2%	L	Q
		0.4	0.8	1.2						
DDM	58.54	51.31	58.01	58.88	12.36	ns	ns	ns	ns	ns
DOM	58.72	54.15	57.51	59.18	18.06	ns	ns	ns	ns	ns
DCP	55.60	50.96	53.59	56.73	19.61	ns	ns	ns	ns	ns
DEE	81.80	77.23	76.72	79.47	12.87	ns	ns	ns	ns	ns
DCHO	58.65	54.13	57.74	59.11	20.28	**	ns	ns	**	ns
DNDF	26.80	26.24	26.26	23.63	51.15	ns	ns	ns	ns	ns
DNFC	79.82	77.31	78.03	86.55	23.13	ns	ns	ns	ns	ns
TDN_{obs}	59.28	54.72	57.98	59.69	17.58	**	ns	ns	**	ns

[a] FS vs. U, L and Q = contrasts of the comparison between soybean meal and different levels of urea and linear and quadratic effects associated with the level of urea, respectively.
[ns] not significant.
** P <0.05.

Consumption a crude protein was lower (P<0.05) for the diet with sugar cane supplemented with a concentrate based on soybean meal compared to diets with urea, at all levels of inclusion studied. Probably the difference in crude protein consumption was due to the source of dietary protein, among other things the diet with sugar cane supplemented with a concentrate of bran had soybean meal as the main source of nitrogen, while diets based on urea had this ingredient added to sugar cane. The addition of urea to sugar cane seems to contribute to the greater consumption of nitrogen, probably due to sugar cane representing a larger volume of natural matter in the cows' diets.

The average NDF consumption, expressed as a percentage of body weight was lower than recommended by Mertens (1985) of 1.25 ± 0.1%, to optimize DM ingestion and energy of lactating cows receiving mixed diets. Lower but similar to the values found in this study were observed by Valvasori et al. (1995), Costa et al. (2005), and Mendonca et al. (2004), and Santos et al. (2011) attributed this to the high lignin content in diets with sugar cane. The largest share of this component in the sugar cane reduces the rate and extent of NDF digestion, giving an increase in digestion retention time in the reticulum-rumen, negatively affecting the NDF consumption (Magalhães et al., 2006; Menezes et al., 2011; Oliveira et al., 2011; Santos et al., 2011). Both, the possibility to include urea in diets based on sugar cane supplemented with a concentrate of soybean meal. When urea was added to sugar cane, the concentrates did not include soybean meal, but cottonseed meal. Lascano et al. (2012) observed an increase in consumption of ADF when dairy cows were fed urea associated with sugar cane compared to cows fed soybean meal associated with the same volume.

The estimated values of protein and energy requirements are presented for lactating cows shown in the table 4 a deficit of dietary protein, which used the concentrate based on soybean meal for supplementation of sugar cane. This can be explained by the lower CP consumption in the soybean meal diet compared to treatments with different levels of urea.

In this research, the variation of body weight (BW) was 0.270, 0.373, 0.321, and 0.311 for the diet with soybean meal, and 0.4, 0.8, and 1.2% urea, respectively. Positive body variation has been observed in cows fed with diets based on sugar cane, used in concentrate of 40:60, similar to that used in this work (Costa et al., 2005; Oliveira et al., 2007). Mendonça et al. (2004) observed no increase in DNDF to raise the level of urea from 0.35% to 1% in diets based on sugar cane using concentration ratio of 60:40. Costa et al. (2005), working with diets based on sugar cane supplemented with 1% urea and ammonium sulphate (9:1), found DNDF higher than those found in this work for a 40:60 diet.

In diets based on sugar cane supplemented with urea, the increase in TDN_{obs} consumption can be explained by the improved food quality, as the level of urea increased, as verified by numerical improvement in the digestibility of all the non-fiber components, which goes against the suggestion of Pereira and Collao-Saenz (2004), according to which nothing is gained in the flow of absorbed nutrients nor available to the animal organism when supplementation exceeds the equivalent of 300 g of urea per 100 kg of raw sugar cane, with 30% DM.

CONCLUSION

The soybean meal in the concentrate is not necessary in diets based on sugarcane supplemented with urea. The inclusion of urea is not necessary in diets based on sugar cane supplemented with a base concentrate of soybean meal for dairy cows producing 20 kg/day.

CONFLICT OF INTEREST

The authors declare that they have no conflict of interest.

REFERENCES

Carvalho, G.G.P., Garcia, R., Pires, A.J.V., Detmann, E., Silva, R.R., Ribeiro, L.S.A., Chagas, D.M.T., Pinho, B.D., Domiciano, E.M.B. 2011. Metabolismo de nitrogênio em novilhas alimentadas com dietas contendo cana-de-açúcar tratada com óxido de cálcio. Revista Brasileira de Zootecnia, 40:622-629.

Cochran, R.C., Adams, D.C., Walace, J.D., Galyean, M. L. 1986. Predicting digestibility of different diets with internal markers: evaluation of four potential markers. Journal of Animal Science, 63:1476-1483.

Correa, C.E.S., Pereira, M.N., Oliveira, S.G., Ramos, M.H. 2003. Performance of Holstein cows fed sugarcane or corn silages of different grain textures. Scientia Agricola, 60:621-529

Costa, M. G., Campos, J.M., Valadares Filho, S. C., Valadares, R. F. D., Mendonça, S. S., Souza, D. P., Teixeira, M. P. 2005. Desempenho produtivo de vacas leiteiras alimentadas com diferentes proporções de cana-de-açúcar e concentrado ou silagem de milho na dieta. Revista Brasileira de Zootecnia, 34:2437-2445.

FNP (2004). Disponível em: <http://www.fnp.com.br>. Acesso em: ago de 2014.

Freitas, A.W.P., Rocha, F.C., Zonta, A., Fagundes, J.L., Fonseca, R., Zonta, M.C.M. 2011. Desempenho de novilhos recebendo dietas à base de cana-de-açúcar *in natura* ou hidrolisada. Revista Brasileira de Zootecnia, 40:2532-2537.

IDF – International Dairy Federation. 1996. Whole milk determination of milkfat, protein and lactose content. Guide fir the operation of mid-infra-red instuments. Bruxelas: 1996. 12p. (IDF Standard 141 B).

Landell, M.G.A., Campana, M.P., Rodrigues, A.A., Cruz, G.M., Batista, L.A.R., Figueiredo, P., Silva, M.A., Bidoia, M.A.P., Rossetto, R., Martins, A.L.M., Gallo, P.B., Kanthack, R.A.D., Cavichioli, J.C., Vasconcelos, A.C.M., Xavier, M.A. 2002. A variedade IAC86-2480 como nova opção de cana-de-açúcar para fins forrageiros: manejo de produção e uso na alimentação animal. Boletim Técnico IAC, n. 193, 36 p.

Lascano, G.J., Velez, M., Tricarico, J.M., Heinrichs, A.J. 2012. Nutrient utilization of fresh sugarcane-based diets with slow-release nonprotein nitrogen addition for control-fed dairy heifers. Journal of Dairy Science, 95:370–376.

Magalhães, A.L.R., Campos, J.M.S., Cabral, L.S., Mello, R., Freitas, J.A., Torres, R.A., Valadares Filho, S.C., Assis, A.J. 2006. Cana-de-açúcar em substituição à silagem de milho em dietas para vacas em lactação: parâmetros digestivos e ruminais. Revista Brasileira de Zootecnia, 35:591-599.

Marin, F., Nassif, D.S.P. 2013. Mudanças climáticas e a cana-de-açúcar no Brasil: Fisiologia, conjuntura e cenário futuro. Revista Brasileira de Engenhria Agricola e Ambiental, 17:232–239.

Mendonça, S.S., Campos, J.M.S., Valadares Filho, S.C., Valadares, R.F.D., Soares, C.A., Lana, R.P., Queiroz, A.C., Assis, A.J., Pereira, M.L.A. 2004. Consumo, produção e composição de leite, variáveis ruminais de vacas leiteiras alimentadas com dietas à base de cana-de-açúcar. Revista Brasileira de Zootecnia, 33:481-492.

Menezes, G.C.C., Valadares Filho, S.C., Magalhães, F.G., Valadares, R.F.D., Pardos, L.F., Detmann, E., Pereira, O.G., Leão, M.I. 2011. Intake and performance of confined bovine fed fresh or ensilaged sugar cane based diets

and corn silage. Revista Brasileira de Zootecnia, 40:1095-1103.

Mertens, D.R. 1985. Factors influencing feed intake in lactating cows: From theory to application using neutral detergent fiber. In: GA Nutrition Conference, 46, 1985, Athens. Proceedings... Athens: University of Georgia. 1-18.

National Research Council – NRC. 2001. Nutrient requirements of dairy cattle. 7. ed. National Academic Press, Washington, DC, USA.

Oliveira, A.S., Campos, J.M.S., Valadares Filho, S.C., Assis, A.J., Texeira, R.M.A., Valadares, R.F.D., Pina, D.S., Oliveira, G.S. 2007. Substituição do milho por casca de café ou de soja em dietas para vacas leiteiras: consumo, digestibilidade dos nutrientes, produção e composição do leite. Revista Brasileira de Zootecnia, 36:1172-1182 (supl.).

Oliveira, A.S., Detemann, E., Campos, J.M.S., Pina, D.S., Souza, S.M., Costa, M.G. 2011. Meta-análise do impacto da fibra em detergente neutro sobre o consumo, a digestibilidade e o desempenho de vacas leiteiras em lactação. Revista Brasileira de Zootecnia, 40:1587-1595.

Pereira, M.N., Collao-Saenz, E.A. 2004. Algumas considerações sobre a velha cana com ureia. Disponível em: <http://www.milkpoint.com.br>. Acesso em: maio de 2004.

Preston, T.R. 1977. Nutritive value of sugarcane for ruminants. Tropical Animal Production., v. 2, p. 125-142, 1977.

Santos, S.A., Valadares Filho, S.C., Detmann, E., Valadares, R.F.D., Ruas, J.R.M., Amaral, P.M. 2011. Different forage sources for F1 Holstein×Gir dairy cows. Livestock Science, 142:48–58.

Schimidt, P., Nussio, L.G. 2005. Produção e utilização de cana-de-açúcar para bovinos leiteiros: novas demandas. Anais... Bovinocultura de Leite: Nutrição, Reprodução e Fertilidade em Bovinos, 2005.

Silva, J.D., Queiroz, A.C. 2002. Análise de alimentos (Métodos químicos e biológicos). 3. ed. Viçosa: Editora UFV, 2002. 235 p.

Siqueira, G.R., Roth, M.T.P., Moretti, M.H., Benatti, J.M.B., Resende, F.D. 2012. Uso da cana-de-açúcar na alimentação de ruminantes. Revista Brasileira de Saúde e Produção Animal, 13:991-1008.

Sklan, D.; Ashkenazi, R.; Braun, A., Devorin, A., Tabori, K. 1992. Fatty acids, calcium soaps of fatty acids and cottonseeds fed to high yielding cows. Journal Dairy Science., 75:2463-2472.

Sniffen, C.J.; O'Connor, J.D.; Van Soest, P.S., Fox, D. G., Russell, J. B. 1992. A net carbohydrate and protein system for evaluating cattle diets. II. Carbohydrate and protein availability. Journal of Animal Science, 70:3562-3577.

Universidade Federal de Viçosa. 1997. Departamento de Engenharia Agrícola. Estação meteorológica. Dados climáticos. Viçosa, MG:UFV..

Valadares Filho, S.C.; Rocha Júnior, V.R.; Cappelle, E.R. 2002. Tabelas brasileiras de composição de alimentos para bovinos. Viçosa: UFV/DZO/DPI, 297 p.

Valvasori, E., Lucci, C.S.L., Pires, F.L., Arcaro, J.R.P., Arcaro Jr, I. 1995. Avaliação da cana-de-açúcar em substituição à silagem de milho para vacas leiteiras. Brazilian Journal of Veterinary Research Animal Science, 32:224-228.

Weiss, W.P. 1999. Energy prediction equations for ruminant feeds. In: Cornell Nutrition Conference for Feed Manufacturers, 61, 1999, Proceeding, Ithaca: Cornell University, p.176-185

PERMISSIONS

All chapters in this book were first published in TSA, by Universidad Autónoma de Yucatájn; hereby published with permission under the Creative Commons Attribution License or equivalent. Every chapter published in this book has been scrutinized by our experts. Their significance has been extensively debated. The topics covered herein carry significant findings which will fuel the growth of the discipline. They may even be implemented as practical applications or may be referred to as a beginning point for another development.

The contributors of this book come from diverse backgrounds, making this book a truly international effort. This book will bring forth new frontiers with its revolutionizing research information and detailed analysis of the nascent developments around the world.

We would like to thank all the contributing authors for lending their expertise to make the book truly unique. They have played a crucial role in the development of this book. Without their invaluable contributions this book wouldn't have been possible. They have made vital efforts to compile up to date information on the varied aspects of this subject to make this book a valuable addition to the collection of many professionals and students.

This book was conceptualized with the vision of imparting up-to-date information and advanced data in this field. To ensure the same, a matchless editorial board was set up. Every individual on the board went through rigorous rounds of assessment to prove their worth. After which they invested a large part of their time researching and compiling the most relevant data for our readers.

The editorial board has been involved in producing this book since its inception. They have spent rigorous hours researching and exploring the diverse topics which have resulted in the successful publishing of this book. They have passed on their knowledge of decades through this book. To expedite this challenging task, the publisher supported the team at every step. A small team of assistant editors was also appointed to further simplify the editing procedure and attain best results for the readers.

Apart from the editorial board, the designing team has also invested a significant amount of their time in understanding the subject and creating the most relevant covers. They scrutinized every image to scout for the most suitable representation of the subject and create an appropriate cover for the book.

The publishing team has been an ardent support to the editorial, designing and production team. Their endless efforts to recruit the best for this project, has resulted in the accomplishment of this book. They are a veteran in the field of academics and their pool of knowledge is as vast as their experience in printing. Their expertise and guidance has proved useful at every step. Their uncompromising quality standards have made this book an exceptional effort. Their encouragement from time to time has been an inspiration for everyone.

The publisher and the editorial board hope that this book will prove to be a valuable piece of knowledge for researchers, students, practitioners and scholars across the globe.

LIST OF CONTRIBUTORS

H. M. Lawal, J. O. Ogunwole and E. O. Uyovbisere
Department of soil science, Institute for Agricultural Research Ahmadu Bello University, Zaria –Nigeria

Godsteven P. Maro and James M. Teri
TaCRI, Lyamungu Moshi, Tanzania

Jerome P. Mrema and Balthazar M. Msanya
Department of Soil Science, Faculty of Agriculture, Sokoine University of Agriculture. Morogoro, Tanzania

M. da C. Mendonça
Emdagro/Embrapa Tabuleiros Costeiros Av. Beira Mar, n. 3250, 13 de Julho, CEP 49025-040, Aracaju, Sergipe, Brazil

M. da F. Santos
Doutoranda do Departamento de Genética da Escola de Superior de Agricultura Luiz de Queiroz, Av. Pádua Dias, 11 CP 9, CEP 13418-900, Piracicaba, São Paulo, Brazil

R. Silva-Mann
Departamento de Engenharia Agronômica da Universidade Federal de Sergipe, Av. Marechal Rondon s/n, Jardim Rosa Elze, CEP 49100-000, Aracaju, Sergipe, Brazil

J. M. Ferreira
Embrapa Tabuleiros Costeiros Av. Beira Mar, n. 3250, 13 de Julho, CEP 49025-040, Aracaju, Sergipe, Brazil

Adrian Guzmán Sánchez and Ana María Rosales-Torres
Departamento de producción Agrícola y Animal Universidad Autónoma Metropolitana Xochimilco. Calzada del Hueso 1100, 04960. México Distrito Federal

Carlos G. Gutiérrez Aguilar
Departamento de Reproducción Facultad de Medicina Veterinaria y Zootecnia. Universidad Nacional Autónoma de México. Ciudad Universitaria 04510 México Distrito Federal

José Andrés Reyes-Gutiérrez and Oziel Dante Montañez-Valdez
Centro Universitario del Sur de la Universidad de Guadalajara. Departamento de Desarrollo Regional. Prolongación Colón S/N. Ciudad Guzmán, Jalisco. CP 49000. México

Cándido Enrique Guerra-Medina
División de Desarrollo Regional, Centro Universitario de la Costa Sur, Universidad de Guadalajara, Autlán de Navarro. Jalisco, México

Hilda G. García-Núñez, E. Gabino Nava-Bernal and A. Roberto Martínez-Campos
Univ. Autónoma del Estado de México. Instituto de Ciencias Agropecuarias y Rurales. Km. 14.5 Autopista Toluca-Atlacomulco. San Cayetano de Morelos. Toluca, Estado de México. C.P. 50295

Sergio de J. Romero-Gómez
Univ. Autónoma de Querétaro. Fac. Química. Av. Hidalgo S/N, Col. Niños Héroes. Querétaro, Qro. C.P. 76010

Carlos. E. González-Esquivel
UNAM. Centro de Investigaciones en Ecosistemas. Antigua Carretera a Pátzcuaro 8701. Col. Ex-Hacienda de San José de la Huerta. CP 58190,Morelia, Michoacán

José Ramos-Zapata, Denis Marrufo-Zapata, Uriel Solís-Rodríguez and Luis Salinas-Peba
Departamento de Ecología Tropical, Campus de Ciencias Biológicas y Agropecuarias, Universidad Autónoma de Yucatán, km 15.5 de la Carretera Mérida-Xmatkuil AP 4-116, Itzimná Mérida, Yucatán, México

Patricia Guadarrama-Chávez
Unidad Multidisciplinaria de Docencia e Investigación, Facultad de Ciencias, Universidad Nacional Autónoma de México, Sisal, Yucatán 97356, México

Pious Soris Tresina and Veerabahu Ramasamy Mohan
Ethnopharmacology Unit, Research Department of Botany, V.O.Chidambaram College, Tuticorin-628008, Tamil Nadu, India

N. Zamarripa, A. M. Patterson, I. Sánchez-Gallen and J. Álvarez-Sánchez
Departamento de Ecología y Recursos Naturales, Facultad de Ciencias, Universidad Nacional Autónoma de México

Jaime Ruiz-Vega, Rafael Pérez-Pacheco, Teodulfo Aquino-Bolaños and María Eugenia Silva-Rivera
Centro de Investigación Interdisciplinaria para el Desarrollo Integral Regional (CIIDIR-IPN-OAXACA). Calle Hornos 1003, Santa Cruz Xoxocotlán, Oax. CP 71 230, México

J. Mondragón-Ancelmo, I. A. Domínguez-Vara and J. L Bórquez-Gastélum
Departamento de Nutrición Animal. Facultad de Medicina Veterinaria y Zootecnia. Universidad Autónoma del Estado de México (UAEMéx.).Campus Universitario "El Cerrillo" Toluca, México. C.P. 50090

S. Rebollar-Rebollar and J. Hernández-Martínez
Centro Universitario UAEM Temascaltepec. Universidad Autónoma del Estado de México (UAEMéx.)

Hortensia Brito-Vega, José Manuel Salaya-Domínguez and Edmundo Gómez-Méndez
División Académica de Ciencias Agropecuarias, Universidad Juárez Autónoma de Tabasco, Km 25.5 Carretera VHA-Teapa, Rancheria 3er. Sección, Vhsa, Tabasco, Mexico. C.P. 86040

David Espinosa-Victoria
Colegio de Postgraduados, Campus Montecillo, Carretera México-Texcoco Km 36.5, Montecillo, Texcoco, C.P. 56230 Estado de Mexico

D. Lepe-Soltero, M. A. García-Neria, D. González de León and L. Silva-Rosales
Laboratorio de Interacciones Planta-Virus del Depto. de Ing. Genética. Cinvestav Unidad Irapuato. Km. 9.6. Lib. Nte. Carr. Irapuato-León. CP 36821

B. M. Sánchez-García, Y. Jiménez-Hernández and J. A. Acosta-Gallegos
Programa de Frijol CEBAJ-INIFAP Km 6.5 Carretera Celaya a San Miguel de Allende Celaya, Gto. Mexico. CP 38110

N. E. Becerra-Leor
Programa de Frijol CECOT-INIFAP Km 34.5 Carretera Federal Veracruz-Córdoba, Medellín de Bravo CP 94270

R. A. Salinas-Perez
Programa de Frijol CEVAF-INIFAP Km 1609 Carretera Internacional Mexico-Nogales, CP 81100 Juan Jose Rios, Los Mochis, Sin

Doug Jackson and Kate Zemenick
University of Michigan, Department of Ecology and Evolutionary Biology, 2077 Kraus Natural Science, 830 North University, Ann Arbor, Michigan, U.S.A., 48109

Graciela Huerta
El Colegio de La Frontera Sur, Departamento de Entomología Tropical, Carretera Antiguo Aeropuerto km 2.5, Tapachula, Chiapas, Mexico

Guillermo M. Carrillo-Castañeda and Francisco Bautista-Calles
Genética Colegio de Postgraduados-Campus Montecillo Km. 36.5 Carretera México-Texcoco Montecillo, Texcoco, México 56230, México

Angel Villegas-Monter
Fruticultura Colegio de Postgraduados-Campus Montecillo Km. 36.5 Carretera México-Texcoco Montecillo, Texcoco, México 56230, México

Danielle Medina Rosa, Lúcia Helena Pereira Nóbrega, Márcia Maria Mauli and Gislaine Piccolo de Lima
Universidade Estadual do Oeste do Paraná (UNIOESTE), CCET – PGEAGRI, Caixa Postal, 711, (85819-110) Cascavel – PR – Brazil

Walter Boller
FAMV/UPF/Passo Fundo – RS-Brazil

Isabel Jiménez-García, Carlos N. Castro-José, Salim Pavón-Suriano, Fabiola Lango-Reynoso and Ma. del Refugio Castañeda-Chávez
Instituto Tecnológico de Boca del Río. Km 12 Carretera Veracruz-Córdoba, CP. 94290, Boca del Río, Veracruz, Mexico

Carlos R. Rojas-García
Universidad Católica de Temuco, Casilla 15D, Temuco, Chile
CCMAR Universidade do Algarve, Campus de Gambelas, 8005-139 Faro – Portugal

Susan A. Nguku, Nashon K. R. Musimba, Dorothy Amwata and Eric M. Kaindi
South Eastern Kenya University, Kitui, Kenya

Donald N. Njarui
Kenya Agriculture and Livestock Research Organisation, Katumani, Machakos, Kenya

P. V. D. Molin, L. Rampim, F. Fávero, M. do Carmo Lana, M. V. M. Sarto, J. S. Rosset, D. Mattei, P. S. Diel and R. N. D. Molin
State University of West Paraná, Unioeste, CCA/PPGA, Pernambuco Street No. 1777, Zip Code 85960-000, city of Marechal Cândido Rondon, Paraná state, Brazil

S. A. Okunade, O. A. Isah, R. Y. Aderinboye and O. A. Olafadehan
Animal Production Technology Department, Federal College of Wildlife Management, P.M.B 268, New Bussa, Nigeria
Department of Animal Nutrition, Federal University of Agriculture, Abeokuta, Nigeria
Department of Animal Science, University of Abuja, Abuja, Nigeria

A. H. N. Rangel, G. S. Difante, L. P. Novaes and M. G. Costa
Unidade Acadêmica Especializada em Ciências Agrárias, Universidade Federal do Rio Grande do Norte – UFRN, 59000-000, Macaíba, RN, Brazil

J. M. S. Campos
Unidade Acadêmica de Garanhuns, Universidade Federal Rural de Pernambuco –UFRPE, Garanhuns, PE, Brazil

S. C. Valadares Filho
Universidade Federal de Viçosa - UFV, Viçosa, MG, Brazil

D. M. Lima Júnior
Campus Arapiraca, Universidade Federal de Alagoas - UFAL, Arapiraca, AL, Brazil

Index